Lecture Notes in Computer Science   11179

Commenced Publication in 1973
Founding and Former Series Editors:
Gerhard Goos, Juris Hartmanis, and Jan van Leeuwen

Wil M. P. van der Aalst · Vladimir Batagelj
Goran Glavaš · Dmitry I. Ignatov
Michael Khachay · Sergei O. Kuznetsov
Olessia Koltsova · Irina A. Lomazova
Natalia Loukachevitch · Amedeo Napoli
Alexander Panchenko · Panos M. Pardalos
Marcello Pelillo · Andrey V. Savchenko (Eds.)

Analysis of Images, Social Networks and Texts

7th International Conference, AIST 2018
Moscow, Russia, July 5–7, 2018
Revised Selected Papers

 Springer

Editors
Wil M. P. van der Aalst (ID)
RWTH Aachen University
Aachen, Germany

Vladimir Batagelj (ID)
University of Ljubljana
Ljubljana, Slovenia

Goran Glavaš (ID)
University of Mannheim
Mannheim, Germany

Dmitry I. Ignatov (ID)
National Research University Higher School
of Economics, Moscow, Russia

Michael Khachay (ID)
Institute of Mathematics and Mechanics
Yekaterinburg, Russia

Sergei O. Kuznetsov (ID)
National Research University Higher School
of Economics, Moscow, Russia

Olessia Koltsova (ID)
National Research University Higher School
of Economics, Saint Petersburg, Russia

Irina A. Lomazova (ID)
National Research University Higher School
of Economics, Moscow, Russia

Natalia Loukachevitch (ID)
Moscow State University
Moscow, Russia

Amedeo Napoli (ID)
Loria
Vandoeuvre lès Nancy, France

Alexander Panchenko (ID)
University of Hamburg
Hamburg, Germany

Panos M. Pardalos (ID)
University of Florida
Gainesville, FL, USA

Marcello Pelillo (ID)
Ca Foscari University of Venice
Venice, Italy

Andrey V. Savchenko (ID)
National Research University Higher School
of Economics, Nizhny Novgorod, Russia

ISSN 0302-9743   ISSN 1611-3349 (electronic)
Lecture Notes in Computer Science
ISBN 978-3-030-11026-0   ISBN 978-3-030-11027-7 (eBook)
https://doi.org/10.1007/978-3-030-11027-7

Library of Congress Control Number: 2018966508

LNCS Sublibrary: SL3 – Information Systems and Applications, incl. Internet/Web, and HCI

This Springer imprint is published by the registered company Springer Nature Switzerland AG
The registered company address is: Gewerbestrasse 11, 6330 Cham, Switzerland

Preface

This volume contains the refereed proceedings of the 7th International Conference on Analysis of Images, Social Networks, and Texts (AIST 2018)[1]. The previous conferences during 2012–2017 attracted a significant number of data scientists – students, researchers, academics, and engineers working on interdisciplinary data analysis of images, texts, and social networks.

The broad scope of AIST made it an event where researchers from different domains, such as image and text processing, exploiting various data analysis techniques, can meet and exchange ideas. We strongly believe that this may lead to the cross-fertilisation of ideas between researchers relying on modern data analysis machinery.

Therefore, AIST brought together all kinds of applications of data mining and machine learning techniques. The conference allowed specialists from different fields to meet each other, present their work, and discuss both theoretical and practical aspects of their data analysis problems. Another important aim of the conference was to stimulate scientists and people from industry to benefit from the knowledge exchange and identify possible grounds for fruitful collaboration.

The conference was held during July 5–7, 2018. The conference was organised in Moscow, the capital of Russia, on the campus of Moscow Polytechnic University (MPU)[2].

This year, the key topics of AIST were grouped into six tracks:

1. General Topics of Data Analysis chaired by Sergei Kuznetsov (Higher School of Economics, Russia) and Amedeo Napoli (LORIA, France)
2. Natural Language Processing chaired by Natalia Loukachevitch (Lomonosov Moscow State University, Russia) and Goran Glavaš (University of Mannheim, Mannheim, Germany)
3. Social Network Analysis chaired by Vladimir Batagelj (University of Ljubljana, Slovenia) and Olessia Koltsova (Higher School of Economics, Russia)
4. Analysis of Images and Video chaired by Marcello Pelillo (University of Venice, Italy) and Andrey Savchenko (Higher School of Economics, Russia)
5. Optimization Problems on Graphs and Network Structures chaired by Panos Pardalos (University of Florida, USA) and Michael Khachay (IMM UB RAS and Ural Federal University, Russia)
6. Analysis of Dynamic Behavior Through Event Data chaired by Wil van der Aalst (RWTH Aachen University, Germany) and Irina Lomazova (Higher School of Economics, Russia)

---

[1] http://aistconf.org/.

[2] http://mospolytech.ru/?eng.

The Program Committee and the reviewers of the conference included 154 well-known experts in data mining and machine learning, natural language processing, image processing, social network analysis, and related areas from leading institutions of 32 countries including Argentina, Australia, Austria, Bangladesh, Canada, China, Croatia, Czech Republic, Denmark, Egypt, Finland, France, Germany, Greece, India, Italy, Kazakhstan, Lithuania, Norway, Qatar, Romania, Russia, Slovenia, Spain, Switzerland, Vietnam, Taiwan, Turkey, Ukraine, The Netherlands, UK, and USA. This year we received 107 submissions mostly from Russia but also from Austria, Bangladesh, Finland, Germany, Norway, Switzerland, Taiwan, Turkey, Ukraine, USA, and Vietnam.

Out of 107 submissions, only 29 papers were accepted as regular oral papers. Thus, the acceptance rate of this volume was around 36% (not taking into account 26 automatically rejected papers). In order to encourage young practitioners and researchers, we included 28 papers in the supplementary proceedings after their poster presentation at the conference. Each submission was reviewed by at least three reviewers, experts in their fields, in order to supply detailed and helpful comments.

The conference featured several invited talks and an industry session dedicated to current trends and challenges.

The topical opening talk was presented by Rostislav Yavorskiy on "Visualization of Data Science Community in Russia."

The invited talks were:

- Anne Rozinat (Fluxicon), "Process Mining: Discovering Process Maps from Data"
- Goran Glavaš (University of Mannheim), "Cars, Drivers, Vehicles, and Wheels: Specializing Distributional Word Vectors for Lexico-Semantic Relations"
- Gleb Gusev (Yandex and Moscow Institute of Physics and Technology), "Machine Learning Research in Yandex and CatBoost"
- Ekaterina Chernyak of HSE and Sberbank presented a tutorial on "Supervised Methods in Natural Language Processing"

The business speakers also covered a wide variety of topics:

- Iosif Itkin (Exactpro Systems), "The World's Most Valuable Commodity"
- Maksim Balashevich (Santiment), "Crypto Assets Valuation"
- Ilya Utekhin (European University at St. Petersburg), "On Understanding Users for Human-Centered Design"
- Denis Stepanov (JetBrains), "Workshop: Data Visualization with Datalore.plot in Research, Education, and Just for Fun"
- Eugeny Malyutin (OK.ru), "Trend Detection at OK"
- Oleg Klepikov (Center for Applied Neuroeconomics), "Psychometric Behavorial Classification of Banking Clients – A Case Study"
- Dmitriy Skougarevskiy (European University at St. Petersburg), "Semantic Representation of Accident Reports to Understand the Dark Figure of Crime"

Under a special agreement between the responsible program co-chairs and the proceedings chair, due to its interdisciplinary nature, one paper was assigned to a special section on Innovative Systems (outside of the main tracks).

We would like to thank the authors for submitting their papers and the members of the Program Committee for their efforts in providing exhaustive reviews.

According to the program chairs, and taking into account the reviews and presentation quality, the best paper awards were granted to the following papers:

Track 1. General Topics of Data Analysis: "Lookup Lateration: Mapping of Received Signal Strength to Position for Geo-Localization in Outdoor Urban Areas" by Andrey Shestakov, Attila Kertesz-Farkas, Dmitri Shmelkin, and Danila Doroshin

Track 2. Natural Language Processing: "Learning Representations for Soft Skill Matching" by Luiza Sayfullina, Eric Malmi, and Juho Kannala

Track 3. Social Network Analysis: "Organizational Networks Revisited: Predictors of Headquarters–Subsidiary Relationship Perception" by Antonina Milekhina, Valentina Kuskova, and Elena Artyukhova

Track 4. Analysis of Images and Video: "Application of Fully Convolutional Neural Networks to Mapping Industrial Oil Palm Plantations" by Artem Baklanov, Mikhail Khachay, and Maxim Pasynkov

Track 5. Optimization Problems on Graphs and Network Structures: "Efficient PTAS for the Euclidean CVRP with Time Windows" by Michael Khachay and Yuri Ogorodnikov

Track 6. Analysis of Dynamic Behavior Through Event Data: "Neural Approach to the Discovery Problem in Process Mining" by Timofey Shunin, Natalia Zubkova, and Sergey Shershakov

We would also like to express our special gratitude to all the invited speakers and industry representatives.

We deeply thank all the partners and sponsors. Our golden sponsor was Exactpro. Exactpro, a fully owned subsidiary of the London Stock Exchange Group, specialises in quality assurance for exchanges, investment banks, brokers, and other financial sector organisations. Our special thanks go to Springer for their help, starting from the first conference call to the final version of the proceedings. JetBrains covered important conference expenses while Springer sponsored the best paper awards. Last but not least, we are grateful to the dean of Information Technologies Faculty of MPU, Andrey Phillipovich, and all the organisers, especially to Andrey Novikov and Marina Danshina, and the volunteers, whose endless energy saved us at the most critical stages of the conference preparation.

Here, we would like to mention the Russian word "aist" is more than just a simple abbreviation (in Cyrillic) – it means a "stork." Since it is a wonderful free bird, a symbol of happiness and peace, this stork gave us the inspiration to organise the AIST conference. So we believe that this young and rapidly growing conference will likewise be bringing inspiration to data scientists around the world!

October 2018

Wil van der Aalst
Vladimir Batagelj
Goran Glavaš
Dmitry Ignatov
Michael Khachay
Olessia Koltsova
Sergei Kuznetsov
Irina Lomazova
Natalia Loukachevitch
Amedeo Napoli
Alexander Panchenko
Panos Pardalos
Marcello Pelillo
Andrey Savchenko

Organisation

Program Committee Chairs

Wil van der Aalst	RWTH Aachen University, Germany
Vladimir Batagelj	University of Ljubljana, Slovenia
Goran Glavaš	University of Mannheim, Germany
Michael Khachay	Krasovskii Institute of Mathematics and Mechanics of Russian Academy of Sciences, Russia and Ural Federal University, Yekaterinburg, Russia
Sergei Kuznetsov	National Research University Higher School of Economics, Moscow, Russia
Olessia Koltsova	National Research University Higher School of Economics, Saint Petersburg, Russia
Amedeo Napoli	LORIA – CNRS, University of Lorraine, and Inria, Nancy, France
Irina Lomazova	National Research University Higher School of Economics, Moscow, Russia
Natalia Loukachevitch	Computing Centre of Lomonosov Moscow State University, Russia
Alexander Panchenko	University of Hamburg, Germany and Université catholique de Louvain, Belgium
Panos Pardalos	University of Florida, USA
Marcello Pelillo	University of Venice, Italy
Andrey Savchenko	National Research University Higher School of Economics, Nizhny Novgorod, Russia

Proceedings Chair

Dmitry I. Ignatov	National Research University Higher School of Economics, Russia

Business Day Chair

Rostislav Yavorskiy	National Research University Higher School of Economics, Moscow

Steering Committee

Dmitry I. Ignatov	National Research University Higher School of Economics, Russia

Michael Khachay	Krasovskii Institute of Mathematics and Mechanics of Russian Academy of Sciences, Russia and Ural Federal University, Yekaterinburg, Russia
Alexander Panchenko	University of Hamburg, Germany and Université catholique de Louvain, Belgium
Rostislav Yavorskiy	National Research University Higher School of Economics, Russia

Program Committee

Ekaterina Arkhangelskaya	St. Petersburg Department of Steklov Institute of Mathematics of Russian Academy of Sciences, Russia
Aleksey Artamonov	Neuromation, USA
Xiang Bai	Huazhong University of Science and Technology, China
Jaume Baixeries	Universitat Politècnica de Catalunya, Spain
Amir Bakarov	National Research University Higher School of Economics, Russia
Artem Baklanov	International Institute for Applied Systems Analysis, Austria
Lamberto Ballan	University of Padova, Italy
Vladimir Batagelj	Institute of Mathematics, Physics and Mechanics, Slovenia
Timo Baumann	University of Hamburg, Germany
Malay Bhattacharyya	Indian Statistical Institute, Kolkata
Chris Biemann	University of Hamburg, Germany
Battista Biggio	University of Cagliari, Italy
Elena Bolshakova	Lomonosov Moscow State University, Russia
Samuel Bulò	Fondazione Bruno Kessler, Italy
Andrea Burattin	Technical University of Denmark, Denmark
Evgeny Burnaev	Skolkovo Institute of Science and Technology, Russia
Aleksey Buzmakov	National Research University Higher School of Economics, Perm, Russia
Ignacio Cassol	Universidad Austral, Argentina
Artem Chernodub	Institute of Mathematical Machines and Systems Problems of National Academy of Sciences of Ukraine, Ukraine
Mikhail Chernoskutov	Krasovskii Institute of Mathematics and Mechanics of the Ural Branch of the Russian Academy of Sciences and Ural Federal University, Russia
Bonaventura Coppola	University of Trento, Italy
Hernani Costa	University of Malaga, Spain
Massimiliano de Leoni	Eindhoven University of Technology, The Netherlands
Boris Dobrov	Research Computing Center of Lomonosov Moscow State University, Russia
Sofia Dokuka	National Research University Higher School of Economics, Russia
Dirk Fahland	Eindhoven University of Technology, the Netherlands

Stefano Faralli	University of Mannheim, Germany
Victor Fedoseev	Samara National Research University, Russia
Elena Filatova	City University of New York, USA
Olga Gerasimova	National Research University Higher School of Economics, Russia
Edward Kh. Gimadi	Sobolev Institute of Mathematics of the Siberian Branch of the Russian Academy of Sciences, Russia
Goran Glavaš	University of Mannheim, Germany
Ivan Gostev	National Research University Higher School of Economics, Russia
Natalia Grabar	STL CNRS Université Lille 3, France
Artem Grachev	Samsung R&D, Russia
Dmitry Granovsky	Yandex, Russia
Anna Gromova	Exactpro, Russia
Marianne Huchard	LIRMM, Université Montpellier 2 and CNRS, France
Dmitry Ignatov	National Research University Higher School of Economics, Russia
Dmitry Ilvovsky	National Research University Higher School of Economics, Russia
Vladimir Ivanov	Innopolis University, Russia
Pei Jun	Hefei University of Technology, China
Alex Jung	Aalto University, Finland
Anna Kalenkova	National Research University Higher School of Economics, Russia
Ilia Karpov	National Research University Higher School of Economics, Russia
Nikolay Karpov	National Research University Higher School of Economics, Nizhniy Novgorod, Russia
Egor Kashkin	Vinogradov Russian Language Institute of Russian Academy of Sciences, Russia
Alexander Kelmanov	Sobolev Institute of Mathematics of Siberian Branch of Russian Academy of Sciences, Russia
Attila Kertesz-Farkas	National Research University Higher School of Economics, Russia
Michael Khachay	Krasovskii Institute of Mathematics and Mechanics, Ural Branch of Russian Academy of Sciences and Ural Federal University, Russia
Sergey Khamidullin	Sobolev Institute of Mathematics, Siberian Branch of Russian Academy of Sciences, Russia
Oleg Khamisov	Melentiev Energy Systems Institute, Siberian Branch of the Russian Academy of Sciences, Russia
Vladimir Khandeev	Sobolev Institute of Mathematics, Siberian Branch of Russian Academy of Sciences, Russia
Alexander Kharlamov	National Research University Higher School of Economics, Russia

Nizar Messai	Université François Rabelais, France
Tristan Miller	Technische Universität Darmstadt, Germany
Olga Mitrofanova	St. Petersburg State University, Russia
Evgeny Myasnikov	Samara National Research University, Russia
Federico Nanni	University of Mannheim, Germany
Amedeo Napoli	Université de Lorraine, Nancy, France
Kirill Nikolaev	National Research University Higher School of Economics, Russia
Damien Nouvel	Institut National des Langues et Civilisations Orientales, France
Evgeniy M. Ozhegov	Higher School of Economics
Alina Ozhegova	National Research University Higher School of Economics
Anna Panasenko	Novosibirsk State University, Russia
Alexander Panchenko	University of Hamburg, Germany
Panos Pardalos	University of Florida, USA
Marcello Pelillo	University of Venice, Italy
Georgios Petasis	National Centre for Scientific Research Demokritos, Greece
Stefan Pickl	Universität der Bundeswehr München
Lidia Pivovarova	University of Helsinki, Finland
Vladimir Pleshko	RCO LLC, Russia
Alexander Porshnev	National Research University Higher School of Economics, Nizhniy Novgorod, Russia
Surya Prasath	Cincinnati Children's Hospital Medical Center, USA
Andrea Prati	University of Parma, Italy
Artem Pyatkin	Novosibirsk State University and Sobolev Institute of Mathematics, Siberian Branch of Russian Academy of Sciences, Russia
Evgeniy Riabenko	Facebook, UK
Martin Riedl	University of Stuttgart, Germany
Alexey Romanov	University of Massachusetts Lowell, USA
Yuliya Rubtsova	Ershov Institute of Informatics Systems, Siberian Branch of Russian Academy of Sciences, Russia
Alexey Ruchay	Chelyabinsk State University, Russia
Eugen Ruppert	University of Hamburg, Germany
Christian Sacarea	Babes-Bolyai University, Romania
Mohammed Abdel-Mgeed M. Salem	German University in Cairo, Egypt
Andrey Savchenko	National Research University Higher School of Economics, Nizhniy Novgorod, Russia
Friedhelm Schwenker	Ulm University, Germany
Oleg Seredin	Tula State University, Russia
Andrey Shcherbakov	University of Melbourne, Australia
Henry Soldano	Université Paris-Nord, France
Alexey Sorokin	Lomonosov Moscow State University, Russia
Tobias Staron	University of Hamburg, Germany

Dmitry Stepanov	Ailamazyan Program Systems Institute of Russian Academy of Sciences, Russia
Vadim Strijov	Dorodnicyn Computing Centre of Russian Academy of Sciences, Russia
Diana Sungatullina	Skolkovo Institute of Science and Technology, Russia
Yonatan Tariku	University of Central Florida, USA
Irina Temnikova	Qatar Computing Research Institute, Qatar
Rocco Tripodi	University of Venice, Italy
Diana Troanca	Babes-Bolyai University, Romania
Christos Tryfonopoulos	University of Peloponnese, Greece
Elena Tutubalina	Kazan Federal University, Russia
Dmitry Ustalov	University of Mannheim, Germany
Wil van der Aalst	RWTH Aachen University, Germnay
Sebastiano Vascon	University Ca' Foscari of Venice, Italy
Natalia Vassilieva	Hewlett Packard Labs, Russia
Dmitry Vetrov	Lomonosov Moscow State University and National Research University Higher School of Economics, Russia
Ekaterina Vylomova	University of Melbourne, Australia
Gui-Song Xia	Wuhan University, China
Roman Yangarber	University of Helsinki, Finland
Marcos Zampieri	University of Wolverhampton, UK
Alexey Zobnin	National Research University Higher School of Economics, Russia
Nikolai Zolotykh	University of Nizhniy Novgorod, Russia

Additional Reviewers

Mikhail Batsyn
Marc Chaumont
Ivan Grechikhin
Alexander Rassadin
Yuri Rykov
François Suro
Arsenis Tsokas

Organising Committee

Rostislav Yavorskiy (Conference Chair)	National Research University Higher School of Economics, Russia
Andrey Novikov (Head of Organisation)	National Research University Higher School of Economics, Russia
Marina Danshina (Venue Organisation and Management)	Moscow Polytechnic University, Russia

Anna Ukhanaeva
(Information
Partners and
Communications)

National Research University Higher School of Economics,
Russia

Anna Kalenkova
(Visa Support and
International
Communications)

National Research University Higher School of Economics,
Russia

Alexander Gnevshev
(Venue Organisation
and Management)

Moscow Polytechnic University, Russia

Volunteers

Daniil Bannyh	Moscow Polytechnic University, Russia
Ksenia Belkova	Moscow Polytechnic University, Russia
Tatiana Mishina	Moscow Polytechnic University, Russia
Zahar Kuhtenkov	Moscow Polytechnic University, Russia
Maxim Pasynkov	Krasovsky Institute of Mathematics and Mechanics of RAS, Russia, Yekaterinburg
Ivan Poylov	Moscow Polytechnic University, Russia

Sponsors

Golden sponsor

Exactpro

Bronze sponsors

JetBrains
Springer

Invited Talks and Tutorials

Process Mining: Discovering Process Maps From Data

Anne Rozinat

Eindhoven University of Technology and Fluxicon BV, Eindhoven,
The Netherlands
anne@fluxicon.com

Abstract. Data scientists deftly move through a whole range of technologies. They know that 80% of the work consists of the processing and cleaning of data. They know how to work with SQL, NoSQL, ETL tools, statistics, scripting languages such as Python, data mining tools, and R. But for many of them Process Mining is not yet part of the data science toolbox.

We explain what Process Mining is, how it works, and what data science teams will be able to do with it.

Keywords: Process mining · Data science · Data processing

Tutorial: Supervised Methods in Natural Language Processing

Ekaterina Chernyak

National Research University Higher School of Economics and Sberbank,
Moscow, Russia
ek.chernyak@gmail.com

Abstract. In this talk we are going to address several aspects of applying supervised methods to natural language processing tasks. We will start by describing several types and tools for text annotation and evaluation of annotation quality. Next we will approach a few supervised tasks, specially, text classification and sequence labelling, describe them formally and discuss both traditional machine learning approaches as well as deep learning approaches to them.

Keywords: Natural language processing · Text annotation · Text classification Sequence labelling

Cars, Drivers, Vehicles, and Wheels: Specializing Distributional Word Vectors for Lexico-Semantic Relations

Goran Glavaš

University of Mannheim, Germany
goran@informatik.uni-mannheim.de

Abstract. In recent years, prediction-based distributional word vectors (i.e., word embeddings) have become ubiqiutous in natural language processing. While word embeddings robustly capture existence of semantic associations between words, they fail to reflect – due to the distributional nature of embedding models – the exact type of the semantic link that holds between the words, that is, the exact semantic relation (e.g., synonymy, antonymy, hypernymy). This talk presents an overview of recent models that fine-tune distributional word spaces for specific lexico-semantic relations, using external knowledge from lexico-semantic resources (e.g., WordNet) for supervision. I will analyze models for specializing embeddings for semantic similarity (vs. other types of semantic association) as well as models that specialize word vectors for detecting particular relations, both symmetric (e.g., synonymy, antonymy) and asymmetric (e.g., hypernymy, meronymy). The talk will also examine evaluation procedures and downstream tasks that benefit from specializing embedding spaces. Finally, I will demonstrate how to transfer embedding specializations to resource-lean languages, for which no external lexico-semantic resources exist.

Keywords: Natural language processing · Word embeddings
Lexico-semantic relations

Contents

Analysis of Images and Video

General Topics of Data Analysis

Analysis of Dynamic Behavior Through Event Data

Optimization Problems on Graphs and Network Structures

Innovative Systems

Opening Talk

Visualization of Data Science Community in Russia

Rostislav Yavorskiy[✉], Tamara Voznesenskaya, and Kirill Rudakov

Faculty of Computer Science, Higher School of Economics,
Myasnitskaya 20, Moscow 101000, Russia
{ryavorsky,tvoznesenskaya}@hse.ru, rudakovkirillx@gmail.com

Abstract. We present in the form of two visualizations some preliminary results of the ongoing study of data science community in Russia. The first visualization aggregates data about top researches and their fields of interest according to the Google Scholar service. The second graph is a map of largest on-line communities on data science on VKontakte platform.

Keywords: Scientometric · Data science · Professional community
On-line community · Visualization

1 Overview of the Project

Importance and high demand for data science competencies have been widely discussed in literature, see e.g. [4,5]. It was shown in [1] that the ascendancy of market and hierarchy principles in the organization of professional work has not diminished the role of community. Important role of community in education, research and professional development has been studied for many years, see [2,3,10,11].

Systematic study of regional community of data science professionals provides useful information to:

- students, who build their career path and search for scientific supervisor or an active research group to join;
- scientist, who are looking for new collaboration and cooperation opportunities;
- research managers who want to know the current situation in the field, trends, strong and weak spots, and better plan development programs and projects.

This project continues our previous research on measuring and analysis of the structure of computer science community in Russia, see [7–9]. An overview of scientometric research in Russia could be found in [6].

© Springer Nature Switzerland AG 2018
W. M. P. van der Aalst et al. (Eds.): AIST 2018, LNCS 11179, pp. 3–9, 2018.
https://doi.org/10.1007/978-3-030-11027-7_1

2   Education

Good education is the key factor in facilitating research, so one of our goals is to measure data science education opportunities across the country. We are collecting data from the official web sites of Russian universities. Some bachelor and master programs have vary clear focus on data science and include that by using specific keywords in their titles, e.g. "data analysis", "big data", "data science" etc. For other the relation to data science is not that clear, that is why we manually inspected all the programs on applied mathematics and informatics.

The following Table 1 presents some of the numbers. In order to make the data for different cities better comparable we have added a column with figures for enrollments per 100k citizens rate.

Table 1. The numbers of annual enrollments in data science related bachelor and master programs in major Russian cities

City	Enrolment per year	City population x1000	Enrollments per 100k citizens
Moscow	2946	12381	23.79
St. Petersburg	892	5282	16.89
Nizhniy Novgorod	463	3248	14.25
Samara	342	3204	10.67
Tomsk	308	1079	28.54
Kazan	212	3885	5.46
Irkutsk	185	2409	7.68
Ekaterinburg	180	4329	4.16
Chelyabinsk	151	3502	4.31
Krasnoyarsk	145	2875	5.04
Perm	131	1007	13.01
Novosibirsk	119	1511	7.88

3   Data Science Researchers

In order to get data about data science researcher working in Russia Google scholar service have been used.

Table 2 lists top 10 scientists with respect to the total number of citations indexed by this service. Among researchers with confirmed Russian affiliations we identified 148 with at least one hundred citations. All they are depicted on Fig. 1 by circles. Size of each circle node is proportional to the total number of citations in Google Scholar. Fill color indicates the region of the affiliation institute or university.

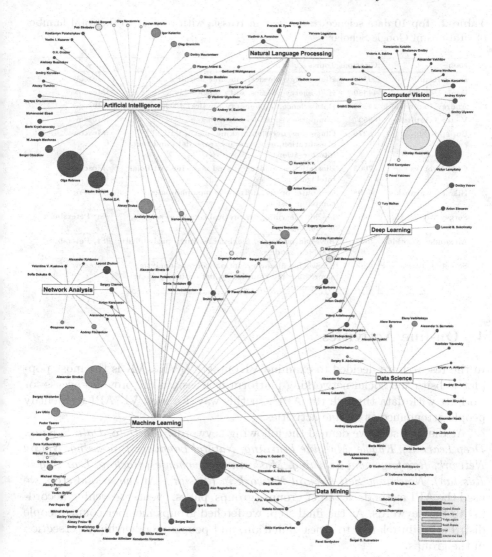

Fig. 1. Map of data science researchers in Russia by field. Size of the circle nodes is proportional to the total number of citations in Google Scholar (minimum is 100). Fill color indicates the region of the affiliation institute or university. (Color figure online)

Table 2. Top 10 data science researchers in Russia with respect to the total number of citations in Google Scholar

Name	Citations	Topics	City
Fedor Ratnikov	99339	High energy physics, machine learning	Moscow
Denis Derkach	44480	High energy physics, data science	Moscow
Andrey Ustyuzhanin	14238	Data science, high energy physics, machine learning	Moscow
Victor Lempitsky	8597	Computer vision, deep learning	Moscow
Boris Mirkin	8276	Clustering, decision making, mathematical classification, evolutionary trees, data and text interpretation	Moscow
Olga Rebrova	7003	Biostatistics, evidence based medicine, artificial intelligence	Moscow
Nikolay Kazanskiy	6408	Computer optics, diffractive nanophotonics, computer vision	Samara
Sergey Nikolenko	5499	Machine learning, theoretical computer science, networking	St. Petersburg
Alexander Sirotkin	5479	Bayesian inference, machine learning, probabilistic graphical models	St. Petersburg
Sergei O. Kuznetsov	4360	Knowledge discovery, formal concept analysis, algorithms and complexity, data mining, discrete mathematics	Moscow

4   On-line Activities

In the research we decided to examine VKontakte[1] (VK). VK is the most popular website and social networking platform in Russia[2]. VK provides access for working with data through the API[3]. In this way, we used VK API to identify people attendance at on-line activities.

At the first step, we defined the following keywords: *Big Data, Data Science, Deep Learning, Kaggle, Machine Learning, Natural Language Processing, Neural Network, Reinforcement Learning, Mashinnoe Obuchenie (Machine Learning in Russian), Nejronnye Seti (Neural Network in Russian)*. At the second step, we used the API methods at target groups (groups, pages, events) searching according to the keywords. At the final step, we fetched groups member count, people distribution in relative to cities affiliation, and people membership intersection in the groups.

The following Table 3 presents top five data science on-line communities on VK platform (those which have 5000 or more members).

We found 19 on-line communities with 1000 members or more (as of June 2018). They are depicted on Fig. 2. Size of a node is proportional to the number of involved users. Edges are used to indicate popularity of these communities in major Russian cities.

---

[1] https://vk.com.
[2] https://www.alexa.com/siteinfo/vk.com.
[3] https://vk.com/dev/methods.

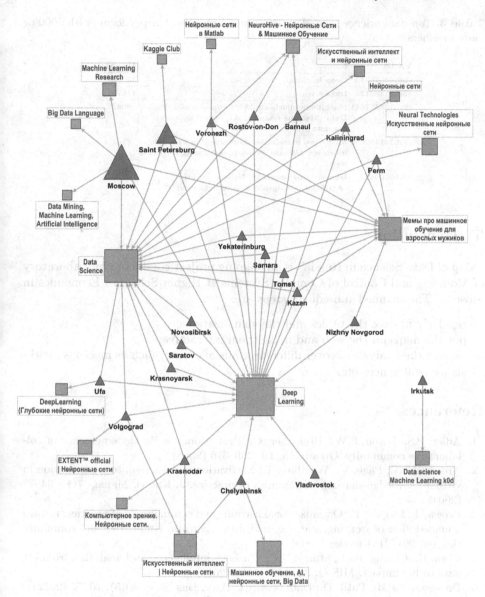

Fig. 2. Map of on-line communities on data science with minimum 1000 members on VK platform. Size of a node is proportional to the number of involved users. Edges indicate popularity in major Russian cities.

Table 3. Top data science related on-line communities on VK platform (with 5000 or more members)

VK ID	Name	Members
club44016343	Deep learning	14098
club91453124	Data science	11678
club138477641	Mashinnoe obuchenie, AI, nejronnye seti, Big Data (Machine learning, AI, neural networks, big data)	8042
club131489096	Memy pro mashinnoe obuchenie dlja vzroslyh muzhikov (Memes on machine learning for adult men)	7141
club55604921	Iskusstvennyj intellekt, Nejronnye seti (Artificial intelligence, neural networks)	6812

5   Conclusion

"Map of Data Science in Russia" is an ongoing project conducted by Laboratory of Modeling and Control of Complex Systems at Higher School of Economics in Moscow. The planned subsequent steps are:

- regularly update the tables and the visualizations;
- put the maps on the web and make them interactive;
- expand the analysis to cover different kinds of events, such as meetups, workshops, conferences etc.

References

1. Adler, P.S., Kwon, S.W., Heckscher, C.: Professional work: the emergence of collaborative community. Organ. Sci. **19**, 359–376 (2008)
2. Ardichvili, A., Page, V., Wentling, T.: Motivation and barriers to participation in virtual knowledge-sharing communities of practice. J. Knowl. Manag. **7**(1), 64–77 (2003)
3. Brown, J., Duguid, P.: Organizational learning and communities of practice: toward a unified view of working, learning, and innovation. In: Knowledge and communities, pp. 99–121. Elsevier (2000)
4. Chen, H., Chiang, R.H., Storey, V.C.: Business intelligence and analytics: from big data to big impact. MIS Q. **36**, 1165–1188 (2012)
5. Davenport, T.H., Patil, D.: Data scientist. Harv. Bus. Rev. **90**(5), 70–76 (2012)
6. Guskov, A., Kosyakov, D., Selivanova, I.: Scientometric research in Russia: impact of science policy changes. Scientometrics **107**(1), 287–303 (2016)
7. Krasnov, F., Vlasova, E., Yavorskiy, R.: Connectivity analysis of computer science centers based on scientific publications datafor major Russian cities. Procedia Comput. Sci. **31**, 892–899 (2014)
8. Krasnov, F., Yavorskiy, R.: Measuring the level of maturity of the professional community. Bus.-Inform. **1**(23), 64–67 (2013). (in Russian)
9. Krasnov, F., Yavorskiy, R.E., Vlasova, E.: Indicators of connectivity for urban scientific communities in Russian cities. In: Ignatov, D.I., Khachay, M.Y., Panchenko, A., Konstantinova, N., Yavorskiy, R.E. (eds.) AIST 2014. CCIS, vol. 436, pp. 111–120. Springer, Cham (2014). https://doi.org/10.1007/978-3-319-12580-0_11

10. Louis, K.S., Marks, H.M.: Does professional community affect the classroom? Teachers' work and student experiences in restructuring schools. Am. J. Educ. **106**(4), 532–575 (1998)
11. Wenger, E.C., Snyder, W.M.: Communities of practice: the organizational frontier. Harv. Bus. Rev. **78**(1), 139–146 (2000)

Social Network Analysis

Echo Chambers vs Opinion Crossroads in News Consumption on Social Media

Sofia Dokuka[1]([⊠]), Sergei Koltcov[2], Olessia Koltsova[2], and Maxim Koltsov[2]

[1] Institute of Education, Higher School of Economics, Moscow, Russia
sdokuka@hse.ru
[2] Laboratory for Internet Studies, Higher School of Economics, Saint-Petersburg, Russia
{skoltsov,ekoltsova,mkoltsov}@hse.ru

Abstract. Social media are often conceived as a mechanism of echo-chamber formation. In this paper we show that in the Russian context this effect is limited. Specifically, we show that audiences of media channels represented in the leading Russian social network VK, as well as their activities, significantly overlap. The audience of the oppositional TV channel is connected with the mainstream media through acceptable mediators such as a neutral business channel. We show this with the data from the VK pages of twelve leading Russian media channels and seven millions users.

Keywords: Social media · Social network analysis · TV · Echo-chamber

1 Introduction and Related Work

Nowadays, the process of information gathering and sharing by media consumers has become simpler than it ever was in the world history. As a result, media audiences can seemingly easily obtain and compare different types of information from different sources.

Despite this, some research has found that social media force users to select social and political content which is not too different from their views (Colleoni 2014), this phenomenon being known as *echo chamber effect*. It refers to users' inclination to select sources of information with similar orientation and to avoid sources with different slants and even agendas. Festinger (1954) explains echo chambers in terms of cognitive dissonance and selective exposure theories. Likewise, Colleoni (2014) claims that people experience positive feelings when they obtain information that supports their opinion and stress when they get different points of views.

The hypothesis about the echo chamber effect facilitated by social media is supported in (Adamic and Glance 2005; Conover et al. 2011; Guerra et al. 2013). Both studies show that social media users tend to link with alters with similar views – an effect also termed *homophily* if this attachment occurs voluntarily (McPherson et al. 2001). Adamic (Adamic and Glance 2005) and Guerra with colleagues (Guerra et al. 2013) provide the evidence of left-right communities with very little mutual contacts. Likewise, a Facebook research (Bakhshi et al. 2015) finds that individuals' choices

W. M. P. van der Aalst et al. (Eds.): AIST 2018, LNCS 11179, pp. 13–19, 2018.
https://doi.org/10.1007/978-3-030-11027-7_2

play a greater role than algorithmic ranking in limiting users' exposure to cross-cutting content. At the same time, this research finds that Facebook users do get exposed to some discordant content and have some heterogeneous friends.

More detailed research finds that homophily levels vary depending on message content and style, as well as across social groups (Barberá et al. 2015; Liao et al. 2014). Thus, lower homophily was observed for non-political and moderately discordant content and among users with high accuracy motives. A study of the Russian blogosphere (Etling et al. 2010) found it to be very connected across all political clusters and very dissimilar to the bi-polar American structure described in Adamic and Glance (2005). The situation opposite to the echo chamber effect has been termed "*national conversation*" (Barberá et al. 2015) and "*opinion crossroads*" (Bodrunova et al. 2015) and is the main alternative theory to explain online political landscapes.

In this study, we seek to test the echo chamber effect in news-related behavior in the largest Russian social networking site VK based on five different measures. Namely, we seek to determine how news consumption on the VK pages of the twelve leading TV channels overlaps in terms of common subscribers, likers, commenters and likes on comments. If individual TV channels or politically homogeneous groups of channels have unique sets of users that do not overlap with other channels or groups of channels, the echo chamber effect hypothesis will hold. If otherwise, the opinion crossroads hypothesis will be more plausible.

2   Data

We perform our analysis on a dataset from VK social networking site (vk.com, a replica of Facebook in terms of its functionality) that has more than 450 millions accounts. We select twelve key Russian media outlets that have the most visible presence in VK, and they happen to be ten TV channels, one news agency and one media organization that is both. We consider only channels with pronounced political agenda. In terms of political landscape, Russia is now divided into a conservative pro-government majority, and an oppositional minority that is mostly liberal, with both "liberal" and "conservative" having different meanings than usually in the "West". Accordingly, we find that only TV Rain is oppositional and liberal (which is why it was deprived of its air frequency and now is an entirely Internet TV channel). RBC is a business channel that tries to keep objective and politically neutral. All the rest may be classified as pro-government. RT was initially an English language TV aimed only at international audience, before it started producing Russian language content in VK in 2017.

For each VK page belonging to a channel we have collected the following information: the lists of users who follow the page, as well as those who have at least once reposted a page's post, or commented on one, or liked either a page's post or a comment to a page's post. The overall dataset embraces the period of one year (2017) and consists of 7,323,983 users. Descriptive statistics for each channel is in Table 1. Such dataset gives us an option to consider different forms of individual participation, both active (reposting, liking and commenting) and passive (following the pages).

After collection, user data were disconnected from their VK IDs and thus anonymized.

Table 1. User activity and subscribers distribution over media pages in VK.

Media outlet	Page audience	Number of comments	Number of reposts	Number of likes to posts	Number of likes to comments
RIA Novosti	2001980	36161	37742	182270	54649
Channel 1	1559147	10654	14145	64612	17887
RT	1001380	26729	15172	74828	36590
RBC	590878	5611	17513	69230	9107
Rain	373963	9360	9327	41647	15305
NTV	299177	5162	6572	23903	7017
Vesti	267568	7752	6856	27572	10639
Channel 5	91809	3248	3919	13461	3537
Russia 1	87573	6203	6349	20153	6846
Kultura	43155	384	2519	6387	464
Mir	23005	227	704	2312	146
TVC	9274	135	255	772	83

Descriptive statistics shows that TV channels are unevenly represented in VK. Some TV channels (Channel 1 or RT) have more than one million subscribers, while other (TVC, Mir or Kultura) have only several thousands.

3 Methods and Results

A database with information about users and their interactions with the channels' pages can be viewed as an adjacency matrix of two-mode (or affiliation) network with two different types of nodes: social network users and public pages. To investigate the structure of channels' similarity in terms of their audiences, we project such two-mode networks into one-mode page-page social networks. As the result we obtain five full weighted graphs with 12 nodes and 66 ties each where the weight of a tie indicates, respectively, the number of subscribers, content sharers, commenters, post likers, or comment likers shared by a given pair of pages. The strength of the ties varies from 0.1% to 5%.

3.1 Relation Between Different Types of Overlaps Between Channels

Theoretically, any two channels may overlap in some aspects (e.g. likes), but not in others (e.g. subscribers). To check this assumption, we test whether the structures of our five networks are correlated. For this, we perform quadratic assignment procedure (QAP-test) (Krackhardt 1988; Simpson 2001) that allows to correlate sets of non-independent observations (Table 2). The empirical estimates show that the different networks are significantly interconnected (correlation is in range 0.84–0.98) which means that if any two channels overlap in one aspect they will also tend to overlap in all others with similar strength.

Table 2. Correlations between the five types of networks according to QAP-test.

	Co-membership	Reposts	Comments	Likes	Likes to comments
Co-membership	–	0.89***	0.84**	0.90***	0.85***
Reposts		–	0.92**	0.96***	0.91***
Comments			–	0.94***	0.98***
Likes				–	0.91*
Likes to comments					–

*p-value < 0.05, **p-value < 0.01, ***p-value < 0.001

3.2 Communities of the Most Overlapping Media Channels

We use Jaccard index to calculate the strength of pairwise intra-channel overlap, while we perform community detection to find clusters of the most overlapping channels. Jaccard index is calculated as the number of common elements in a pair of sets divided by the total number of unique elements in both sets. It is thus sensitive to the difference in the sizes of the compared sets, which is why we, instead of looking at its highest absolute values, find the most similar (overlapping) pair for each channel. First, we confirm the results of the QAP test finding out that such a pair is the same across all or most aspects (likes, shares etc.) for each given channel: Channel 1 and TVC – 3/5, Mir – 4/5, others except Kultura – 5/5. Kultura's similarity pattern is inconsistent. Second, when measured by the sum of Jaccard indices across all aspects, the most similar pair is nearly always much more similar to the given channel than all the rest. An exception is Channel 1 which overlaps with NTV nearly as strongly as it does with its most similar pair (RT). By pointing an arrow from each given channel (except Kultura) to its most overlapping pair we obtain the following scheme:

1. [Rain ↔RBC]
2. [RIA Novosti ↔ RT] ← Channel 1
3. [Channel 5 → NTV ⇆ Vesti ↔ Russia 1]
4. [TVC ↔ Mir]

We thus observe two closed dyads, as well as a dyad and a quartet linked by Channel 1 for which we also show its second strongest link. When other second strongest links are added, more ties emerge between neighboring lines, but never between lines 1 & 3, 2 & 4 or 1 & 4 (Kultura gets linked to lines 3 and 4). Also, mean values of Jaccard index are low in all networks (0.016–0.022, max = 0.1) which indicates overall modest level of audience intersection. Before interpreting these results, we report the results of community detection performed with Louvain algorithm for weighted networks (Blondel et al. 2008). The cells in Table 3 contain labels of the clusters detected in each of five networks. If the labels for different social networks are the same, it means that these two channels are from the same clusters in both social networks.

Table 3. Clusters of the most overlapping media channels in VK.

Channel	Subscribers	Reposts	Comments	Likes to posts	Likes to comments
RIA Novosti	1	1	1	1	1
Rain	1	1	1	1	1
Mir	2	2	3	2	3
Vesti	2	2	2	2	2
Russia 1	2	2	2	2	2
RBC	1	1	1	1	1
Channel 1	1	2	2	2	2
NTV	2	2	2	2	2
Channel 5	2	2	2	2	2
RT	1	1	1	1	1
Kultura	2	2	3	2	3
TVC	2	2	3	2	3

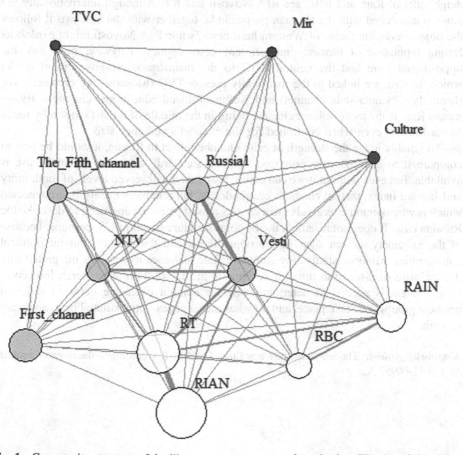

Fig. 1. Community structure of the likes to comments network projection. The size of the node is proportional to the number of likes to the comments of the corresponding channel. Same colored nodes are from the same cluster. The strength of the tie is proportional to the number of users who liked comments on both channels. (Color figure online)

We see that, first and foremost, cluster structure is weakly expressed (modularity values are: co-membership network 0.14; comments and likes networks 0.1; reposts and likes on comments networks 0.08). At the same time, cluster composition is relatively stable across different networks, except that a third cluster occurs in comments and in likes to comments networks (Fig. 1), but again with the same set of members.

4 Discussion and Conclusion

In this paper we investigate the overlap of news consumption between the leading media channels represented in VK social network and try to relate those overlaps with media political stances. A difficulty of this research is that media political stances are not as easily defined in Russia as they are e.g. in the US. Nevertheless, it is obvious that oppositional Rain is reliably connected to RBC that tries to serve business audience irrespective of its political views. The two channels that emerge as the nearest neighbours of Rain and RBC are RIA Novosti and RT. Although internationally the latter is associated with the Russian propaganda, together with the first two it follows the objective-looking style of Western hard news, while RIA Novosti might exploit its lasting reputation of formerly independent news agency. Anyway, both link the oppositional Rain and the neutral RBC to the mainstream pro-government media which, in turn, are linked to the marginally present TVC (Moscow city channel), Mir (formerly CIS inter-state channel) and Kultura (official educational channel). By no means Rain is the most isolated channel being in the middle of the list (sorted by mean Jaccard index) even when controlled for the channel's audience size.

To reliably judge the strength of echo chamber effect in Russia, it would be best to compare it to an obviously polarized "golden standard" case, but no such case is available. But even without it we can claim, based on the observed levels of modularity and Jaccard index, that all channels are modestly, but relatively evenly interconnected which is why opinion crossroads theory looks a most plausible approach to describe the Russian case. It does not resemble the bi-polar structure found in US, perhaps, because of the extremely uneven sizes of the oppositional and the pro-government political communities. Minority status may stimulate Rain audience to be more integrated into the mainstream discussion through acceptable mediators. Further research into comment content is needed to determine whether opinion exchange between different political groups does take place, and on what media pages - oppositional or mainstream or both.

Acknowledgement. The reported study was funded by RFBR according to the research project № 18-011-00997 A.

References

Adamic, L.A., Glance, N.: The political blogosphere and the 2004 US election: divided they blog. In: Proceedings of the 3rd International Workshop on Link Discovery, pp. 36–43. ACM, August 2005

Bakhshi, S., Shamma, D.A., Kennedy, L., Gilbert, E.: Why we filter our photos and how it impacts engagement. In: ICWSM, pp. 12–21, April 2015

Barberá, P., Jost, J.T., Nagler, J., Tucker, J.A., Bonneau, R.: Tweeting from left to right: is online political communication more than an echo chamber? Psychol. Sci. **26**(10), 1531–1542 (2015)

Blondel, V.D., Guillaume, J.L., Lambiotte, R., Lefebvre, E.: Fast unfolding of communities in large networks. J. Stat. Mech.: Theory Exp. **2008**(10), P10008 (2008)

Bodrunova, S.S., Litvinenko, A.A., Gavra, D.P., Yakunin, A.V.: Twitter-based discourse on migrants in Russia: the case of 2013 bashings in Biryulyovo. Int. Rev. Manag. Mark. **5** (Special Issue), 97–104 (2015)

Colleoni, E., Rozza, A., Arvidsson, A.: Echo chamber or public sphere? Predicting political orientation and measuring political homophily in Twitter using big data. J. Commun. **64**(2), 317–332 (2014)

Conover, M., Ratkiewicz, J., Francisco, M.R., Gonçalves, B., Menczer, F., Flammini, A.: Political polarization on twitter. In: ICWSM, vol. 133, pp. 89–96 (2011)

Etling, B., Alexanyan, K., Kelly, J., Faris, R., Palfrey, J., Gasser, U.: Public discourse in the Russian blogosphere: mapping RuNet politics and mobilization. Berkman Center Research Publication, 19 October 2010 http://cyber.law.harvard.edu/publications/2010/Public_Discourse_Russian_Blogosphere

Festinger, L.: A theory of social comparison processes. Hum. Relat. **7**(2), 117–140 (1954)

Guerra, P.H.C., Meira Jr, W., Cardie, C., Kleinberg, R.: A measure of polarization on social media networks based on community boundaries. In: ICWSM, July 2013

Krackhardt, D.: Predicting with networks: nonparametric multiple regression analysis of dyadic data. Soc. Netw. **10**(4), 359–381 (1988)

Liao, Q.V., Fu, W.-T.: Can you hear me now?: mitigating the echo chamber effect by source position indicators. In: Proceedings of the 17th ACM Conference on Computer Supported Cooperative Work & Social Computing, CSCW 2014, 15–19 February 2014, pp. 184–196 (2014)

McPherson, M., Smith-Lovin, L., Cook, J.M.: Birds of a feather: homophily in social networks. Ann. Rev. Sociol. **27**(1), 415–444 (2001)

Simpson, W.: QAP: the quadratic assignment procedure. In: North American Stata Users' Group Meeting, pp. 12–13, March 2001

Joint Node-Edge Network Embedding
for Link Prediction

Ilya Makarov[1,2](✉) ⓘ, Olga Gerasimova[1] ⓘ, Pavel Sulimov[1] ⓘ,
Ksenia Korovina[1] ⓘ, and Leonid E. Zhukov[1] ⓘ

[1] National Research University Higher School of Economics, Moscow, Russia
{iamakarov,ogerasimova,psulimov,lzhukov}@hse.ru, ks.korovina@gmail.com
[2] Faculty of Computer and Information Science, University of Ljubljana,
Ljubljana, Slovenia

Abstract. In this paper, we consider new formulation of graph embedding algorithm, while learning node and edge representation under common constraints. We evaluate our approach on link prediction problem for co-authorship network of HSE researchers' publications. We compare it with existing structural network embeddings and feature-engineering models.

Keywords: Graph embedding · Link prediction · Node2vec
Machine learning

1 Introduction

In the last several years, the world has seen drastic development in the number and quality of conferences and journal articles that are available. In order to get acquainted with all of the relevant recent advances, a researcher would need to look through hundreds of papers, which takes up a lot of time. As a consequence, finding co-authors with similar interests may be a demanding task.

The solution to this problem lies in devising a recommender system that would be able to suggest co-authors. We formulate this task as a link prediction problem [1], in which the model should predict the likelihood of an edge appearing between each pair of nodes, as well as the parameters of such an edge. Many applications can be formulated as link prediction problems, including real-world friend search in social networks [2], web hyper-link prediction [3,4], social dating [5], genomics [6,7], and digital libraries [8,9]. For a complete survey on link prediction applications, we refer the reader to [10].

One of the major advances in graph representation learning was the transfer of vectorized language models to graphs through node sequence sampling [11]. These models [11–14] demonstrate state-of-art performance on link prediction, multi-class node classification, and community detection tasks.

The work was supported by the Russian Science Foundation under grant 17-11-01294 and performed at National Research University Higher School of Economics, Russia.

W. M. P. van der Aalst et al. (Eds.): AIST 2018, LNCS 11179, pp. 20–31, 2018.
https://doi.org/10.1007/978-3-030-11027-7_3

In this article we construct graph-based co-authorship recommendation system that takes into account not only the similarities, but also constructed embeddings. In what follows, we describe in detail the process of evaluating the recommender system based on co-authorship network and information on the staff units profiles of each author from HSE (Higher School of Economics).

2   Related Work

2.1   Link Prediction

The link prediction problem was first formulated in [15], which suggested the approach based on measures for analyzing the nodes proximity in a network. The described experiments on large co-authorship networks showed that information on future links could be extracted from network structure alone.

In [16], the authors presented a survey on link prediction methods categorized on three types of models: feature extraction for binary classification, Bayesian graphical models, and an approach of dimensionality reduction from linear algebra matrix operations. The survey on classical link prediction was made in [17].

In the next section we describe two most studied approaches for link prediction task for large-scale networks, providing feasible solutions based on state-of-the-art techniques for Homophily and Graph Embedding, while leaving the up-to-date survey on link prediction to reader [18].

2.2   Actor Homophily and Graph-Based Recommender Systems

The simplest baseline solution is based on Common Neighbors or other network similarity scores [19]. A number of coded solutions was developed in a standard part of NetworkX Python package [20]. In [21], the authors showed that the similarity measures are highly influenced by the type of network, and could give a wide spread in the results depending on network properties, thus being an inaccurate standalone feature.

The impact of the attribute-based formation of social networks rather than only self-organizing (according to power-law and preferential attachment) was considered in [22]. The main idea is to consider attribute properties such as attribute homophily, one of the most known empirical observations from the real world [23]. The different assortative mixing patterns based on age, sex, group interest, and even romance relations, have been studied as a part of emerging social network analysis [24]. Mostly, all these approaches require manually engineered feature selection from the problem domain. A modern approach for computationally efficient link prediction via feature engineering was suggested in [25].

A series of works on recommender systems via link prediction problem [1,26,27] started the theory of graph recommender, which showed state-of-art performance based on extracting structure information from the networks.

In [28], the authors studied the effect of homophily in a university community, considering temporal co-authorship network accompanied by staff and affiliation information. The authors concluded that not only structural proximity, but also an orientation to form links with actors possessing similar attributes highly impacts the social network structure over time.

We follow this idea while studying the large HSE research community, creating staff units and researcher interests' attributes as a foundation for actor-based homophily.

2.3 Graph Embeddings

In supervised machine learning problems, one has to construct a feature vector in order to represent network nodes and edges. Usually hand-engineering features based on domain expert knowledge would be required. Their applicability would be highly influenced by particular tasks and by quality of an expert and size of the data. Nowadays, the development of digital technologies leads to aggregating big data that could not be manually engineered, that is why the theory of finding hidden representations (see [29]) has impacted the whole field of machine learning and artificial intelligence community. The challenge in feature learning via optimization problem consists of defining loss function while simultaneously supporting robustness of inner representation and domain generalization. Local Linear Embedding [30], IsoMAP [31], Laplacian Eigenmap [32], Spectral Clustering [33], MFA [34] and GraRep [35] were the first attempts to embed the graphs into vector space based on several locality measures. However, development of representation learning for networks has major drawbacks when considering dimensionality reduction techniques such as PCA [36], SVD [37], MDS [38], LDA [39], etc. Based on eigenvalue-based decomposition of the corresponding network matrix, which is not computationally efficient for large-scale networks and have low accuracy in many classification tasks, these methods fail to preserve robustness of the model for noised edge data of the network.

Recent papers on network embeddings concentrate mostly on improving performance on several typical machine learning tasks based on describing network in terms of random walking. In DeepWalk [11] and node2vec [14] algorithms, the authors use Skip-gram [40] based model to simulate breadth-first sampling (BFS) and depth-first sampling (DFS). More weak representations, based on the first-order and second-order nodes proximity, were suggested in LINE [12] and SDNE [41] models.

In [42], research group make more thorough investigation of node2vec [14] graph embedding, finding certain drawbacks in the suggested fine-tuning of hyper-parameters for balancing BFS and DFS. In [43], the authors argue that node2vec model should be accompanied by a vectorized model of the whole network.

While the connections between node and edge embedding are still to be studied (as mentioned in [14]), we concentrate on applying graph and feature embedding methods to a link prediction task on a particular network of HSE researchers co-authorship network with many additional attributes on actors. Purely network-based methods fail to include the information from actors obtained from other sources, resulting in decrease in efficiency of network embeddings. In what follows, we consider several approaches aimed to combine structure and attribute information of network actors.

2.4 Combined Deep Learning Architectures

Several approaches were suggested recently to incorporate actor's label and text information, such as text-associated DeepWalk model TADW [44] and more advanced label informed attributed network embedding LANE [45]. In tri-party deep network representation TriDNR [46], the authors proposed to separately learn embeddings from DeepWalk [11] and label and content embedding modeled by Doc2Vec [47] model. On the contrary, social network embedding framework ASNE [48] learns representations for social actors by preserving both the structural equivalence and node attribute proximity simultaneously in end-to-end neural network architecture. The semi-supervised learning for network embedding has been studied in [49,50].

3 Data Processing and Feature Generation

In this section we present information on data extraction from the HSE digital publications archive and feature selection process according to author homophily and structural graph embeddings.

3.1 HSE Publications Digital Library

We started with the database of all the records from the NRU HSE publication portal [51]. The data source is the electronic library of scientific papers published in co-authorship with HSE researchers and placed on this site by one of the co-authors.

The database contains information on 6996 HSE authors who published 30668 research papers, including journals, conference reports, proceedings, monographs, books, preprints and scientific reviews. The portal contains tools for uploading data for a fixed time period. The fields of the database contain information on a title, a list of authors, keywords, abstract, data, place, journal/conference and publishing agency, and several attributes of the paper and journal issue, as well as information on indexing in Scopus, Web of Science (WoS) Core Collection, and Russian Science Citation Index (RSCI).

Unfortunately, the database has not been integrated with any research paper digital libraries and does not support automatically uploading of bibliography descriptions, such as *.ris, *.enw or *.bib files, which results in noisy data representation with heterogeneous descriptions, abbreviated author homonimy, and incomplete information on the publications. The full texts are usually unavailable under security properties chosen by the corresponding authors, while the paper abstracts are not sufficient to be used for topic modelling. When obtaining the necessary information from the article was impossible, we included the information on research interests from the Scopus subject categories of the journals in which authors have published research articles, and manually input the research interest list according to RSCI categorization.

In what follows, we briefly describe the process of cleaning the original database:

1. All the duplicate records were merged under assumption that any conflict could be resolved by choosing the data from verified record at the portal.
2. All the missing fields were omitted during computational part or filled with a median over respective category of articles and authors.

3.2 Actor Attributes and Similarity Scores for Pairs of Vertices

We consider the problem of finding authors with similar interests to a selected one as a foundation for collaboration search. We choose well-known similarity scores described in [19]. The most popular are represented in Table 1.

Table 1. Similarity score for a pair of nodes u and v with local neighborhoods N(u) and N(v) correspondingly, and for vectorized representations of two authors' attributes and research interests X and Y.

Similarity metric	Definition
Common neighbors	$|N(u) \cap N(v)|$
Jaccard coefficient	$\frac{|N(u) \cap N(v)|}{|N(u) \cup N(v)|}$
Adamic-Adar score	$\sum_{w \in N(u) \cap N(v)} \frac{1}{\ln |N(w)|}$
Preferential attachment	$|N(u)| \cdot |N(v)|$
Graph distance	Length of shortest path between u and v
Metric score	$\frac{1}{1+||x-y||}$
Cosine score	$\frac{(x,y)}{||x||||y||}$
Pearson coefficient	$\frac{cov(x,y)}{\sqrt{cov(x,x)} \cdot \sqrt{cov(y,y)}}$
Generalized jaccard	$\frac{\sum min(x_i,y_i)}{\sum max(x_i,y_i)}$

In order to define similarity of actors by their known features from the HSE staff information, publication activity represented by centralities of co-authorship network, and publication activity descriptive statistics, we use additional content-based and graph-based features and corresponding similarity score.

Considering edge representation, firstly we obtain features from similarity scores, and then we pass into the vectorized space (see Table 2) with operators inspired by [14].

We computed the number of publications between two authors and their weighted sum of quartiles for each respective edge (for negative sampling we make it equal to zero) and combined several feature representations for edge

Table 2. Symmetric binary functions for computing vectorized (u, v)-edge representation based on node attribute embeddings $f(x)$ for ith component for $f(u, v)$

Symmetry operator	Definition
Average	$\frac{f_i(u)+f_i(v)}{2}$
Hadamard	$f_i(u) \cdot f_i(v)$
Weighted-L_1	$\lvert f_i(u) - f_i(v) \rvert$
Weighted-L_2	$(f_i(u) - f_i(v))^2$
Neighbor Weighted-L_1	$\left\lvert \frac{\sum\limits_{w \in N(u) \cup \{u\}} f_i(w)}{\lvert N(u) \rvert + 1} - \frac{\sum\limits_{t \in N(v) \cup \{v\}} f_i(t)}{\lvert N(v) \rvert + 1} \right\rvert$
Neighbor Weighted-L_2	$\left(\frac{\sum\limits_{w \in N(u) \cup \{u\}} f_i(w)}{\lvert N(u) \rvert + 1} - \frac{\sum\limits_{t \in N(v) \cup \{v\}} f_i(t)}{\lvert N(v) \rvert + 1} \right)^2$

embedding as a number of similarity scores (compact representation) and the nodes embedding pairwise representation via operators from Table 2.

3.3   Models Using Dual Embeddings

2-GAE Model Description

2-GAE is a model based on several ideas, namely, learning metafeatures with sequence-based methods, and utilizing dual information.

Assume we receive only an adjacency matrix as an input (the graph can be either directed or undirected), and no features for nodes. We may then attempt to generate helper features (meta-features) for nodes in some way, and then apply a master algorithm to obtain final representations. One possible way to do this is to use a first-level algorithm for node embedding as well. Moreover, it is reasonable to choose a light-weight first-level algorithm for learning primary representations, and a more expensive algorithm above it that is capable of making the most use of added features.

Another novel idea we will employ is utilizing the dual graph and representations for its nodes. This way, we not only incorporate neighboring nodes information, but also account for edge similarities.

The model *2-GAE* is as follows. On the first stage, we take either a diffusion-based embedding algorithm *diff2vec* or a biased random walk-based embedding algorithm *node2vec* as the first-level algorithm, and learn both embeddings $f_{seq,primal} : V \to \mathbb{R}^{d_1}$ for the original graph and for its dual graph $f_{seq,dual} : E \to \mathbb{R}^{d_2}$, separately.

On the subsequent stage, we construct features $f : V \to \mathbb{R}^{d_1+d_2}$ for the original graph by concatenating these sequence-based node representations from the former together with average edge representation over edges adjacent to a given node from the latter (see the formula below).

Then the structure data, augmented with these features, is supplied to a variational autoencoder, which is composed of a graph convolutional encoder and an inner product decoder.

Prior to all learning, some edges are withheld from the training data and not used in any computations, to be used later as positive testing samples on link prediction.

$$g_{seq,dual}(u) = \sum_{(u,v)\in\delta(u)} f_{seq,dual}((u,v))$$

$$f(u) = [f_{seq,primal}(u),\ g_{seq,dual}(u)]$$

For directed graphs, one may account for directions by using signed representations and altering the dual component as follows:

$$g_{seq,dual}(u) = \sum_{(u,v)\in\delta_{out}(u)} f_{seq,dual}((u,v)) - \sum_{(v,u)\in\delta_{in}(u)} f((v,u))$$

Results on HSE Dataset

Option 1. Diffusion: More Visible Clustering
Using the model described, we can learn representations only from graph structure and still obtain high quality results (see Table 3).

Table 3. Diffusion model results

Params	Test F1 score	Test Log-Loss	Test ROC-AUC score
400 epochs, 20 diffusions	0.85269	0.45298	0.90189
400 epochs, 40 diffusions	0.85135	0.43415	0.92061

Note again that these results are achieved without using any additional features or edge weights, at the additional cost of having to learn such features.

One more notable characteristic of the learned embeddings is that they form tighter clusters. The following graphic was produced by fitting a two-component TSNE to the resulting embeddings (see Fig. 1).

Option 2. Node2Vec: Uses Edge Weights, Slightly Better Quality
The biased random walk procedure of node2vec admits incorporation of edge weights. To do this weighted random walk on the dual graph as well, we have defined the connection between two dual nodes (that is, two edges) to be the average of their weights. From experiments it can be seen that this approach proves superior on our data:

The can be seen in the Fig. 2, random walk based 2-GAE is less prone to cluster in the embedding space (see Table 4).

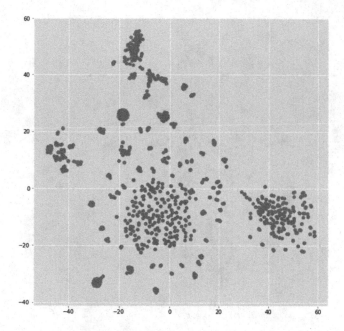

Fig. 1. Diffusion-based 2-GAE induces clustering

Table 4. Embedding model for different random walks

Dual graph	Params	F1 score	Log-Loss	ROC-AUC score
Uniform	200, p $=0.8$, q $=1.2$	0.82943	0.45047	0.91099
Weighted	200, p $=0.8$, q $=1.2$	**0.83697**	**0.44299**	**0.92202**
Uniform	400, p $=0.8$, q $=1.2$	**0.85185**	0.44174	0.91659
Weighted	400, p $=0.8$, q $=1.2$	0.83980	**0.43564**	**0.92537**

Discussion. In Table 5 we could see that the suggested local proximity operator for link embedding that we call Neighbor Weighted-L2 link embedding outperforms all the other approaches for embedding edges based on node vector representations.

While GAE approach showed promising results compared to simple node2vec, it still lacks the ideas on how to incorporate edge information.

There still may be new ideas on how to make use of weights in diffusion or replace inner product decoder $\hat{A} = \sigma\left(ZZ^T\right)$ with neighbor-weighted-L2 expression that we leave for the future work.

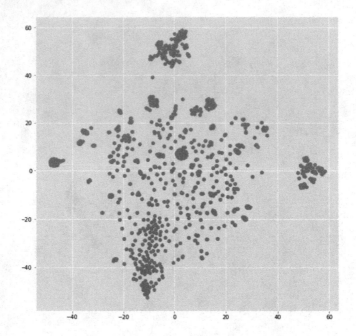

Fig. 2. Random walk based 2-GAE

Table 5. Comparing link embeddings for link prediction problem on HSE dataset

	Precision	Accuracy	F1-score (macro)	F1-score (micro)	Logloss	ROC-AUC
Average	0.628	0.611	0.693	0.738	1.092	0.641
Hadamard	0.721	0.728	0.673	0.687	0.703	0.771
Weighted-L1	0.776	0.765	0.727	0.759	0.376	0.817
Weighted-L2	0.786	0.782	0.751	0.782	0.355	0.827
Neighbor Weighted-L1	0.816	0.834	0.762	0.801	0.214	0.839
Neighbor Weighted-L2	**0.836**	**0.871**	**0.774**	**0.831**	**0.193**	**0.888**

4   Conclusion

We have improved recommender systems [52,53] based on various feature engineering techniques for co-authorship search, with a new model including learned graph embeddings with dual representations.

We are looking forward to the evaluation of our system for several tasks inside the NRU HSE (though, it could be applied to any other research community), such as:

– finding an expert based on text evaluation;
– matchmaking for co-authored research papers with novice researchers;
– searching for scientific adviser based on co-authorship network and the probability of publication in co-authorship with a student;
– searching for collaborators on specific grant proposal.

An application of this system may support novice researchers and increase their publishing activity or be used to estimate the future collaboration of staff units. The combined deep learning approaches for integrating label, text, and network features for advanced graph embedding is left to the future work.

References

1. Li, X., Chen, H.: Recommendation as link prediction: a graph kernel-based machine learning approach. In: Proceedings of the 9th ACM/IEEE-CS Joint Conference on Digital Libraries. JCDL 2009, pp. 213–216. ACM, New York (2009)
2. Backstrom, L., Leskovec, J.: Supervised random walks: predicting and recommending links in social networks. In: Proceedings of the Fourth ACM International Conference on Web Search and Data Mining. WSDM 2011, pp. 635–644. ACM, New York (2011)
3. Adafre, S.F., de Rijke, M.: Discovering missing links in wikipedia. In: Proceedings of the 3rd International Workshop on Link Discovery. LinkKDD 2005, pp. 90–97. ACM, New York (2005)
4. Zhu, J., Hong, J., Hughes, J.G.: Using markov models for web site link prediction. In: Proceedings of the Thirteenth ACM Conference on Hypertext and Hypermedia. HYPERTEXT 2002, pp. 169–170. ACM, New York (2002)
5. Fiore, A.T., Donath, J.S.: Homophily in online dating: when do you like someone like yourself? In: CHI 2005 Extended Abstracts on Human Factors in Computing Systems. CHI EA 2005, pp. 1371–1374. ACM, New York (2005)
6. Vazquez, A., Flammini, A., Maritan, A., Vespignani, A.: Global protein function prediction from protein-protein interaction networks. Nature Biotechnol. **21**(6), 697–700 (2003)
7. Freschi, V.: A graph-based semi-supervised algorithm for protein function prediction from interaction maps. In: Stützle, T. (ed.) LION 2009. LNCS, vol. 5851, pp. 249–258. Springer, Heidelberg (2009). https://doi.org/10.1007/978-3-642-11169-3_18
8. Malin, B., Airoldi, E., Carley, K.M.: A network analysis model for disambiguation of names in lists. Comput. Math. Organ. Theory **11**(2), 119–139 (2005)
9. Elmagarmid, A.K., Ipeirotis, P.G., Verykios, V.S.: Duplicate record detection: a survey. IEEE Trans. Knowl. Data Eng. **19**(1), 1–16 (2007)
10. Srinivas, V., Mitra, P.: Applications of link prediction. Link Prediction in Social Networks. SCS, pp. 57–61. Springer, Cham (2016). https://doi.org/10.1007/978-3-319-28922-9_5
11. Perozzi, B., Al-Rfou, R., Skiena, S.: DeepWalk: online learning of social representations. In: Proceedings of the 20th ACM SIGKDD International Conference on Knowledge Discovery and Data Mining. KDD 2014, pp. 701–710. ACM, New York (2014)
12. Tang, J., Qu, M., Wang, M., Zhang, M., Yan, J., Mei, Q.: LINE: large-scale information network embedding. In: Proceedings of the 24th International Conference on World Wide Web. WWW 2015, pp. 1067–1077. International World Wide Web Conferences Steering Committee, Republic and Canton of Geneva (2015)
13. Chang, S., Han, W., Tang, J., Qi, G.J., Aggarwal, C.C., Huang, T.S.: Heterogeneous network embedding via deep architectures. In: Proceedings of the 21th ACM SIGKDD International Conference on Knowledge Discovery and Data Mining. KDD 2015, pp. 119–128. ACM, New York (2015)

14. Grover, A., Leskovec, J.: Node2vec: scalable feature learning for networks. In: Proceedings of the 22nd ACM SIGKDD International Conference on Knowledge Discovery and Data Mining. KDD 2016, pp. 855–864. ACM, New York (2016)
15. Liben-Nowell, D., Kleinberg, J.: The link prediction problem for social networks. In: Proceedings of the Twelfth International Conference on Information and Knowledge Management. CIKM 2003, pp. 556–559. ACM, New York (2003)
16. Hasan, M.A., Zaki, M.J.: A survey of link prediction in social networks. In: Aggarwal, C. (ed.) Social Network Data Analytics, pp. 243–275. Springer, Boston (2011)
17. Lü, L., Zhou, T.: Link prediction in complex networks: a survey. Phys. A: Stat. Mech. Appl. **390**(6), 1150–1170 (2011)
18. Wang, P., Xu, B., Wu, Y., Zhou, X.: Link prediction in social networks: the state-of-the-art. Sci. China Inf. Sci. **58**(1), 1–38 (2015)
19. Liben-Nowell, D., Kleinberg, J.: The link-prediction problem for social networks. J. Assoc. Inf. Sci. Technol. **58**(7), 1019–1031 (2007)
20. Network Developers: Link prediction algorithms (2017). https://networkx.github.io/documentation/latest/reference/algorithms/link_prediction.html. Accessed 17 Jan 2018
21. Gao, F., Musial, K., Cooper, C., Tsoka, S.: Link prediction methods and their accuracy for different social networks and network metrics. Sci. Program. **2015**, 1 (2015)
22. Robins, G., Snijders, T., Wang, P., Handcock, M., Pattison, P.: Recent developments in exponential random graph (p*) models for social networks. Soc. Netw. **29**(2), 192–215 (2007)
23. McPherson, M., Smith-Lovin, L., Cook, J.M.: Birds of a feather: homophily in social networks. Annu. Rev. Sociol. **27**(1), 415–444 (2001)
24. Wasserman, S., Faust, K.: Social Network Analysis: Methods and Applications, vol. 8. Cambridge University Press, Cambridge (1994)
25. Fire, M., Tenenboim-Chekina, L., Puzis, R., Lesser, O., Rokach, L., Elovici, Y.: Computationally efficient link prediction in a variety of social networks. ACM Trans. Intell. Syst. Technol. **5**(1), 10:1–10:25 (2014)
26. Chen, H., Li, X., Huang, Z.: Link prediction approach to collaborative filtering. In: Proceedings of the 5th ACM/IEEE-CS Joint Conference on Digital Libraries (JCDL 2005), pp. 141–142, June 2005
27. Liu, Y., Kou, Z.: Predicting who rated what in large-scale datasets. SIGKDD Explor. Newsl. **9**(2), 62–65 (2007)
28. Kossinets, G., Watts, D.J.: Origins of homophily in an evolving social network. Am. J. Sociol. **115**(2), 405–450 (2009)
29. Bengio, Y., Courville, A., Vincent, P.: Representation learning: a review and new perspectives. IEEE Trans. Pattern Anal. Mach. Intell. **35**(8), 1798–1828 (2013)
30. Roweis, S.T., Saul, L.K.: Nonlinear dimensionality reduction by locally linear embedding. Science **290**(5500), 2323–2326 (2000)
31. Tenenbaum, J.B., Silva, V.D., Langford, J.C.: A global geometric framework for nonlinear dimensionality reduction. Science **290**(5500), 2319–2323 (2000)
32. Belkin, M., Niyogi, P.: Laplacian eigenmaps and spectral techniques for embedding and clustering. In: Advances in Neural Information Processing Systems, pp. 585–591 (2002)
33. Tang, L., Liu, H.: Leveraging social media networks for classification. Data Min. Knowl. Discov. **23**(3), 447–478 (2011)
34. Yan, S., Xu, D., Zhang, B., Zhang, H.J., Yang, Q., Lin, S.: Graph embedding and extensions: a general framework for dimensionality reduction. IEEE Trans. Pattern Anal. Mach. Intell. **29**(1), 40–51 (2007)

35. Cao, S., Lu, W., Xu, Q.: GraRep: learning graph representations with global structural information. In: Proceedings of the 24th ACM International on Conference on Information and Knowledge Management. CIKM 2015, pp. 891–900. ACM, New York (2015)
36. Karl Pearson, F.R.S.: LIII. on lines and planes of closest fit to systems of points in space. Lond. Edinb. Dublin Philos. Mag. J. Sci. **2**(11), 559–572 (1901). https://www.tandfonline.com/doi/abs/10.1080/14786440109462720?journalCode =tphm17
37. Eckart, C., Young, G.: The approximation of one matrix by another of lower rank. Psychometrika **1**(3), 211–218 (1936)
38. Torgerson, W.S.: Theory and Methods of Scaling. Wiley, Hoboken (1958)
39. Blei, D.M., Ng, A.Y., Jordan, M.I.: Latent dirichlet allocation. J. Mach. Learn. Res. **3**(Jan), 993–1022 (2003)
40. Mikolov, T., Sutskever, I., Chen, K., Corrado, G.S., Dean, J.: Distributed representations of words and phrases and their compositionality. In: Advances in Neural Information Processing Systems, pp. 3111–3119 (2013)
41. Wang, D., Cui, P., Zhu, W.: Structural deep network embedding. In: Proceedings of the 22nd ACM SIGKDD International Conference on Knowledge Discovery and Data Mining. KDD 2016, pp. 1225–1234. ACM, New York (2016)
42. Carstens, B.T., Jensen, M.R., Spaniel, M.F., Hermansen, A.: Vertex similarity in graphs using feature learning (2017). https://projekter.aau.dk/projekter/files/259997796/mi109f17___Vertex_Similarity.pdf
43. Wu, H., Lerman, K.: Network vector: distributed representations of networks with global context. arXiv preprint arXiv:1709.02448 (2017)
44. Yang, C., Liu, Z., Zhao, D., Sun, M., Chang, E.Y.: Network representation learning with rich text information. In: IJCAI, pp. 2111–2117 (2015)
45. Huang, X., Li, J., Hu, X.: Label informed attributed network embedding. In: Proceedings of the Tenth ACM International Conference on Web Search and Data Mining. WSDM 2017, pp. 731–739. ACM, New York (2017)
46. Pan, S., Wu, J., Zhu, X., Zhang, C., Wang, Y.: Tri-party deep network representation. Network **11**(9), 12 (2016)
47. Le, Q., Mikolov, T.: Distributed representations of sentences and documents. In: Proceedings of the 31st International Conference on Machine Learning (ICML 2014), pp. 1188–1196 (2014)
48. Liao, L., He, X., Zhang, H., Chua, T.S.: Attributed social network embedding. arXiv preprint arXiv:1705.04969 (2017)
49. Weston, J., Ratle, F., Mobahi, H., Collobert, R.: Deep learning via semi-supervised embedding. In: Montavon, G., Orr, G.B., Müller, K.-R. (eds.) Neural Networks: Tricks of the Trade. LNCS, vol. 7700, pp. 639–655. Springer, Heidelberg (2012). https://doi.org/10.1007/978-3-642-35289-8_34
50. Yang, Z., Cohen, W.W., Salakhutdinov, R.: Revisiting semi-supervised learning with graph embeddings. arXiv preprint arXiv:1603.08861 (2016)
51. Powered by HSE Portal: Publications of HSE (2017). http://publications.hse.ru/en. Accessed 9 May 2017
52. Makarov, I., Bulanov, O., Zhukov, L.E.: Co-author recommender system. In: Kalyagin, V., Nikolaev, A., Pardalos, P., Prokopyev, O. (eds.) Models, Algorithms, and Technologies for Network Analysis. PROMS, vol. 197, pp. 251–257. Springer, Cham (2017). https://doi.org/10.1007/978-3-319-56829-4_18
53. Makarov, I., Bulanov, O., Gerasimova, O., Meshcheryakova, N., Karpov, I., Zhukov, L.E.: Scientific matchmaker: collaborator recommender system. In: van der Aalst, W.M., et al. (eds.) AIST 2017. LNCS, vol. 10716, pp. 404–410. Springer, Cham (2018). https://doi.org/10.1007/978-3-319-73013-4_37

Co-authorship Network Embedding
and Recommending Collaborators
via Network Embedding

Ilya Makarov[1,2]([✉]) [iD], Olga Gerasimova[1] [iD], Pavel Sulimov[1] [iD],
and Leonid E. Zhukov[1] [iD]

[1] School of Data Analysis and Artificial Intelligence,
National Research University Higher School of Economics,
3 Kochnovskiy Proezd, 125319 Moscow, Russia
{iamakarov,ogerasimova,psulimov,lzhukov}@hse.ru
[2] Faculty of Computer and Information Science, University of Ljubljana,
Ljubljana, Slovenia

Abstract. Co-authorship networks contain invisible patterns of collaboration among researchers. The process of writing joint paper can depend of different factors, such as friendship, common interests, and policy of university. We show that, having a temporal co-authorship network, it is possible to predict future publications. We solve the problem of recommending collaborators from the point of link prediction using graph embedding, obtained from co-authorship network. We run experiments on data from HSE publications graph and compare it with relevant models.

Keywords: Co-authorship network · Graph embedding
Machine learning · Recommender system

1 Introduction

In this paper, we construct a co-authorship recommender system using information on co-authors from the Higher School of Economics (HSE). While preprocessing data, we try to extract similarity features as well as word [1,2] and graph [3,4] embeddings from network authors, their papers and attributes. We aim to show that combined methods, using not only ordinary features, but also embeddings, can outperform state-of-the-art methods in link prediction problem.

Such a system could be used in expert search [5], constructing a system for matching collaborators [6,7], and also building a system for finding relevant publications.

I. Makarov—The work was supported by the Russian Science Foundation under grant 17-11-01294 and performed at National Research University Higher School of Economics, Russia.

W. M. P. van der Aalst et al. (Eds.): AIST 2018, LNCS 11179, pp. 32–38, 2018.
https://doi.org/10.1007/978-3-030-11027-7_4

2   Preprocessing and Feature Generation

In this section we describe the process of generating data and preparing it for training. A part of HSE network representing the ability to generalize corporate relations by visualizing co-authorship subgraph of the most publishable person (rector of HSE, in fact) and his co-authors (representing leading research laboratories and administrative units) could be seen at Fig. 1.

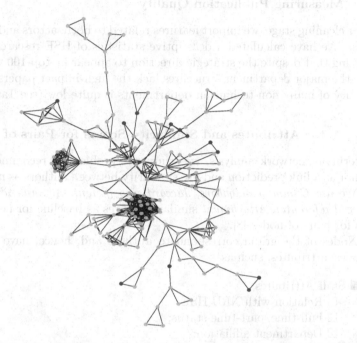

Fig. 1. Visualizing co-authorship network for understanding administrative structure solely based on co-authorship network topology. Green color denotes for HSE workers, while blue color represents other co-authors not belonging to HSE. (Color figure online)

2.1   Author Disambiguation

World-wide, many researchers with the same name or initials work in the same organization, that is why we need to resolve the conflicts of potential author name multiple matching to the existing workers entries, which is called disambiguation. In certain cases when the author is mentioned as HSE worker the database contain the link to the personal web page, which is unique to all the staff members. A number of ambiguous author descriptions was approximately 2% of the whole database. The authors, which have published their work in English and Russian languages, have been transliterated into English equivalent and matched under assumption of small Levenshtein distance measure for Last name and initials.

The disambiguation of the authors with the same initials was solved using simple logistic regression model based on the number of common co-authors in a given dataset, similarity score on the vectors of research interests from both candidates and the category of the published article with ambiguous author description. Finally, we removed all non-HSE authors due to lack of information on their activity and status outside of HSE collaboration.

2.2 Measuring Publication Quality

After cleaning stage, we import features related to both, actors and links between them. We have calculated a descriptive statistics of HSE research publications showing that despite the strategic direction to appear in Top-100 world universities, the major department structures lack the high-impact papers, and research activity of many non-technician departments is quite low (see Table 1).

2.3 Actor Attributes and Similarity Scores for Pairs of Vertices

In terms of network analysis, we study the problem of recommending similar author as a link prediction and use similarity between authors as model features.

We use *Common neighbors, Jaccard's coefficient, Adamic-Adar, Graph distance, Preferential Attachment* similarity scores as baseline for network descriptors for pairs of nodes [8].

Nodes of the graph correspond to authors and, hence, have binarized and numeric attributes, such as

- Staff Attributes
  - ☐ Relation with NRU HSE;
  - ☐ Full-time/part-time status;
  - ☐ Department affiliation;
  - ☐ Administrative, research or lecturer position;
  - ☐ Type of position (lecturer, professor, senior researcher, etc.)
- Network Attributes
  - ☐ Degree centrality;
  - ☐ Betweenness centrality;
  - ☐ Closeness centrality;
  - ☐ Clustering;
- Publication Activity
  - ☐ Number of publications in total;
  - ☐ Number of publications in total in Scopus;
  - ☐ Weighted sum of quartiles of the current publications;
  - ☐ Number of publications in three years;
  - ☐ Number of publications for three years in Scopus;
  - ☐ Number of publications for three years in Scopus Q1, Q2;
  - ☐ Number of publications total in Scopus Q1, Q2;
- Unified vector of Subject Categories as a union of all the categories corresponding to the author papers;

Table 1. HSE research activity on 2015 year

Features\Departments	Education	Computer Science and Math	Economics	Social	Management	Humanities	Law	Media & Design
Number of authors	19.38	19.45	14.15	14.74	12	20.24	12.41	11.36
Number of publications	442.15	434.52	231.63	402.01	187.33	256.88	264.71	146.86
Average publications per author from department	27.66	26.55	21.3	30.51	15.44	13.67	22.3	10.94
Average department authors per publication	0.09	0.06	0.07	0.05	0.09	0.12	0.07	0.1
Average publications per HSE authors	1.71	1.89	1.61	1.76	1.63	1.63	1.56	1.25
Average HSE authors per publications	0.66	0.54	0.65	0.62	0.65	0.71	0.67	0.82
Papers indexed by Scopus	5.08	47	4.5	4.39	2.13	2.79	0.24	1
Papers indexed by Scopus per author	0.21	2.14	0.36	0.29	0.16	0.16	0.03	0.08
Papers indexed by Scopus for the latter 3 years	3.77	13	2.71	2.12	1.41	1.7	0.24	0.57
Papers indexed by Scopus for the latter 3 years per author	0.16	0.59	0.21	0.15	0.12	0.09	0.03	0.07
Papers in Q1, Q2 quartiles in Scopus	2.31	36.07	2.88	2	1.63	2.48	0	1
Papers in Q1, Q2 quartiles in Scopus per author	0.12	1.64	0.22	0.13	0.12	0.13	0	0.03
Papers in Q1, Q2 quartiles in Scopus for the latter 3 years	1.69	9.41	1.4	0.91	1.06	1.7	0	0.57
Papers in Q1, Q2 quartiles in Scopus for the latter 3 years per author	0.1	0.4	0.1	0.06	0.1	0.09	0	0.07
Average co-author number	12.91	13.1	12.71	15.98	9.29	8.21	14	8.89
Number of connected components	9	10.14	7.33	6.7	6.31	9.73	5.29	7.57
Size of the greatest connected component	9.62	7.43	6.65	8.28	5.97	9.91	7.47	4.43
Average distance	1.51	1.43	1.43	1.55	1.35	1.52	1.42	1.19
Diameter	6.31	33.21	7.4	7.97	5.31	6.33	6.18	4.57
Clustering coefficient	0.33	0.34	0.36	0.49	0.44	0.46	0.44	0.31
Density	0.2	0.12	0.15	0.19	0.16	0.14	0.29	0.09
Number of lecturers	3.29	3.75	3.23	2.62	1.81	3.13	2.33	2.6
Number of papers by lecturers	3.3	4.37	2.41	3.7	3.43	4.73	6.17	2.03
Average grade of lecturers (0–30) as 15* (number of papers per 3 years)	25.71	19.99	16.09	17.91	15.69	18.85	19.06	18.25
Number of senior lecturers	3.3	4.76	2.98	4.71	5.43	5.18	7.05	2.03
Number of papers by senior lecturers	25.71	21.81	39.93	22.8	25.1	20.64	21.79	18.25
Average grade of senior lecturers (0–30) as 15* (number of papers per 3 years)	1.43	1.55	2.71	1.67	0.6	6.29	1.29	0.2
Number of assistant professors	25.71	21.81	19.93	22.8	25.1	20.64	21.79	13.25
Number of papers by assistant professors	1.43	1.55	2.71	1.67	0.6	6.29	1.29	0.2
Average grade of assistant professors (0–30) as 10* (number of papers per 3 years)	3.45	4.63	3.44	8	0.5	6.46	0.77	6
Number of professors	1.86	2	2.41	2.19	1.89	6.36	2.8	1
Number of papers by professors	10.81	5.96	4.29	9.72	1.34	6.65	2.24	10
Average grade of professors (0–30) as 6* (number of papers per 3 years)	25.71	19.52	17.06	21.35	10.56	19.2	15.05	10

■ node2vec embedding with tuned p, q hyper-parameters during validation step.

As for edge representation, we use two-sided approach. At first, we compute several similarity scores based on separate node feature representations and obtain exact numerical features. Secondly, similar to graph embedding edge representation, we use the operations for representing elements of Cartesian product of feature spaces as some symmetric function applied by each component of independent vectorized feature space with operators from [3].

We are building 4 different models with following features included:

1. Edge Embedding based on node2vec actor embeddings
2. Edge Embedding based on node2vec embedding for keywords co-occurrence network (only for Scopus dataset)
3. Network Similarity Score
   - Common Neighbors;
   - Jaccard's Coefficient;
   - Adamic-Adar Score;
   - Preferential Attachment;
   - Graph Distance;
4. Author Similarity Scores
   - Cosine Similarity;
   - Common Neighbors;
   - Jaccard's Generalized Coefficient;
   - Pearson's Correlation Coefficient;
   - Metric Score;

For each paper, we choose the most frequent keyword in the co-occurrence network and then construct general embedding using node2vec with automatically chosen parameters.

3   Training Model

In link prediction problem for co-authorship network, we consider the problem of using historical data to predict the future links and their weights according to the suggested publication quality measurement via regression models and neural networks with MSE loss function. We consider machine learning based approaches for evaluating regression/classification models based on regularized linear/logistic regression, SVR/SVM, XGBoost and Neural Networks with fully-connected layers and ELU activators. We have compared training models for the several combinations of edge training data attributes.

The resulting measures are placed in Table 2. The combined approach incorporating information on structural properties, author attributes and keywords embedding outperforms the other, while only structural embedding and node similarity features give less precise results but presenting better micro F1-score.

Table 2. Prediction of publication quality for 2017 year on HSE dataset as a regression problem

	MSE	MAE	MAPE
(1)	0.205	0.191	28.313
(3)	0.407	0.254	35.74
(4)	0.408	0.257	36.82
(1) + (3)	**0.105**	0.143	29.11
(1) + (4)	0.113	**0.133**	**27.65**
(3) + (4)	0.407	0.249	35.25
(1) + (3) + (4)	**0.106**	0.148	28.99

Table 3. Prediction of publications for 2017 year based on previous years information from Scopus dataset

	Precision	Accuracy	F1-score (macro)	F1-score (micro)	Logloss	ROC-AUC
(1)	0.712	0.784	0.761	0.754	0.634	0.816
(3)	0.652	0.745	0.731	0.719	0.682	0.767
(1) + (2)	0.735	0.822	0.775	0.759	0.593	0.836
(1) + (3)	0.728	0.796	0.781	**0.789**	0.618	0.827
(1) + (4)	0.722	0.791	0.772	0.751	0.611	0.819
(3) + (4)	0.683	0.762	0.782	0.781	0.671	0.762
(1) + (2) + (3)	0.742	0.863	0.787	0.751	0.582	0.855
(1) + (2) + (4)	0.738	0.852	0.791	0.731	0.604	0.842
(1) + (2) + (3) + (4)	**0.798**	**0.866**	**0.793**	0.786	**0.573**	**0.878**

In what follows, we evaluate our model on binary classification task, similar to link prediction problem but future graph state. The results of Table 3 show that combined approach also outperform network- or author- only feature spaces.

The combined approach for graph embedding of co-authorship network, author features and research interests lead to drastical improvement in the ROC AUC metric for link prediction task for both, regression and classification tasks.

4   Conclusion

We improved recommender systems presented in [6,7] using different feature engineering techniques including new model of co-authorship graph embeddings. We have compared several machine learning solutions for link prediction task interpreted as binary classification problem and regression task for predicting the quality of possible joint publication. This system may be considered as a recommender system for searching candidates for collaboration based on HSE co-authorship network and database of publications. The recommender system demonstrates promising results on predicting new collaborations between existing authors and the accuracy of the system was improved by adding graph embedding component for extracting structural features from the network. The recommendations could also be made for a new author, who should state research interests and/or load his research papers for topics extraction.

References

1. Wang, C., Blei, D.M.: Collaborative topic modeling for recommending scientific articles. In: Proceedings of the 17th ACM SIGKDD International Conference on Knowledge Discovery and Data Mining, pp. 448–456. ACM (2011)
2. Le, Q., Mikolov, T.: Distributed representations of sentences and documents. In: Proceedings of the 31st International Conference on Machine Learning (ICML-2014), pp. 1188–1196 (2014)
3. Grover, A., Leskovec, J.: Node2vec: scalable feature learning for networks. In: Proceedings of the 22nd ACM SIGKDD International Conference on Knowledge Discovery and Data Mining, KDD 2016, pp. 855–864. ACM, New York (2016)
4. Wu, H., Lerman, K.: Network vector: distributed representations of networks with global context. arXiv preprint arXiv:1709.02448 (2017)
5. Mimno, D., McCallum, A.: Expertise modeling for matching papers with reviewers. In: Proceedings of the 13th ACM SIGKDD International Conference on Knowledge Discovery and Data Mining, pp. 500–509. ACM (2007)
6. Makarov, I., Bulanov, O., Zhukov, L.E.: Co-author recommender system. In: Kalyagin, V., Nikolaev, A., Pardalos, P., Prokopyev, O. (eds.) Models Algorithms and Technologies for Network Analysis. Springer Proceedings in Mathematics and Statistic. Springer, Cham (2017). https://doi.org/10.1007/978-3-319-56829-4_18
7. Makarov, I., Bulanov, O., Gerasimova, O., Meshcheryakova, N., Karpov, I., Zhukov, L.E.: Scientific matchmaker: collaborator recommender system. In: van der Aalst, W.M.P., et al. (eds.) AIST 2017. LNCS, vol. 10716, pp. 404–410. Springer, Cham (2018). https://doi.org/10.1007/978-3-319-73013-4_37
8. Liben-Nowell, D., Kleinberg, J.: The link-prediction problem for social networks. J. Assoc. Inf. Sci. Technol. 58(7), 1019–1031 (2007)

Organizational Networks Revisited: Predictors of Headquarters-Subsidiary Relationship Perception

Antonina Milekhina, Elena Artyukhova[✉], and Valentina Kuskova

National Research University Higher School of Economics,
Moscow, Russian Federation
evartyukhova@hse.ru

Abstract. The problem of effective management of company subsidiaries has been on the forefront of strategic management research since early mid-1980s. Recently, special attention is being paid to the effect of headquarters - subsidiary conflicts on the company performance, especially in relation to the subsidiaries' resistance, both active and passive, to following the directives of the headquarters. A large number of theoretical approaches have been used to explain the existence of intraorganizational conflicts. For example, Strutzenberger and Ambos (2013) examined a variety of ways to conceptualize a subsidiary, from an individual up to a network level. The network conceptualization, at present, is the only approach that could allow explaining the dissimilarity of the subsidiaries' responses to headquarters' directives, given the same or very similar distribution of financial and other resources, administrative support from the head office to subsidiaries, and levels of subsidiary integration. This is because social relationships between different actors inside the organization, the strength of ties and the size of networks, as well as other characteristics, could be the explanatory variables that researchers have been looking for in their quest to resolve varying degrees of responsiveness of subsidiaries, and – in fact – headquarters' approaches – to working with subsidiaries. The purpose of this study is to evaluate the variety of characteristics of networks formed between actors in headquarters and subsidiaries, and their effects on a variety of performance indicators of subsidiaries, as well as subsidiary-headquarters conflicts. Data is being collected in two waves at a major Russian company with over 200,000 employees and several subsidiaries throughout the country.

Keywords: Network relationships · Headquarters
Subsidiaries relations

The article was prepared within the framework of the Basic Research Program at the National Research University Higher School of Economics (HSE) and supported within the framework of a subsidy by the Russian Academic Excellence Project '5-100'.

© Springer Nature Switzerland AG 2018
W. M. P. van der Aalst et al. (Eds.): AIST 2018, LNCS 11179, pp. 39–50, 2018.
https://doi.org/10.1007/978-3-030-11027-7_5

1 Introduction

In today's management studies, a lot of attention is devoted to the subsidiary autonomy and its impact on the interaction with the headquarters of large companies [1,2]. Beginning from the 1990s, it was on the forefront of strategic management studies [3,4], and the wide variety of studies have examined the questions of conflict between parts of organization (e.g., [5]), and more importantly, the refusal of the subsidiaries to comply with headquarters' requests [5].

The importance of this topic cannot be underestimated, as subsidiary performance has been shown to have an impact on the overall company's strategic development [6,7], and the subsidiary autonomy especially was shown as one of the most important factors affecting that performance. In turn, there are a number of factors, shown to have an impact on the subsidiary autonomy: access to resources and long-term commitments [8], level of headquarters' formalization and related control mechanisms [8,9], level of innovativeness, and subsidiary size. While multiple factors of the relationship between subsidiary autonomy and overall company performance were examined, most studies examined the effects of tangible attributes, such as finances (e.g., profitability [10], level of competition [11], etc.), very little is known about the impact of individual employees and their relationships on subsidiary effectiveness. Some studies have shown that an important factor in the subsidiary-headquarters relationship is the social networks built between the main office and its branches [12], but at this point, the exact mechanisms of subsidiary performance that individual relationships affect, remain unclear.

Especially interesting is the idea that social networks built within a subsidiary may help or harm its relationship with headquarters. Several studies have looked at the role of social capital in subsidiary-headquarters conflict (e.g., [3,5]). Other studies looked at the relationship between social networks and knowledge transfer [13]. Yet another group of studies have examined the role of human resource management practices [7]. To the best of our knowledge, however, previous studies have not looked at the employee perceptions of relationships with headquarters, formed within social networks, and how these perceptions affect the headquarters-subsidiary relationships. Understanding the roots of conflict - where they start on personal relationship level - can help the management build more effective professional networks, and as a result, more effective companies.

This paper reports on a study designed to fill an important gap in our understanding of the relationship between organizational social structure subsidiary-headquarters relationship. The study was conducted in a large energy distribution company in Russia and all of its subsidiaries throughout the country. Employees provided data on their networks and important organizational and job characteristics, previously shown to be related to perception of conflict and justice. This is the first study of several, designed to address the important gaps in understanding listed above.

2 Headquarters-Subsidiary Relationship: Theoretical Considerations

According to the broad body of literature on the relationship between company divisions, headquarters (HQ) play a number of important roles, identified by Ghoshal and Bartlett [4]: *implemementor, black hole, contributor, and strategic leader*. These roles are developed based on the dependence of the subsidiary on its environment and access to resources. While this approach is well-established and researched, it leaves very little in terms of the understanding of how the HQ and the subsidiaries formulate their unique relationships.

Another approach describes the closeness of the relationship between HQ and subsidiaries: the more they are dependent on each other, the more of an advantage the company has in the uncertain industry environment. HQ are responsible for creating a unified vision of the company's goals and objectives among its subsidiaries and establish a system of organizational learning [14]. The correspondence between HQ and subsidiaries is established on both the formal and the informal levels. On the formal level, it's the collection of company's rules and procedures; on the informal - trust, personal development, and other intangible human relations constructs that establish the unique organizational culture. Kostova and Roth [14] also establish each subsidiary as a unique network, where there are certain people, called *boundary spanners*, who act as bridges between headquarters and subsidiaries. At first they form their individual social capital, which later, through the mechanisms of knowledge transfer, turn into *public social capital* [14]. This shared vision of the company's goals and objectives allows to increase the level of personal motivation and the resulting company performance.

The "headquarters-subsidiary relations" were also examined from the standpoint of the agency theory [15], where the HQ is the principal and the subsidiary is the agent. This theory is mostly concerned with assymetry of information, where formal rules may create lack of proper communication, and lack of control can generate moral hazard. Several methods are recommended as measures of control from the HQ, taking into account both the formal and the informal relationships between company parts. However, this theory completely ignores the personal relations between managers and employees and the role of social capital.

A large body of literature is devoted to another approach to the HQ-subsidiary relationship, explained by role theory [16]. From this approach, every part of the organization has people who disseminate information in the network. Also, there are network brokers, who coordinate the actions of others in the network; then - innovation sponsors, who generate and distribute new ideas, network structuring agents and many others. This theory makes it possible to explain why and how the organizational network can be managed, and the HQ can use the knowledge generated by testing this theory in order to manage the subsidiary more effectively.

Social networks play an important role in understanding the structure of social exchange [17]. For example, Burt [18] looked at competitive advantage that is created by social capital, and both the positive effects of such capital and negative effects (in case of gossip). In the relationship between headquarters and subsidiaries, the impact of individual networks has not been thoroughly investigated. According to Harzing and Feely [19] only a handful of studies have examined various network-related communication barriers, information distribution problems, and corporate culture differences. However, the impact of networks between individual employees of different organizational units on the organizational effectiveness remains largely unknown. Therefore, this analysis is largely exploratory, designed to look at the relationship between individuals' network characteristics and various measures of organizational outcomes.

3   Method and Analysis

3.1   Sample

For our research sample was gathered within a Russian electricity supplying firm with 8 regional departments and 41 subsidiaries. Unique feature of the firm is its geographic spread within Russia. The company operates only within one country which minimizes the cultural difference of people in subsidiaries. However its geographical spread is very large due to the fact that company is represented in 74 regions of Russian Federation. Such geographical spread is typical of multinational companies (MNC).

Respondents were key employees of manager level and above in subsidiaries. We resulted with 201 usable responds. 104 of respondents were female, average age of 37.3 years. Data collection was not anonymous, but respondents were assured of full confidentiality of data. Questionnaire consisted of 3 parts: sociodemographic (or introductory) part, network information part and survey on perception of headquarter by subsidiary employees. In network part respondents were asked to indicate employees of the firm with whom they have a connection. They were also asked to indicate type (friendship, professional, support and boss-subordinate) and strength (on scale from 1 to 7) of connection. In order to induce the respondents to mention not only coworkers from the same subsidiary, but also from headquarters general questions on working with headquarters were asked in introductory part. As third part of questionnaire we used survey on perception of headquarters based on approach suggested by Roth and Nigh [3]. The survey was conducted in Russian language.

3.2   Dependent Variables

Questionnaire covered a large number of constructs related to perception of conflict between headquarters and subsidiaries by employees. In our final model 5 constructs were included: Perception of rules, Manager interactions, Non-contingent reward and punishment, Feedback and Feedback speed.

Perception of Rules. We measured it on a three item scale. This concept reflects the subsidiary employees' perception of independence level of their actions. The more rules are induced by headquarters the less independent employee feel. This may also lead to the perception of the firm structure as bureaucratic.

Manager Interactions. We measured it on a three item scale. This construct measures the perception of managers' interconnection within the whole firm. The more interconnection leads to higher level of successful collaboration.

Non-Contingent Reward and Punishment. We measured it on a three item scale. This construct reflects the subsidiary employees' perception of fair treatment by headquarters. The more reward or punishment is connected to actual effort, the higher is perception of fair treatment.

Feedback Quality. We measured it on a three item scale. This construct reflects the perception of receiving appropriate feedback from headquarters. Receiving appropriate feedback leads to behaviour correction and performance improvement [20].

Feedback Speed. We measured it on a two item scale. This construct is also related to feedback as previous one, but it reflects different aspect of feedback, namely timeliness of feedback.

All scale items are provided in Appendix.

3.3 Independent Variables

Network characteristics were used as independent variables. Networks characteristics were calculated for each respondent on both four types of relationship and whole network. Examples of extracted egonetworks are presented in Fig. 1. Egonetwork characteristics (Krackhardt's indices, degree, strength of tie) are measured on extracted network of each respondent. Example of a full network on professional level is presented in Fig. 2. Characteristics of full networks (closeness) was measured for each respondent based on constructed full network. Five characteristics included into final model are described below.

Krackhardt-Stern E-I Indices for Network of Friendship and for Network of Support Relationship. Krackhardt-Stern E-I Ratio is a measure of homophily [21]. It measures relationship between external and internal ties. In our research we regard a person from the same region as respondent as internal tie (within-group) and a person from different region as external tie (between-group).

$$Index = \frac{E - I}{E + I}$$

where E is number of external ties (between different regions) and I is number of internal ties (within the respondent's region).

Strength of Ties for Network of Boss-Subordinate Relationship. It measures the weighted degree of nodes in boss-subordinate relationship. Tie weights were measured on 7-point scale of Likert-type. For any node i strength was calculated as:

$$s_i = \sum_{j}^{N} w_{ij}$$

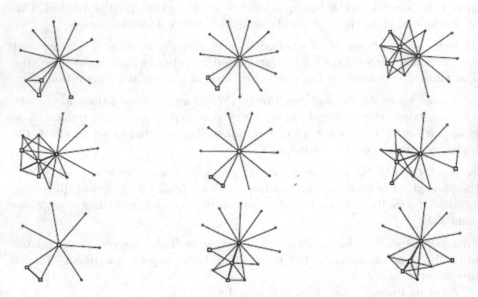

Fig. 1. Example of egonetworks of respondents with professional type of relationship

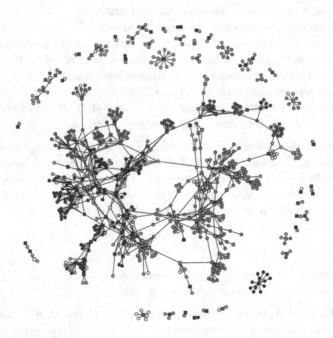

Fig. 2. Example of full network of professional relationships. Squares indicate respondents, circles indicate alters.

where N is the total number of nodes, w_{ij} is the element of the weights matrix of the boss-subordinate relationship network, $j \in [1, N]$, $j \neq i$.

Closeness to High-Ranked Nodes in Network of Friendship. Closeness centrality was developed by Bavelas [22]. Closeness centrality is a measure of shortest paths between the node and all other nodes in a graph. For any node i the closeness centrality was calculated as:

$$c_i = \left[\sum_{j=1}^{N} d(i,j) \right]^{-1}$$

where N is the total number of nodes, $d(i,j)$ is the shortest distance between nodes i and j for all $j \in [1, N]$, $j \neq i$.

Number of Ties (Degree) in Network of Support. Degree is number of ties a person has. For any node i degree was calculated as:

$$d_i = \sum_{j}^{N} x_{ij}$$

where N is the total number of nodes, x_{ij} is the element of the adjacency matrix of the support network, $j \in [1, N]$, $j \neq i$,

$$x_{ij} = \begin{cases} 1, \text{if exists the tie between persons } i \text{ and } j \\ 0, \text{otherwise.} \end{cases}$$

3.4 Analysis

All data were analyzed using structural equation modeling (SEM) in Lisrel 8.8 and network structure was analyzed using R. Network characteristics were then used as inputs into the structural model. This approach has been successfully used in the large number of studies, and description below demonstrates the general approach to SEM. The resulting general estimation procedure for the structural model followed the standard SEM algorithm [23]. The general matrix form for model with latent variables is defined as follows:

$$\Sigma = \left[\frac{\Sigma_{yy} | \Sigma_{yx}}{\Sigma_{xy} | \Sigma_{xx}} \right] = \Sigma(\Phi), \tag{1}$$

which can be further specified as follows:

$$\frac{\Lambda_y (I-B)^{-1} (\Gamma \Phi \Gamma' + \Psi) \left[(I-B)^{-1} \right]' \Gamma'_y + \Theta_\epsilon \quad \Lambda_y (I-B)^{-1} \Gamma \Phi \Lambda'_x}{\Lambda_x \Phi \Gamma' \left[(I-B)^{-1} \right]' \Lambda'_y \quad \Lambda_x \Phi \Lambda'_x + \Theta_\delta}$$

Where the following parameters are present:

- Λs are the matrices of individual indicators xs for exogenous factors or ys for endogenous factor indicators;
- B and Γ are the matrices of structural relationship parameters;
- Φ is the matrix of exogenous factor intercorrelations;
- Ψ is the matrix of endogeneous factor errors;
- Θs are matrices of random errors on x-variables (δs) and y-variables (ϵs).

The model is based on general laws of variances, covariances, and means of the observed variables.

For the observed variables (V), resulting equations can be written as follows:

$$V_i = \lambda_{ij} F_i + E_i \tag{2}$$

Or, equivalently,

$$X_i = \lambda_{ij} \Xi_i + \delta_i \tag{3}$$

where i is the number of the indicator, j - latent factor indicator, F and Ξ are factors, E and δ - errors. With two indicator variables (V_1 and V_2), loading on the same factor F_1, their implied covariances are expressed as follows:

$$
\begin{aligned}
Cov(V_1, V_2) &= Cov(\lambda_{11} F_1 + E_1, \lambda_{21} F_1 + E_2) \\
&= Cov(\lambda_{11} F_1, \lambda_{21} F_1) + Cov(\lambda_{11} F_1, E_2 + Cov(E_1, \lambda_{21} F_1) + Cov(E_1, E2) \\
&= \lambda_{11} \lambda_{21} Cov(F_1, F_1) + \lambda_{11} Cov(F_1, F_2) + \lambda_{21} Cov(E_1, F_1) + Cov(E_1, E_2) \\
&= \lambda_{11} \lambda_{21} Cov(F_1, F_1) \\
&= \lambda_{11} \lambda_{21} Var(F_1) \\
&= \lambda_{11} \lambda_{21}
\end{aligned}
\tag{4}
$$

When two indicator variables (V_1 and V_4) load on different factors (F_1 and F_2), the picture is calculated similarly, though slightly more complicated:

$$
\begin{aligned}
Cov(V_1, V_4) &= Cov(\lambda_{11} F_1 + E_1, \lambda_{4,2} F_2 + E_4) \\
&= Cov(\lambda_{11}, F_1, \lambda_{42} F_2) + Cov(\lambda_{11} F_1, E_4) + Cov(E_1, \lambda_{42} F_2) + Cov(E_1, E_4) \\
&= \lambda_{11} \lambda_{42} Cov(F_1, F_2) + \lambda_{11} Cov(F_1, E_4) + \lambda_{4,2} Cov(E_1, F_2) + Cov(E_1, E_4) \\
&= \lambda_{11} \lambda_{42} Cov(F_1, F_2)
\end{aligned}
\tag{5}
$$

The rest of the calculations follow the same logic, though relationships become much more complicated with increased number of indicators and variables. For the model presented in this paper, we estimated the model for 5 exogenous latent factors (all of them network characteristics) with 5 indicators and 5 endogenous latent factors with 14 indicators.

Diagram representing the overall model is presented in Fig. 3, with only the statistically significant results shown. Model was subjected to the standard tests

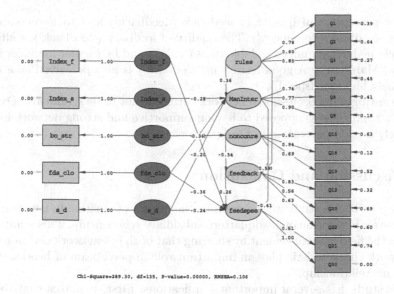

Chi-Square=289.30, df=135, P-value=0.00000, RMSEA=0.106

Fig. 3. Structural model of the hypothesized relationships

of fit; measurement model was built before structural model, and only the significant factor loadings (above 0.7 per generally accepted standards) were retained. Then, hypothesized structural relationships were tested one at a time, and only the significant relationships were retained in the final model.

4   Results

Our research is of exploratory nature. As our model shows having more friends in external network than in internal network ('Index f') enhances speed of feedback ('feedspeed'). However, having a lot of contacts in support network ('s d') results in reducing speed of feedback ('feedspeed'). This may indicate the difference in nature of support and friendship. Support of all kind is not aimed at providing feedback, it is aimed at helping and finding solution, whether friendship might mean being upright with your friend and thus giving feedback of all kind. Interestingly that closeness in friendship network ('fds clo') also affects speed of feedback ('feedspeed') negatively. This might indicate that having close access to other people is not enough for actually receiving feedback. Closeness in friendship network ('fds clo') also impacts negatively interactions between managers ('ManInter'). To enhance manager interactions ('ManInter') clear and strict rules ('rules') are necessary. Also interesting is that strong connection to your boss ('bo str') could lead to higher perception of reward and punishment as non-contingent ('nonconre'). If your boss and/or subordinates are strongly connected this means higher group cohesion in a subsidiary as opposed to headquarters and thus perceiving actions of headquarters as less fair. Higher perception of reward and punishment as non-contingent ('nonconre') also leads to lower perception of quality of feedback ('feedback') and to lower speed of receiving feedback ('feedspeed'). The quality of feedback ('feedback') is affected positively by having understandable rules ('rules') and by regular manager interactions ('ManInter'). A good feedback ('feedback') is also perceived as a quick and timely one ('feedspeed').

Interestingly the professional network turns out not to be significant. Overall, having a good rules as opposed to having supportive and strong network impact positively a good perception of headquarters.

5   Discussion and Conclusion

The purpose of this study was to examine the impact of position in network of an employee on perception of headquarters-subsidiary relationship. This study filled a gap in the field of management by showing that both full network centrality and egonetwork characteristic play an important role in perception of headquarters-subsidiary relationship.

This study has several important implications. First, it makes contribution to understanding what role network position plays in perceiving headquarters-subsidiary relationship. It also has practical implications. Understanding that certain position in organizational network and design of one's egonetwork impacts perception of employee of processes in organization is important for management of the organization. Managing the whole network structure and egonetwork structure of key employees by both formal and informal methods can lead to reducing conflicts and resistance to organizational changes and, in the end, to enhancing the overall performance of the firm.

This study has some limitations. First, it was conducted on a limited set of employees. Though we've applied full network analysis we are aware that the network we were able to build is not a full network of all employees of the organization. However when dealing with this limitation we take into consideration that among respondents there were key contacts in subsidiaries with most influence on other employees. Thus omitting non-influential nodes will not impact the results of research. Secondly, Krackhardt index is applied only to geographical characteristic of employees. However, group withinness and group betweenness could be percepted not only from geographical point of view, but also from job level, department etc. Third, the model explores the relationship at one time period. For more generalized results time series research is essential.

Appendix

Questionnaire Items

Perception of Rules:

1. In organization there are strict rules and procedures to control and coordinate personnel working in subsidiaries

2. Rules and procedures that are applied to personnel in subsidiaries are developed by headquarters
3. Rules and procedures for personnel in subsidiaries are controlled strictly.

Interactions of Managers:

1. Managers are often transferred from one subsidiary to another
2. Managers with same functional roles in different subsidiaries and headquarter often work together
3. Managers are regularly transferred between subsidiaries and headquarters.

Non-Contingent Reward and Punishment:

1. Subsidiary is punished regardless of actual work and results
2. Employees of headquarters often express dissatisfaction with work of subsidiary without particular reason
3. Headquarters often criticizes subsidiary even if subsidiary does work well.

Feedback Quality:

1. Headquarters always express approval if subsidiary works well
2. Subsidiary receives reward in case its results exceed expectation
3. Subsidiary receives regular feedback on its performance from headquarters.

Feedback Speed:

1. Subsidiary receives answers from headquarters to its requests as soon as possible.
2. It takes a lot of time to make decisions in headquarters even for a simple task.

References

1. Ferner, A., et al.: Dynamics of central control and subsidiary autonomy in the management of human resources: case-study evidence from US MNCs in the UK. Organ. Stud. **25**, 363–391 (2004). https://doi.org/10.1177/0170840604040041
2. Edwards, R., Ahmad, A., Moss, S.: Subsidiary autonomy: the case of multinational subsidiaries in Malaysia. J. Int. Bus. Stud. **33**, 183–191 (2002). https://doi.org/10.1057/palgrave.jibs.8491011
3. Roth, K., Nigh, D.: The effectiveness of headquarters-subsidiary relationships: the role of coordination, control, and conflict. J. Bus. Res. **25**, 277–301 (1992). https://doi.org/10.1016/0148-2963(92)90025-7
4. Ghoshal, S., Bartlett, C.A.: The multinational corporation as an interorganizational network. Acad. Manag. Rev. **15**, 603–626 (1990). https://doi.org/10.5465/amr.1990.4310825
5. Schotter, A., Beamish, P.W.: Performance effects of MNC headquarters-subsidiary conflict and the role of boundary spanners: the case of headquarter initiative rejection. J. Int. Manag. **17**, 243–259 (2011). https://doi.org/10.1016/j.intman.2011.05.006

6. Andersson, U., Forsgren, M., Holm, U.: The strategic impact of external networks: subsidiary performance and competence development in the multinational corporation. Strateg. Manag. J. **23**, 979–996 (2002). https://doi.org/10.1002/smj.267

7. Fey, C.F., Björkman, I.: The effect of human resource management practices on MNC subsidiary performance in Russia. J. Int. Bus. Stud. **32**, 59–75 (2001). https://doi.org/10.1057/palgrave.jibs.8490938

8. Hedlund, G.: Autonomy of subsidiaries and formalization of headquarters-subsidiary relationships in Swedish MNCs. In: Otterbeck, L. (ed.) The Management of Headquarters-Subsidiary Relationships in Multinational Corporations, pp. 27–76. Gower Publishing, Aldershot (1981)

9. Egelhoff, W.G.: Patterns of control in US, UK and European multinational corporations. J. Int. Bus. Stud. **15**, 73–83 (1984)

10. Delios, A., Beamish, P.W.: Survival and profitability: the roles of experience and intangible assets in foreign subsidiary performance. Acad. Manag. J. **44**, 1028–1038 (2001)

11. Birkinshaw, J., Hood, N., Young, S.: Subsidiary entrepreneurship, internal and external competitive forces, and subsidiary performance. Int. Bus. Rev. **14**, 227–248 (2005)

12. Nell, P.C., Ambos, B., Schlegelmilch, B.B.: The MNC as an externally embedded organization: an investigation of embeddedness overlap in local subsidiary networks. J. World Bus. **46**, 497–505 (2011)

13. Inkpen, A.C., Tsang, E.W.K.: Social capital, networks, and knowledge transfer. Acad. Manag. Rev. **30**, 146–165 (2005)

14. Kostova, T., Roth, K.: Social capital in multinational corporations and a micro-macro model of its formation. Acad. Manag. Rev. **28**, 297–317 (2003)

15. O'Donnell, S.W.: Managing foreign subsidiaries: agents of headquarters, or an interdependent network? Strateg. Manag. J. **21**, 525–548 (2000)

16. Knight, L., Harland, C.: Managing supply networks: organizational roles in network management. Eur. Manag. J. **23**, 281–292 (2005)

17. Emerson, R.M.: Social exchange theory. Annu. Rev. Soc. **2**, 335–362 (1976)

18. Burt, R.S.: Bandwidth and echo: trust, information, and gossip in social networks. In: Casella, A., Rauch, J.E. (eds.) Network and Markets, pp. 30–74. Sage, New York (2001)

19. Harzing, A.W., Feely, A.J.: The language barrier and its implications for HQ-subsidiary relationships. Cross Cult. Manag. Int. J. **15**, 49–61 (2008). https://doi.org/10.1108/13527600810848827

20. London, M., Smither, J.W.: Feedback orientation, feedback culture, and the longitudinal performance management process. Hum. Res. Manag. Rev. **12**, 81–100 (2002). https://doi.org/10.1016/S1053-4822(01)00043-2

21. Krackhardt, D., Stern, R.N.: Informal networks and organizational crises: an experimental simulation. Soc. Psychol. Q. **51**, 123–140 (1988)

22. Bavelas, A.: Communication patterns in task-oriented groups. J. Acoust. Soc. Am. **22**, 725–730 (1950). https://doi.org/10.1121/1.1906679

23. Bollen, K.A., Long, J.S.: Testing Structural Equation Models, vol. 154. SAGE Publications, Newbury Park (1993)

Natural Language Processing

Vector Space Models in Detection of Semantically Non-compositional Word Combinations in Turkish

Levent Tolga Eren and Senem Kumova Metin[✉]

Faculty of Engineering, İzmir University of Economics,
Sakarya C., No. 156 Balçova, İzmir, Turkey
{tolga.eren, senem.kumova}@ieu.edu.tr

Abstract. The semantic compositionality presents the relation between the meanings of word combinations and their components. Simply, in non-compositional expressions, the words combine to generate a different meaning. This is why, identification of non-compositional expressions (e.g. idioms) become important in natural language processing tasks such as machine translation and word sense disambiguation.

In this study, we explored the performance of vector space models in detection of non-compositional expressions in Turkish. A data set of 2229 uninterrupted two-word combinations that is built from six different Turkish corpora is utilized. Three sets of five different vector space models are employed in the experiments. The evaluation of models is performed using well-known accuracy and F-measures. The experimental results showed that the model that measures the similarity between the vectors of word combination and the second composing word produced higher average F-scores for all testing corpora.

Keywords: Semantic compositionality · Vector space model · Turkish

1 Introduction

The compositionality is described to be the degree to which the features of the con-stituents of a multiword expression combine to predict the feature of the whole [1]. In other words, it is the amount of information that is hold by the constituents that enables or eases the prediction of the expression, especially the meaning of the whole expression.

In this study, the notion of compositionality is limited to semantic compositionality of expressions that is composed of two consecutive words known as bigram. In this per-spective, compositionality is the degree of relation between the meaning of expression and the individual meanings of its constituents. In compositional expressions, the meaning of expression can be predicted from the meanings of its composing words. For example, the two-word expression *trafik ışığı* (Eng. traffic light) is a compositional expression (to some degree). The regarding expression corresponds to signaling devices positioned at road intersections, pedestrian crossings etc. to control the flow of traffic. A person who knows the dictionary-based definitions/meanings of the words *trafik* (Eng.

W. M. P. van der Aalst et al. (Eds.): AIST 2018, LNCS 11179, pp. 53–63, 2018.
https://doi.org/10.1007/978-3-030-11027-7_6

traffic) and *ışığı* (Eng. light) may guess that the expression points to an object/item that includes a lighting item and is related somehow to traffic. On the other hand, in non-compositional expressions, the combined meaning of words is unrelated to individual meanings of its components. For instance, the two-word (idiomatic) expression *kanı bozuk* (Eng. corrupt or evil by nature) is a fully non-compositional expression. Even if a person is a native speaker of Turkish, he may not predict the meaning of multiword expression by the meanings – dictionary-based definitions-of *kanı* (Eng. blood) and *bozuk* (Eng. spoilt). Though it is almost impossible to discover such non-compositional multiword expressions utilizing dictionaries, distributional hypothesis/semantics where each word is represented by a vector of words enables the regarding discovery. Simply in distributional semantics, a word is expressed commonly by a vector of its neighboring words targeted with their occurrence frequency and the context of a target word is defined as the vector of its neighboring words in a fixed window size. As a result, one can tell that given two words are similar if they have a similar distribution of contexts.

The objective of this study is exploring the performance of vector space models (VSM) in detection of non-compositional expressions in Turkish. In line with this objective, a dataset of 2229 bigrams is constructed from 6 different Turkish corpora by the use of occurrence frequency methods (chi square, occurrence frequency counts, point-wise mutual information and t-test). This dataset is annotated by 4 human judges. The dataset is utilized in the experiments of vector space models that are previously proposed to measure the semantic compositionality/non-compositionality in different languages.

In following sections, we are going to explore the related work on semantic compositionality, and then we will present the data set, our proposed models and evaluation measures. Following the experimental results, we will conclude the paper.

2   Related Work

In recent years, there has been a growing awareness in the natural language processing field about the problems related to semantic compositionality/non-compositionality. Such that special interest workshops have been arranged and discussed issues like automatically acquiring semantic compositionality [2]. In Table 1, a set of studies presented in Distributional Semantics and Compositionality Workshop (DiSCo 2011) [2] that not only provide the inspiration but also directed our work are given. As the studies in Table 1 and the other works on distributional semantics are examined, it is clearly seen that the majority of the works on non-compositionality are performed on English and/or German corpora. In DiSCo shared task, the samples in data set are separated into multiple classes based on the compositionality score (low in compositional ($0 < $ score $ < 37$). Medium in compositional ($36 < $ score $ < 75$) and high in compositional ($74 < $ score)). Though no clear winner emerged in the task, it was examined that the approaches based on distributional semantics seemed to outperform those based on statistical association measures [3].

Following the shared task. Krčmář et al. [12] evaluated various distributional semantic approaches in compositionality detection and showed that LSA-based models

Table 1. A set of studies presented in DiSCo 2011 workshop.

#	The study	Approaches
1	Identifying collocations to measure compositionality [4]	Statistical association measures. t-score and pmi
2	Measuring the compositionality of bigrams using statistical methodologies [5]	A mix of statistical association measures
3	Measuring the compositionality of collocations via word co-occurrence vectors [6]	Unsupervised WSM, cosine similarity
4	(Linear) maps of the impossible: capturing semantic anomalies in distributional space [7]	Cosine similarity
5	Frustratingly hard compositionality prediction [8]	Support vector regression with COALS-based endocentricity features
6	Exemplar-based word-space model for compositionality detection [9]	Exemplar-based WSM, prototype-based WSM
7	Distributed structures and distributional meaning [10]	Distributed tree vector, distributed kernel tree vector
8	Two multivariate generalizations of point-wise mutual information [11]	Multi-way co-occurrences

perform quite well. There also exist a few studies in literature that employ parallel corpora or other resources such as Wiktionary to detect the degree of compositionality/ non-compositionality in word combinations, especially in multiword expressions (e.g. [13–15]).

3 Detection of Non-compositional Word Combinations in Turkish

In this section, we will present our work to measure the performance of vector space models in detection of non-compositional word combinations in Turkish texts. To the best of our knowledge, this is the first work for Turkish language in this scope.

In following subsections, firstly the dataset and the procedure to prepare the set will be explained. Secondly the vector space models utilized in this study will be presented. In third subsection, evaluation methods will be given.

3.1 Data Set

In this study, the experiments on non-compositionality/compositionality in Turkish are limited to bigrams that are known as uninterrupted two-word combinations in text. BilCol [16], Bilkent [17], Ege, Leipzig [18], Metu [19] and Muder [20] corpus are utilized to obtain bigrams that will be used in experiments. Briefly, punctuation marks are removed from each corpus and the text is tokenized to obtain bigrams. Table 2 gives total number of tokens (unigrams) and bigrams in regarding corpora.

Table 2. The corpora used in experiments

Corpus	Size (Number of tokens)	Number of unique tokens	Number of unique bigrams
BilCol [16]	42414743	984434	11759532
Bilkent [17]	706443	94552	507758
Ege	2465285	259196	1637055
Leipzig [18]	13389049	745446	7350443
Metu [19]	1987447	212853	1388722
Muder [20]	638547	82145	437826

In order to decrease the number of samples (bigrams), a set of occurrence frequency methods (chi-square, occurrence frequency counts, point-wise mutual information and the t-test) is applied and a sorted list of bigrams is built individually for each method from each corpus. Each method yielded 1200 distinct bigrams from 6 different corpora. The best scoring 200 bigrams of sorted lists are selected to construct the final data set of 2229 bigrams.

The data set is annotated by four human judges and the interrater agreement among judges is measured by Fleiss kappa metric (kappa = ~ 0.738). In this study, assuming that majority of multiword expressions (idioms. technical terms. named entities. some phrasal verbs) are non-compositional expressions as it was done in the study of Bu et al. [21]. Choueka [22] and Almi et al. [3], the human judges are asked to label the samples either as MWE (multiword expression) or non-MWE (non-multiword expression). A guide that directs judges to assign idiomatic expressions, named entities, technical terms, phrasal verbs and multi-word conjunctions in same class and all other word combinations in the other class is provided. Further information/details on the data set preparation may be found in [23]. In our experiments, all MWE labeled samples are accepted to be non-compositional and all non-MWE labeled samples are considered as compositional. As a result, we employed 1194 compositional bigrams (54% of the whole data set) and 1035 (46% of the whole set) non-compositional bigrams in our final set.

3.2 Proposed Method: Vector Space Models

In information retrieval systems, vector space models are commonly employed to represent the text documents as vectors of identifiers, such as, for example index terms. The vectors may be composed by occurrence frequencies of terms (term frequencies-tf), document frequencies (df), inverse document frequencies (idf) or a combination of these such as well-known tf.idf measure.

In this study, a vector of word occurrence frequencies will represent each bigram and/or its constituting words. The words that compose the vector are the ones that reside in same sentence with the target (bigram or a constituents of a bigram). For sentences that are relatively longer, a window size of 5 words is defined, in other words the preceding and/or following 5 words of the target are considered while building the regarding vector. To measure similarity between the vectors, we used cosine similarity as given below

$$sim\left(\vec{V_1} \cdot \vec{V_2}\right) = \frac{\vec{V_1} \cdot \vec{V_2}}{\left\|\vec{V_1}\right\|\left\|\vec{V_2}\right\|} \tag{1}$$

where V_i represents the ith vector. In our experiments, we have used a normalized cosine similarity function that produces values in range [0.2] instead of range [−1.1] (0 indicates the exact similarity between vectors. 2 is vice versa).

The compositionality for a given bigram is measured by 5 different models that can be obtained from the following equation proposed by Reddy et al. [9] in DiSCo 2011 shared task [2]:

$$\alpha\left(\vec{w_1} \cdot \vec{w_2}\right) = a + b * sim\left(\overrightarrow{w_1 w_2} \cdot \vec{w_1}\right) + c * sim\left(\overrightarrow{w_1 w_2} \cdot \vec{w_2}\right) + d$$
$$* sim\left(\overrightarrow{w_1 w_2} \cdot \vec{w_1} + \vec{w_2}\right) + e * sim\left(\overrightarrow{w_1 w_2} \cdot \vec{w_1} * \vec{w_2}\right) \tag{2}$$

where $\overrightarrow{w_1 w_2}$, $\vec{w_1}$ and $\vec{w_2}$ represent the vectors of bigram, first and second words of the bigram respectively. In Reddy et al. [9], it is stated that the models that are derived from Eq. (2) outperform existing prototype-based models in DiSCo 2011 shared task [2].

In our experiments, we have employed 3 different sets of models that are derived from the Eq. 2. The brief definitions of model sets are given below.

Set 1 (S_1): This set includes 5 different models where the vectors include raw frequencies of neighboring words that reside in same window size with the target. The models are

$$(S_1 M_1) : sim\left(\overrightarrow{w_1 w_2} \cdot \vec{w_1}\right)$$
$$(S_1 M_2) : sim\left(\overrightarrow{w_1 w_2} \cdot \vec{w_2}\right)$$
$$(S_1 M_3) : sim\left(\overrightarrow{w_1 w_2} \cdot \vec{w_1} + \vec{w_2}\right)$$
$$(S_1 M_4) : sim\left(\overrightarrow{w_1 w_2} \cdot \vec{w_1} * \vec{w_2}\right)$$
$$(S_1 M_5) : sim\left(\overrightarrow{w_1 w_2} \cdot \vec{w_1}\right)$$
$$+ sim\left(\overrightarrow{w_1 w_2} \cdot \vec{w_2}\right)$$
$$+ sim\left(\overrightarrow{w_1 w_2} \cdot \vec{w_1} + \vec{w_2}\right)$$
$$+ sim\left(\overrightarrow{w_1 w_2} \cdot \vec{w_1} * \vec{w_2}\right)$$

For example, in Model 1 ($S_1 M_1$) the similarity of vectors that belong to bigram and the first word in bigram is measured. On the other hand, in Model 3, the similarity is measured between the vector of bigram and the vector that is obtained by summation of first and second word's vectors. Briefly, in models $S_1 M_1$ and $S_1 M_2$, the semantic similarity between the bigram and its constituents are measured. If the given bigram is non-compositional it is expected that the similarity score of these vectors will not be high. In models $S_1 M_3$ and $S_1 M_4$, the similarity between the bigram and a combined version of vectors (point-wise summation and multiplication) for the constituents is measured as in Mitchell and Lapata [24]. Finally in model $S_1 M_5$, the results of the previous models are summed up.

Set 2 (S_2): This set includes models where the vectors of component words are refined. In this set, the refined vector for each constituent is built by the sentences that include the constituent but not the bigram. As a result the refined vector of $\overrightarrow{w_1}$ is $\overrightarrow{w_1'} = \overrightarrow{w_1} - \overrightarrow{w_1w_2}$ and refined vector of $\overrightarrow{w_2}$ is $\overrightarrow{w_2'} = \overrightarrow{w_2} - \overrightarrow{w_1w_2}$. The models in set 2 are listed as below

$$(S_2M_1) : sim\left(\overrightarrow{w_1w_2}.\,\overrightarrow{w_1'}\right)$$

$$(S_2M_2) : sim\left(\overrightarrow{w_1w_2}.\,\overrightarrow{w_2'}\right)$$

$$(S_2M_3) : sim\left(\overrightarrow{w_1w_2}.\,\overrightarrow{w_1'} + \overrightarrow{w_2'}\right)$$

$$(S_2M_4) : sim\left(\overrightarrow{w_1w_2}.\,\overrightarrow{w_1'} * \overrightarrow{w_2'}\right)$$

$$(S_2M_5) :sim\left(\overrightarrow{w_1w_2}.\,\overrightarrow{w_1'}\right)$$
$$+ sim\left(\overrightarrow{w_1w_2}.\,\overrightarrow{w_2'}\right)$$
$$+ sim\left(\overrightarrow{w_1w_2}.\,\overrightarrow{w_1'} + \overrightarrow{w_2'}\right)$$
$$+ sim\left(\overrightarrow{w_1w_2}.\,\overrightarrow{w_1'} * \overrightarrow{w_2'}\right)$$

Set 3 (S_3): In Set 3, while building the vectors of constituents irrelevant sentences in the corpus are removed. For example, building the vector of w_1, only the sentences that includes both w_1 and a word that is semantically related to w_2 are considered and the other sentences that includes only w_1 are removed.

In [9], it is stated that the composing words of a bigram may be used in a different context that may be unrelated to the regarding bigram. Reddy et al. [9] exemplified this by the bigram *traffic light*. The composing word *light* may occur in different context in corpus. And some of the occurrences may be unrelated to notion of *traffic light*. These unrelated occurrences tend to decrease the semantic relation between the composing words; *light* and *traffic*. In order to decide relevant occurrences of *light*, a group of words that appears in similar context of *traffic* is defined. This group of words will be named as context words from now on. While building the vector of *light,* the sentences where both *light* and at least one of the context words of *traffic* are selected, the other sentences where only *light* is observed are accepted to be in a context that is unrelated to *traffic light*.

In this study, for each composing word a group of context words is determined. The context words group includes the words that are most frequently co-occurring words with the regarding word. Simply, for each word, the most frequently co-occurring words are listed in the corpus, the stop words are removed from the list and finally the first five words are assigned as context words. The list of stop words that is given in [25] is used. Following models in Set 3 are determined:

$$(S_3M_1) : sim\left(\overrightarrow{w_1w_2}.\overrightarrow{w_1}^r\right)$$
$$(S_3M_2) : sim\left(\overrightarrow{w_1w_2}.\overrightarrow{w_2}^r\right)$$
$$(S_3M_3) : sim\left(\overrightarrow{w_1w_2}.\overrightarrow{w_1}^r + \overrightarrow{w_2}^r\right)$$
$$(S_3M_4) : sim\left(\overrightarrow{w_1w_2}.\overrightarrow{w_1}^r * \overrightarrow{w_2}^r\right)$$
$$(S_3M_5) : sim\left(\overrightarrow{w_1w_2}.\overrightarrow{w_1}^r\right)$$
$$+ sim\left(\overrightarrow{w_1w_2}.\overrightarrow{w_2}^r\right)$$
$$+ sim\left(\overrightarrow{w_1w_2}.\overrightarrow{w_1}^r + \overrightarrow{w_2}^r\right)$$
$$+ sim\left(\overrightarrow{w_1w_2}.\overrightarrow{w_1}^r * \overrightarrow{w_2}^r\right)$$

where $\overrightarrow{w_k}^r$ represents the revised vector of kth word in bigram.

3.3 Evaluation

The evaluation of proposed models is performed in 3 steps. For each proposed model below steps are followed:

1. Bigrams are sorted according to the similarity score that is produced by the model in decreasing order. It is accepted that if the similarity score of a bigram is low, then it is non-compositional.
2. F1-score and accuracy are measured in a point-wise manner. Simply they are measured for set size N where N is increased from 1 to the total set size. The resulting N number of regarding evaluation values (F1 or accuracy) are summed up and divided by N to obtain average evaluation score.
3. Average F1-score and accuracy are compared to respectively F1 and accuracy scores of other models.

The F1 measure is given as

$$F1 = \frac{2TP}{2TP + FN + FP} \tag{2}$$

where TP is the number of true positives (samples that are both expected and predicted to belong to the same class (non-compositional or compositional)). FN is the number of false negatives (type 2 error). FP is the number of false positives (type1 error).

Accuracy (A) is measured as follows

$$A = \frac{TP + TN}{TP + TN + FN + FP} \tag{3}$$

where TN is the number of true negatives.

4   Experimental Results

In experiments, three sets of models are tested for 3 corpora of different sizes: Bilkent [17], Muder [20] and Metu [19] corpus. Table 3 presents accuracy and F1 scores obtained from the regarding corpora. In Table 3, bold cells represent the highest scores obtained from each corpus for the regarding evaluation measure.

Table 3. Average F1-measure and accuracy values obtained from Bilkent [17]. Muder [20] and Metu [19] corpora

Model	Bilkent		Muder		Metu	
	A	F1	A	F1	A	F1
S_1M_1	0.626	0.501	0.576	0.521	0.656	0.519
S_1M_2	0.615	0.490	0.565	0.513	0.650	0.513
S_1M_3	0.600	0.476	0.549	0.500	0.633	0.499
S_1M_4	0.591	0.469	0.551	0.503	0.624	0.492
S_1M_5	0.602	0.479	0.560	0.509	0.639	0.503
S_2M_1	0.694	0.564	0.580	0.527	0.709	0.570
S_2M_2	**0.703**	**0.570**	**0.595**	**0.537**	0.710	0.569
S_2M_3	0.694	0.565	0.576	0.524	0.704	0.565
S_2M_4	0.686	0.557	0.573	0.521	0.707	0.568
S_2M_5	0.697	0.567	0.583	0.529	**0.711**	**0.571**
S_3M_1	0.627	0.503	0.585	0.530	0.660	0.523
S_3M_2	0.628	0.504	0.589	0.533	0.662	0.524
S_3M_3	0.613	0.489	0.570	0.517	0.647	0.511
S_3M_4	0.615	0.493	0.576	0.522	0.650	0.516
S_3M_5	0.620	0.496	0.583	0.528	0.657	0.520

The number of bigrams that reside both in data set and Bilkent corpus is 957 in which 63.32% of bigrams is annotated as non-compositional. The highest averaged F1 and accuracy values are obtained by S_2M_2. Considering average F1 values (average $F_{Set1} = 0.483$, $F_{Set2} = 0.565$, $F_{Set3} = 0.497$) it is observed that in Bilkent corpus, Set 2 outperforms the other sets of models. As a result, it is possible to state that refined vectors of composing words; the vectors that are built by the sentences that include the constituents but not the bigram; are better representatives to detect non-compositionality.

Muder corpus includes 798 bigrams (56% non-compositional, 44% compositional) of the data set. The maximum F1 and accuracy values are obtained in model S_2M_2 similar to Bilkent corpus. In addition, though the difference in average F-values is not as high as the values in Bilkent corpus, it is observed that still Set 2 models generate higher F1 performance scores compared to other sets (average $F_{Set1} = 0.509$, $F_{Set2} = 0.528$, $F_{Set3} = 0.526$).

The third corpus in our experiments, Metu, includes 1129 of bigrams in data set. In Metu corpus, it is examined that the models in Set 2 are performing better compared to the other models (average F_{Set1} = 0.505, F_{Set2} = 0.569, F_{Set3} = 0.519), supporting the results in previous corpora. The best performing model is observed to be S_2M_5.

Based on the overall results of 3 corpora, it is examined that models in Set 2 (especially Model 2 – S_2M_2) are succeeding in Turkish corpora in this experimental set up. Though the size of the corpus changes the evaluation scores, the best set of models do not differ according to the corpus size.

To the best of our knowledge, since this is the first study that employs vector space models in detection of non-compositional word combinations in Turkish, there exist no other evaluation results for alternative vector space models. Nevertheless it may be stated that the range of highest average F1 scores ([0.537 0.571]) is promising compared to the performance scores reported in previous frequency-based studies employing different Turkish data sets. For example, in [26] where Google search engine is used as an additional resource, it is reported that considering web-based frequency metrics, highest F scores are observed to be in range [0.570 0.585] and the highest F score when corpus-based frequency is utilized, is given as 0.57.

5 Conclusion

In this study, we analyzed the semantic compositionality/non-compositionality in Turkish by vector space models. We introduced three sets of 5 different VSMs that assess the non-compositionality in Turkish. VSMs of Set 2; the models where the vector of composing words are built by ignoring the sentences that hold the word combination; are observed to provide better performance results compared to other models. It is also examined that as the size of the corpus increases, the difference in performances of successful and unsuccessful methods becomes more significant.

Due to the high time and space complexity of the algorithms that are used to implement models, we were unable to work on larger corpus. As a future work, we are planning to repeat our experiments in larger corpora and with different settings (e.g. windows size. stemmed/surface formed corpus. binary/weighted vectors. unigrams/bigrams/trigrams). Moreover, we plan to evaluate the performance of word-embedding models (e.g. word2vec, skip-gram models) in Turkish data sets [27].

References

1. Baldwin, T.: Compositionality and multiword expressions: six of one, half a dozen of the other? In: COLING/ACL 2006 Workshop on MWEs, Invited Speech (2006)
2. Biemann, C., Giesbrecht, E.: Distributional semantics and compositionality 2011: shared task description and results. In: Proceedings of the Workshop on Distributional Semantics and Compositionality, pp. 21–28 (2011)
3. Almi, P., Snajder, J.: Determining the semantic compositionality of croatian multiword expressions. In: 9th Language Technologies Conference Information Society, IS 2014 (2014)
4. Pedersen, T.: Identifying collocations to measure compositionality: shared task system description. In: Proceedings of the Workshop on Distributional Semantics and Compositionality, pp. 33–37 (2011)

5. Chakraborty, T., Pal, S., Mondal, T., Saikh, T.: Shared task system description: measuring the compositionality of bigrams using statistical methodologies, pp. 38–42. Association for Computational Linguistics (2011)
6. Maldonado-guerra, A., Emms, M.: Measuring the compositionality of collocations via word co-occurrence vectors: shared task system description, pp. 48–53. Association for Computational Linguistics (2011)
7. Vecchi, E.M., Baroni, M., Zamparelli, R.: (Linear) maps of the impossible: capturing semantic anomalies in distributional space. In: Proceedings of the Workshop on Distributional Semantics and Compositionality, pp. 1–9 (2011)
8. Johannsen, A., Alonso, H.M., Rishøj, C., Søgaard, A.: Shared task system description: frustratingly hard compositionality prediction. In: Proceedings of DiSCo, pp. 29–32 (2011)
9. Reddy, S., McCarthy, D., Manandhar, S., Gella, S.: Exemplar-based word-space model for compositionality detection: shared task system description. In: Proceedings of the Workshop on Distributional Semantics and Compositionality, pp. 54–60. Association for Computational Linguistics (2011)
10. Zanzotto, F.M., Dell'Arciprete, L.: Distributed structures and distributional meaning. In: Proceedings of the Workshop on Distributional Semantics and Compositionality, DiSCo 2011, Portland, Oregon, pp. 10–15 (2011)
11. Van De Cruys, T.: Two multivariate generalizations of pointwise mutual information. In: Proceedings of the Workshop on Distributional Semantics and Compositionality, pp. 16–20 (2011)
12. Krčmář, L., Ježek, K., Pecina, P.: Determining compositionality of word expressions using various word space models and measures. In: Proceedings of the Workshop on Continuous Vector Space Models and their Compositionality, Sofia, Bulgaria, 9 August 2013, pp. 64–73 (2013)
13. Salehi, B., Cook, P., Baldwin, T.: Detecting non-compositional MWE components using Wiktionary. In: EMNLP (2014)
14. Salehi, B., Askarian, N., Fazly, A.: Automatic identification of Persian light verb constructions. In: Gelbukh, A. (ed.) CICLing 2012. LNCS, vol. 7181, pp. 201–210. Springer, Heidelberg (2012). https://doi.org/10.1007/978-3-642-28604-9_17
15. de Medeiros Caseli, H., Ramisch, C., das Graças Volpe Nunes, M., Villavicencio, A.: Alignment-based extraction of multiword expressions. Lang. Resour. Eval. 44, 59–77 (2010)
16. Can, F., Kocberber, S., Baglioglu, O., Kardas, S., Ocalan, H.C., Uyar, E.: New event detection and topic tracking in Turkish. J. Am. Soc. Inf. Sci. Technol. 61, 802–819 (2010)
17. Tur, G., Hakkani-Tur, D., Oflazer, K.: A statistical information extraction system for Turkish. Nat. Lang. Eng. 9, 181–210 (2003)
18. Quasthoff, U., Richter, M., Biemann, C.: Corpus portal for search in monolingual corpora. In: Proceedings of 5th International Conference on Language Resources and Evaluation, pp. 1799–1802 (2006)
19. Say, B., Zeyrek, D., Oflazer, K., Umut, Ö.: Development of a corpus and a treebank for present-day written Turkish. In: Proceedings of the Eleventh International Conference of Turkish Linguistics (2002)
20. Dinçer, B.T.: Türkçe için istatistiksel bir bilgi geri-getirim sistemi (2004)
21. Bu, F., Zhu, X., Li, M.: Measuring the non-compositionality of multiword expressions. In: Proceedings of the 23rd International Conference on Computational Linguistics, Coling 2010, pp. 116–124 (2010)
22. Choueka, Y.: Looking for needles in a haystack or locating interesting collocational expressions in large textual databases. In: RIAO, pp. 609–624 (1988)
23. Metin, S.K., Taze, M.: A procedure to build multiword expression data set. In: 2nd International Conference on Computer and Communication Systems, pp. 46–49 (2017)

24. Mitchell, J., Lapata, M.: Vector-based models of semantic composition, pp. 236–244. Association for Computational Linguistics (2008)
25. Turkish Stopwords. http://www.ranks.nl/stopwords/turkish
26. Aka Uymaz, H., Metin, S.K.: A comprehensive analysis of web-based frequency in multiword expression detection. Int. J. Intell. Syst. Appl. Eng. (2017, in print)
27. Salehi, B., Cook, P., Baldwin, T.: A word embedding approach to predicting the compositionality of multiword expressions. In: Proceedings of the 2015 Conference of the North American Chapter of the Association for Computational Linguistics: Human Language Technologies, pp. 977–983 (2015)

Authorship Verification on Short Text Samples Using Stylometric Embeddings

Johannes Jasper[✉], Philipp Berger, Patrick Hennig, and Christoph Meinel

Hasso Plattner Institute, University of Potsdam, Potsdam, Germany
`aist@johannesjasper.de`

Abstract. Given the increasing amounts of textual data published online and our inability to reliably identify a person by their writing style, impersonation in the context of social media applications becomes a real-world problem. This work explores how deep learning and metric learning techniques can be applied to the challenge of *authorship verification*—given a collection of text samples by one author and another document of unknown origin, determine if the new document is written by the same author or not. Using fastText word embeddings, deep LSTMs, and triplet loss, we propose a system that is able to learn *stylometric embeddings* of different documents and measure their stylistic distance. Unlike most approaches that work on entire documents, our system is able to work on very short text samples of 1–3 sentences, which resembles the length of typical social media posts. We successfully evaluated our approach on the PAN 2014 challenge on authorship verification for English text. The presented system outperforms competing approaches in the PAN 2014 challenge when using 10 short text samples or more.

Keywords: Authorship identification · Natural language processing Metric learning · Embedding · Deep learning

1 Introduction

With the advent of social media platforms such as Facebook or Twitter, users share increasing amounts of textual content online. Also communication turns from telephony to text-based messaging using WhatsApp or other instant messaging services. While most users use these platforms only casually, some users build up a reputation and gain large attention from their followers. What such influencers publish online is closely observed by their followers and has immediate real-world implications. At the same time, however, most people struggle with reliably judging the authenticity of textual posts. Most people's intuition for linguistics and writing style is far inferior to their ability to distinguish faces or hand written texts, for instance. Given the growing impact of online media, impersonation on social media platforms becomes a real-world problem. Impersonation, in this context, can manifest in the creation of profiles in another

© Springer Nature Switzerland AG 2018
W. M. P. van der Aalst et al. (Eds.): AIST 2018, LNCS 11179, pp. 64–75, 2018.
https://doi.org/10.1007/978-3-030-11027-7_7

person's name. When security measures fail, existing accounts could be compromised and controlled by a third party. The immediate result of a third party writing posts in another person's name would be a change in writing style. For social media platforms the need for an automated style breach detection system, as depicted in Fig. 1, arises. Given a social media account with a certain number of existing posts, the goal is to verify if new content is authored by the same user. This is a form of the *authorship verification* problem—given a collection of text samples by one author and another sample of unknown origin, determine if they were written by the same author [10]. This challenge is closely related to authorship *attribution*, where the task is to assign a document of unknown origin to an author from a given finite set. While verification is only a binary classification, it is considered more difficult due to its inherent open world assumption [10].

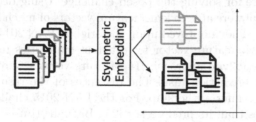

Fig. 1. Using stylometric embedding, documents can be clustered by writing style, thus detecting single samples that do not stem from the suggested author.

This work introduces a system that employs deep and metric learning techniques on the problem of authorship verification using very short text samples as they occur in social media applications. It discusses the challenges that arise with the use of short text samples and presents techniques to mitigate that problem. Finally this work introduces triplet loss as a spatial clustering algorithm to the problem. Using this technique, the system is able to learn stylometric embeddings of text samples that not only capture the author's writing style, but also discriminate it against other authors. The proposed approach is evaluated using the benchmarks provided by the PAN 2014 challenge on authorship verification [24].

2   Related Work

Authorship attribution and verification are problems usually solved with machine learning. Traditionally, large feature vectors are extracted from pairs of documents in order to assess their stylometric similarity. Adequate features are designed manually and have been subject to major change over the past decades [22]. Khonji and Iraqi [11] performed best on the original PAN 2014 challenge by extracting n-grams of characters, words, function words, word shapes, POS tags, and POS-words for varying n resulting in a feature vector of 43,931 dimensions for English novels. They use a variant of the impostor method [12,13], in which the unknown document has to be more similar to the original document than a set of imposter documents–documents that are known to stem from different authors. This assessment is performed with multiple randomly selected feature sets. Other approaches use similar feature sets that are well established in natural language processing such as $tf\text{-}idf$ or vocabulary richness functions [24]. Rocha et al. [19] contributed a series of text pre-processing techniques for social media applications. They especially highlight that details such as dates, times, URLs, or hash tags should be placed by generic placeholders, in order to emphasize the writing style over the specific meaning of a post.

While such approaches proved effective, deep learning techniques have become the de facto standard in pattern recognition tasks, in recent years. Rather than manually designing and combining features, deep neural networks *learn* the features they require for solving the posed challenge. Using deep learning techniques, Ding et al. [6] were able to learn feature vectors of much lower dimensionality (few hundred). They improved upon the original PAN 2014 performance by learning a stylometric feature vector for different modalities respectively. Similar to the original doc2vec approach [14], feature vectors were learned as part of an unsupervised prediction task. The distances of the resulting stylometric vectors were used to infer authorship. For the PAN 2015 challenge, which uses shorter text samples than its predecessor [23], Bagnall [1] used a shallow recurrent neural network in order to read character sequences and train a language model for each candidate author. Each language model is trained to predict the local character flow for a particular author. Authorship was inferred by assessing how well the respective language model predicted the unknown document. Shrestha et al. [20] also focused on shorter text samples and learned embeddings for character-n-grams ($n \in \{1, 2\}$ performed best). They used a convolutional neural network with varying filter sizes and max-over-time pooling to extract the most discriminative features. In contrast to the aforementioned works, however, Shrestha et al. performed authorship attribution rather than verification.

The main obstacle to apply the traditional authorship verification challenge on social media applications is the length of text samples [13,20]. The PAN challenges [23,24] on this problem provide evaluation corpora with documents spanning several hundred to several thousand words. Social media posts, in contrast, are only few sentences short. This work focuses on performing authorship verification with samples spanning 1–3 sentences in order to simulate a social media application while remaining comparable to PAN benchmarks.

Another aspect in that our approach differs from the aforementioned works is the use of spatial clustering. Most approaches create a feature vector that captures the writing style of an author. They do not, however, present the system with negative samples in order to find discriminative features. The system presented in this work applies triplet loss in order to spatially disseminate feature vectors, thus comparing authors amongst each other and differentiating them.

3   Deep Stylometric Embeddings

We propose a system that reads text samples of variable length on a word level and creates a stylometric embedding in the form of a low dimensional vector. This embedding captures the stylometric properties of the given text sample. Text samples of the same author should therefore have a similar embedding. The overall goal is to verify authorship of an unknown document by its distance to a known document. As shown in Fig. 2, the proposed system uses fastText word embeddings [2] to map discrete tokens to continuous vectors, while preserving their semantic and syntactic properties. We use a model that was pre-trained[1] on English Wikipedia articles in an unsupervised manner. While similar in design to the traditional skip-gram approach [16], fastText uses sub-word (character-n-gram) embeddings in order to create an inherent association between words that share the same stem.

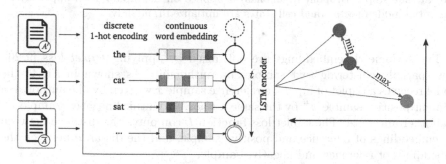

Fig. 2. The proposed system reads short text samples and learns stylometric embeddings. It maps words to vectors and feeds them into an LSTM. The hidden state at the final time step is interpreted as stylometric embedding. Triplet loss is used to cluster the embeddings by author such that samples by the same author (green) are embedded close to each other, and samples of other authors (orange) are moved further away. (Color figure online)

Besides the aforementioned doc2vec [14], one way to embed higher order structures, such as sentences, is the use of recurrent neural networks. Their hidden state can be utilized to transfer information over time and capture relevant aspects of sequential input. Embedding word sequences into a low dimensional vector representation that captures its semantic and syntactic properties proved especially useful in machine translation using RNN-autoencoders [3–5,15,17,25]. Using a stacked LSTM [18], as depicted in Fig. 3, the word embedding at each time step is read as input, thus enriching the LSTM's hidden state over time. The LSTM's hidden state at the last time step S captures information about the entire input sequence and is interpreted as stylometric sequence embedding.

---

[1] https://fasttext.cc/docs/en/pretrained-vectors.html.

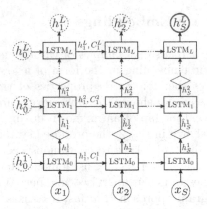

Fig. 3. The sequence embedding model uses $L = 3$ layers of LSTM cells (blocks) to iteratively read word sequences of size S. Each input x_i is the word embedding of the word at time step i. Dropout (diamonds) is applied on the inputs of all upper layers. The initial hidden vectors and cell states are initialized randomly.

The stylometric embeddings are optimized by applying *triplet loss* [9,26], thus spatially clustering text samples by authorship. As shown in Fig. 2, the model receives a triplet of inputs: a reference sample x written by one author A, a different positive sample x^+ by the same author, and a third negative sample x^- of another author $\overline{A}$. The triplet loss function L computes the distance between the embeddings of reference and positive sample, and the distance between the embeddings of reference and negative sample.

$$L(x, x^+, x^-) = \frac{1}{N} \left(\sum_i^N max\{d\left(x_i, x_i^+\right) - d\left(x_i, x_i^-\right) + m, 0\} \right) \qquad (1)$$

where d is the squared euclidean distance

$$d(x, y) = ||x_i - y_i||_2^2 \qquad (2)$$

and where m is a minimum margin. Using triplet loss minimizes the distance between samples of one class and maximizes their relative distance to samples of other classes. When applied on large corpora of text samples, they form clusters, as depicted in Fig. 4. Note, that the text samples in Fig. 4 stem from an authorship verification challenge generated from data that stems from the PAN 2012 dataset on authorship attribution.

Implementation Details. The pre-trained fastText model embeds each token into a vector of length 300. The embedding model, consisting of 3 stacked LSTMs, therefore has an input layer size of 300. Input and output of the remaining LSTM layers have a size of 300, as well. As suggested by Pascanu et al. [18] and depicted in Fig. 3, the output of one LSTM cell is used as input to the next LSTM cell, i.e. the cells are stacked vertically, while the state vector C is transferred horizontally to the same cell in the next iteration. Dropout [8, 21] with a ration of 0.25 is applied on the input of all but the first layer in order to make the model more robust. The LSTMs' hidden state and cell state are initialized randomly.

Data Representation and Augmentation. The corpora provided for the PAN challenges on authorship verification [23, 24] consist of multiple *cases*. Each case provides one or more *known* documents that are guaranteed to be written by one author, and one document of *unknown* origin. The documents each span far more than the 1–3 sentences that we aim for in this work. Multiple shorter samples are therefore extracted from each document. Sentences are treated as atomic units in this work and sentence boundaries are preserved. While approaches such as Bagnall [1] work on fixed size character sequences, thus breaking words at arbitrary positions, operating on word level encourages us to preserve higher order syntactic structures. Since deep learning techniques require more training data than traditional machine learning approaches, we augment our dataset by permuting the sentences within one document. For the sentences in one document we generate all combinations that (a) preserve the original order, (b) do not contain gaps, and (c) do not exceed a maximum length of $k \in \mathbb{N}$. A document consisting of the four sentences $(1, 2, 3, 4)$ and $k = 3$, for example, yields the combinations (1), (2), (3), (4), $(1, 2)$, $(2, 3)$, $(3, 4)$, $(1, 2, 3)$, and $(2, 3, 4)$. It omits $(1, 3)$, however, since it contains a gap, $(2, 1)$ since it does not correspond to the original sentence order, and $(1, 2, 3, 4)$ since it exceeds the length limit $k = 3$. This way, samples of different length containing varying amounts of context can be created from less original content. We preserve sentence order and sentence connections, thus avoiding samples that are semantically or syntactically incorrect. For each case in the PAN corpus, we create two lists of known and unknown samples from the respective documents.

Sample Grouping. As Sect. 4 will discuss in greater detail, single short samples often do not contain sufficient stylometric information to predict authorship reliably. This is a direct result of the self-imposed restriction on sample length. In this case embeddings of text samples could move into clusters representing a different author. The proposed system therefore uses groups of size $j \in \mathbb{N}$ composed of randomly selected samples by one author. All samples within one group are embedded individually and summarized into one spatial centroid. This way, the impact of outliers can be reduced and the overall confidence in a prediction can be raised. Figure 4 shows that the system forms more disjoint clusters once sample grouping is introduced. Using single short samples, the presented system is able to form clusters by authorship, many outliers merge with other clusters, however. Combining multiple text samples, as shown in Fig. 4b, leads to clearer separation. In a social media use case, this means that the system's ability to reliably verify or reject the authenticity of new posts grows, the more new content is available.

We would like to emphasize that using groups of short samples merely mitigates the underlying problem that short text samples carry little stylometric information. Section 4 demonstrates how longer samples lead to much more reli-

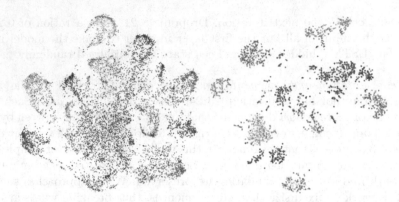

(a) Embeddings of single short text   (b) Embeddings of short text sam-
samples   ples in groups of 5

Fig. 4. 2-dimensional t-SNEs of stylometric embeddings, colored by authorship. Compare the clusters of single text samples and the centroids of multiple embeddings. The exemplary text samples stem from a PAN 2012 dataset on authorship attribution.

able verification results. Nonetheless, taking multiple posts into account for verification is a realistic scenario for an automated style breach detection system.

Distance-Based Authorship Verification. During training, a triplet of samples is generated from the known documents exclusively. The negative sample is picked from the known sample of another author in the dataset. Other resources such as Wikipedia articles could also be used.

At application time, only the documents within one *case* of the PAN challenge are considered. Pairs of known and unknown samples are created. For each pair, both text samples are embedded and their respective distance is computed. If a group of text samples is used as described above, the distance between their respective centroids is computed. Since documents with the same writing style should be embedded close to each other, the distance between two embeddings can be regarded as metric similarity of two text samples. Their similarity is used to infer if the two samples are authored by the same person or not. In order to evaluate the performance of the presented system, we compute the area under the receiver operation curve (AUROC) as an application-independent performance measure. In a real-world application, a threshold would need to be found depending on the requirements at hand.

4   Results

The presented system was evaluated on the English novel section of the PAN 2014 challenge on authorship verification [24]. Its evaluation corpus provides 200 cases, each with one known and one unknown document. Each document contains 6,104 words on average and thus far exceeds the limit on sample size that we imposed

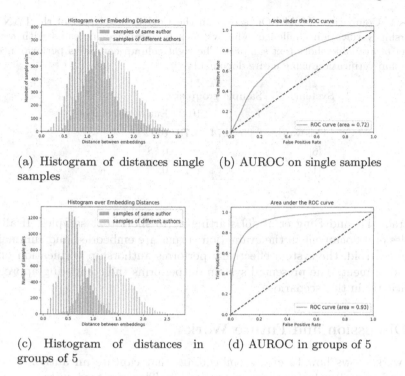

(a) Histogram of distances single samples

(b) AUROC on single samples

(c) Histogram of distances in groups of 5

(d) AUROC in groups of 5

Fig. 5. Histograms and ROC curves for pairwise distances of embeddings. In the histograms, blue indicates the distance between samples by the same author, whereas orange indicates different authors. The data for these exemplary figures stems from the PAN 2012 authorship attribution challenge. (Color figure online)

on our system. As described before, short samples of 1–3 sentences are extracted from each document and embedded using the proposed embedding model. The pairwise distances between embeddings of known and unknown samples form two distributions, as presented in Fig. 5a. The histogram shows that distances between samples that were authored by the same person tend to be smaller than between samples of different authors. The overlap between the two distributions is rather large, however, leading to high false-positive and false-negative rates for any given threshold, which is reflected in the AUROC, shown in Fig. 5b. Introducing sample grouping, as shown in Fig. 5c and d with $j = 5$ in this example, reduces the overlap of the two distributions and increases the discriminative power of the embedding system. Table 1 presents the performance of our system on the English novel section of the PAN 2014 authorship verification challenge, compared to the work of Khonji and Iraqi [11] and Ding et al. [6], as discussed in Sect. 2. When comparing single short text samples of 1–3 sentences, i.e. using a group size of 1, our system is not able to verify authorship with competitive performance. If more than one text sample of unknown origin is available, however, the performance of our classifier increases. It exceeds the performance of Khonji

Table 1. Comparison of AUROC scores on the English novel part of the PAN 2014 authorship verification challenge. The system's performance increases with a rising number of available short text samples. The right column compares performances on authorship verification using entire documents.

System	Sample group size			
	1	5	10	Entire document
(This work)	0.65	0.74	**0.78**	**0.92**
[6]	-	-	-	0.78
[11]	-	-	-	0.75

and Iraqi [11] and Ding et al. [6] starting at 10 short text samples. If all text samples of a document in the evaluation corpus are embedded and summarized in one centroid, the system effectively performs authorship verification on the entire document. The presented system outperforms the competing approaches significantly in this scenario.

5   Discussion and Future Work

This work shows how to create embeddings that capture an author's writing style and spatially cluster them by authorship. The proposed system is able to reliably verify authorship on short text samples with little stylometric value if multiple (10 or more for competitive results) samples are available for comparison. For social media platforms such as Facebook or Twitter this means that, given a number of previous posts by a user, new posts on this user's account can be reliably verified once a larger number of new posts is added to his or her timeline. If the presented approach was combined with other features, such as access time, used IP addresses or hardware identifiers, for instance, the number of posts required for reliable authorship verification could be reduced. Furthermore, a social media provider could verify the identity of an author with minimal user interaction by asking for password confirmation or other means of authentication.

While the presented system performs well on a subset of the PAN 2014 challenge, it struggled on the PAN 2015 dataset, which provides far fewer known text samples for each author. Bagnall [1] as well as Shrestha et al. [20] argue that a focus on character-level features and word morphology is more promising when restricted to short text samples. A future implementation should therefore adapt the concept of character-bi-gram embeddings as presented by Shrestha et al. [20] and combine it with triplet loss in order to perform authorship verification.

Finally, our work performed little preprocessing. As introduced in Sect. 2, it would be valuable to replace specific semantic content such as hashtags, dates, or URLs with placeholders that focus on sentence composition rather than content. Especially regarding social media, special character sequences such as acronyms and emoji should be preserved in any feature extraction process. Eisner et al.

[7] introduced the tool `emoji2vec` to embed emoji into high dimensional space, an approach that could be incorporated into future implementations.

6 Conclusion

This work described the problem of compromised social media accounts and the inability of people to detect the resulting style breaches. In order to solve this problem computationally, this work explored how deep and metric learning techniques can be employed for the authorship verification challenge. It presents a system that learns stylometric embeddings which spatially disseminate authors, thus distinguishing them by writing style. The resulting distance between two text samples was successfully used to infer if they were authored by the same person or not. We described how text samples can be augmented in order to create more training samples from limited data sets. Furthermore, we presented techniques that can mitigate the challenge of very short text samples in social media applications and improve performance to competitive levels. A key contribution of this work is that the presented system uses text samples that range between one and three sentences in length, as opposed to other approaches that used entire essay-length documents. This restrictions was imposed on our system in order to simulate the requirements of an authorship identification system applied on a social media platform. The proposed approach significantly outperformed competitors on the PAN 2014 authorship verification challenge while at the same time using far shorter text samples. We reach competitive performance using 10 or more short text samples and reach AUROC scores of up to 0.91 on entire documents.

References

1. Bagnall, D.: Author identification using multi-headed recurrent neural networks. In: CEUR Workshop Proceedings, vol. 1391 (2015). https://arxiv.org/pdf/1506.04891.pdf
2. Bojanowski, P., Grave, E., Joulin, A., Mikolov, T.: Enriching word vectors with subword information. arXiv:1607.04606v1 [cs.CL] (2016). http://arxiv.org/abs/1607.04606
3. Cho, K., et al.: Learning phrase representations using RNN encoder-decoder for statistical machine translation. In: Proceedings of the 2014 Conference on Empirical Methods in Natural Language Processing (EMNLP), pp. 1724–1734 (2014). http://arxiv.org/abs/1406.1078
4. Cho, K., Merrienboer, B.V., Bahdanau, D., Bengio, Y.: On the properties of neural machine translation: encoder-decoder approaches. In: SSST-2014, pp. 103–111 (2014). http://www.aclweb.org/anthology/W14-4012
5. Dai, A.M., Le, Q.V.: Semi-supervised sequence learning. In: Advances in Neural Information Processing Systems (NIPS 2015), pp. 1–9 (2015). http://arxiv.org/abs/1511.01432
6. Ding, S.H.H., Fung, B.C.M., Iqbal, F., Cheung, W.K.: Learning stylometric representations for authorship analysis (2016). https://arxiv.org/pdf/1606.01219.pdf

7. Eisner, B., Rocktäschel, T., Augenstein, I., Bošnjak, M., Riedel, S.: emoji2vec: learning emoji representations from their description, pp. 48–54 (2016). http://arxiv.org/abs/1609.08359

8. Hinton, G.E., Srivastava, N., Krizhevsky, A., Sutskever, I., Salakhutdinov, R.R.: Improving neural networks by preventing co-adaptation of feature detectors, pp. 1–18 (2012). http://arxiv.org/abs/1207.0580

9. Hoffer, E., Ailon, N.: Deep metric learning using triplet network. Lecture Notes in Computer Science (including subseries Lecture Notes in Artificial Intelligence and Lecture Notes in Bioinformatics), vol. 9370, no. 2010, pp. 84–92 (2015). https://arxiv.org/pdf/1412.6622.pdf

10. Juola, P., Stamatatos, E.: Overview of the author identification task at PAN 2013. In: Forner, P., Navigli, R., Tufis, D. (eds.) CLEF 2013 Evaluation Labs and Workshop - Working Notes Papers, 23–26 September, Valencia, Spain, September 2013

11. Khonji, M., Iraqi, Y.: A slightly-modified GI-based author-verifier with lots of features (ASGALF): notebook for PAN at CLEF 2014. In: CEUR Workshop Proceedings, vol. 1180, no. 1, pp. 977–983 (2014). http://ai2-s2-pdfs.s3.amazonaws.com/cab5/af021b2277c860cb5095c7f29c49084e3ff1.pdf

12. Koppel, M., Seidman, S.: Automatically identifying pseudepigraphic texts. In: EMNLP 2013 (2004), pp. 1449–1454 (2013). http://www.aclweb.org/anthology/D13-1151

13. Koppel, M., Winter, Y.: Determining if two documents are written by the same author. J. Assoc. Inf. Sci. Technol. 65(1), 178–187 (2014)

14. Le, Q., Mikolov, T.: Distributed representations of sentences and documents. In: International Conference on Machine Learning - ICML 2014, vol. 32, pp. 1188–1196 (2014). http://arxiv.org/abs/1405.4053

15. Li, J., Luong, M.T., Jurafsky, D.: A hierarchical neural autoencoder for paragraphs and documents. In: ACL, pp. 1106–1115 (2015)

16. Mikolov, T., Corrado, G., Chen, K., Dean, J.: Efficient estimation of word representations in vector space. CrossRef Listing of Deleted DOIs 1, 1–12 (2013). http://arxiv.org/pdf/1301.3781v3.pdf, http://www.crossref.org/deleted_DOI.html

17. Palangi, H., et al.: Deep sentence embedding using long short-term memory networks: analysis and application to information retrieval. IEEE/ACM Trans. Audio Speech Lang. Process. 24(4), 694–707 (2016). http://arxiv.org/abs/1502.06922%5Cnwww.arxiv.org/pdf/1502.06922.pdf%5Cnieeexplore.ieee.org/lpdocs/epic03/wrapper.htm?arnumber=7389336%5Cnarxiv.org/abs/1502.06922

18. Pascanu, R., Gulcehre, C., Cho, K., Bengio, Y.: How to construct deep recurrent neural networks, pp. 1–13 (2013). http://arxiv.org/abs/1312.6026

19. Rocha, A., et al.: Authorship attribution for social media forensics. IEEE Trans. Inf. Forensics and Secur. PP(99), 1–30 (2016)

20. Shrestha, P., Sierra, S., González, F.A., Rosso, P., Montes-y Gómez, M., Solorio, T.: Convolutional neural networks for authorship attribution of short texts. In: EACL 2017, p. 669 (2017)

21. Srivastava, N., Hinton, G., Krizhevsky, A., Sutskever, I., Salakhutdinov, R.: Dropout: a simple way to prevent neural networks from overfitting. J. Mach. Learn. Res. 15, 1929–1958 (2014). http://www.jmlr.org/papers/volume15/srivastava14a/srivastava14a.pdf?utm_content=buffer79b43&utm_medium=social&utm_source=twitter.com&utm_campaign=buffer

22. Stamatatos, E.: A survey of modern authorship attribution methods. 14(4), 90–103 (2013). http://www.clips.ua.ac.be/stylometry/Lit/Stamatatos_survey2009.pdf

23. Stamatatos, E., amd Ben Verhoeven, W.D., Juola, P., López-López, A., Potthast, M., Stein, B.: Overview of the author identification task at PAN 2015. In: Cappellato, L., Ferro, N., Jones, G., San Juan, E. (eds.) CLEF 2015 Evaluation Labs and Workshop - Working Notes Papers, 8–11 September, Toulouse, France. CEUR-WS.org, September 2015

24. Stamatatos, E., et al.: Overview of the author identification task at PAN 2014. In: Cappellato, L., Ferro, N., Halvey, M., Kraaij, W. (eds.) CLEF 2014 Evaluation Labs and Workshop - Working Notes Papers, 15–18 September, Sheffield, UK. CEUR-WS.org, September 2014

25. Sutskever, I., Vinyals, O., Le, Q.V.: Sequence to sequence learning with neural networks. In: Advances in Neural Information Processing Systems (NIPS), pp. 3104–3112 (2014). http://papers.nips.cc/paper/5346-sequence-to-sequence-learning-with-neural

26. Wang, J., et al.: Learning fine-grained image similarity with deep ranking. In: Proceedings of the IEEE Computer Society Conference on Computer Vision and Pattern Recognition, pp. 1386–1393 (2014). https://arxiv.org/pdf/1404.4661.pdf

Extraction of Hypernyms
from Dictionaries with a Little Help
from Word Embeddings

Maria Karyaeva[1]([✉]), Pavel Braslavski[2], and Yury Kiselev[3]

[1] Yaroslavl State University, Yaroslavl, Russia
mari.karyaeva@gmail.com
[2] Ural Federal University, Yekaterinburg, Russia
pbras@yandex.ru
[3] Yandex, Yekaterinburg, Russia
ykiselev.loky@gmail.com

Abstract. The paper investigates several techniques for hypernymy extraction from a large collection of dictionary definitions in Russian. First, definitions from different dictionaries are clustered, then single words and multiwords are extracted as hypernym candidates. A classification-based approach on pre-trained word embeddings is implemented as a complementary technique. In total, we extracted about 40K unique hypernym candidates for 22K word entries. Evaluation showed that the proposed methods applied to a large collection of dictionary data are a viable option for automatic extraction of hyponym/hypernym pairs. The obtained data is available for research purposes.

Keywords: Hypernymy · Semantic relations · Thesaurus · Word2vec

1 Introduction

Hypernymy relations such as *dog* ≺ *mammal* and *roadster* ≺ *car* are essential for organization of electronic thesauri, ontologies and taxonomies for natural language processing. Building such resources manually is a labor-intensive task. Manually built resources have an inherently narrow scope and need to be constantly maintained and kept up-to-date. Therefore, automatic methods for constructing linguistic resources have been demanded and actively explored over the last two decades. Pioneering experiments on the extraction of hypernymy relations between words from text corpora in early 1990s employed hand-crafted templates. Later, attempts were made to automatically expand the initial set of templates, as well as to combine them with distributional semantics representations. Nowadays neural networks have been actively using to detect hypernym/hyponym word pairs.

In this paper, we address the task of noun hypernym extraction from dictionary definitions in Russian. Earlier work on template-based extraction of hierarchical relations showed that encyclopedic data is best suited for this approach.

© Springer Nature Switzerland AG 2018
W. M. P. van der Aalst et al. (Eds.): AIST 2018, LNCS 11179, pp. 76–87, 2018.
https://doi.org/10.1007/978-3-030-11027-7_8

Template-based methods require that both words (hyponym and hypernym) occur in the same sentence, which is not very common in arbitrary texts. Dictionaries often define a word through a more generic concept, for example:

- *Car*: a wheeled <u>vehicle</u> that moves independently, powered mechanically, steered by a driver.
- *Convertible*: a <u>car</u> whose roof can be removed or folded.[1]

The main advantage of this approach is its simplicity, but it has also additional benefits. When employing dictionary definitions, we naturally solve the problem of distinguishing hypernyms corresponding to different word senses expressed in turn in separate definitions. For example, both pairs $bass \prec fish$ and $bass \prec musical\ instrument$ are expected to be presented in a thesaurus. Dictionaries make it possible to find hypernyms even for infrequent words, which is rather hard in case of purely statistical methods. In addition, we use multiple dictionaries simultaneously, which potentially improves both precision and recall of the resulting pairs. As a supplemental method, we reproduce experiments on hypernym/hyponym pairs detection based on pre-trained word embeddings and supervised classification.

The study resulted in 47,831 (40,298 unique) hypernym candidates extracted for the 21,957-word initial list by three methods. Evaluation showed that the precision of the simple template-based methods lies in the range of 0.57–0.64 (see Sect. 5 for details). Classification-based methods using word embeddings demonstrate lower quality, but still can be considered as a complementary solution.

2 Related Work

There are two main approaches to extraction of hyponymy/hyperonymy relations: (1) lexico-syntactic templates and (2) semantic vector representations. In addition, there are methods for building linguistic resources based on semi-structured data from Wiktionary and Wikipedia, as well as combinations of different approaches.

Researchers have been engaged with the task of automatic detection of hierarchical relations between words since early 1990s. In her pioneering work, Hearst [5] proposed a set of manually crafted templates aimed at extraction of hypernyms and co-hyponyms from a large text corpus, for example: x_1, x_2, x_3 *and other* y, where x_i and y are noun phrases representing co-hyponyms and a hypernym, respectively. Sabirova and Lukanin [17] applied a similar method to a corpus of texts in Russian. The studies showed that this kind of approach works best with encyclopedic data as a source. Kiselev et al. [8] followed this line of research and elaborated extraction of hyponymy/hypernymy relations from dictionary definitions in Russian using lexico-syntactic patterns. In the

---

[1] A different, though less common approach is to define a word trough its synonyms: *car – a motorcar or automobile,* or through cognate words: *running – the action of the verb to run.*

current work, we improve on the approach by using several dictionaries in parallel, extracting multiwords in addition to single-word candidates, as well as compliment the method with a simple approach based on pre-trained semantic vector representations.

In the works [2,3] initial template-based methods were improved by taking into account semantic similarity of words based on conjunctive structures and LSA representations. Snow et al. [20] elaborated a method for automatic extraction of such patterns, followed by classification-based hyponymy/hyperonymy detection.

Word embeddings [12] proved to be a good semantic representation of words in many natural language processing tasks. Thus, attempts have been made to use pre-trained models for the task of hypernymy detection. Several studies [1,16,21] investigated supervised classification and vector representations of words for hypernymy detection task: word pairs are presented either as concatenation or difference of their vector representations, and classifiers are trained on positive and negative word pairs. We reproduce these approaches in the current work. Fu et al. [4] elaborated this approach: first, vector differences are clustered (clusters are expected to reflect different types of hierarchical relations), then a separate projection is learned for each cluster using a training sample from an existing thesaurus. After space partitioning and projections learning, each word pair is assigned to a positive or a negative class. Shwartz et al. [19] combined syntactic and semantic information to address the task of hyponym/hypernym pair extraction: each candidate word pair is represented as concatenation of individual words' vectors and averaged vector representation of the paths in the dependency tree between words.

Many studies explore the possibility of building structured semantic resources from Wikipedia data, see for example [6,13,15].

3   Data

3.1   Dictionaries

We use six dictionaries of Russian as the main source of data in the current study:

(EFR) T. F. Efremova, New dictionary of Russian, 2000.
(BAS) S. A. Kuznetsov, Comprehensive Dictionary of the Russian Language, 1998.
(MAS) A. P. Evgenyeva, Small Academic Dictionary, 1957.
(BAB) L. Babenko, Dictionary of synonyms of the Russian Language, 2011.
(USH) D. N. Ushakov, Explanatory Dictionary of the Russian Language, 1935.
(OZH) S. I. Ozhegov, Explanatory Dictionary of the Russian Language, 1949 (1992).

These dictionaries span 75 years and reflect well the diversity of Russian lexicography, in particular – different approaches to definitions. Five of them are explanatory dictionaries, BAB is a relatively small vocabulary of synonyms, with definitions given for the whole synsets. EFR differs through its more 'analytical' definitions. As shown in [7], MAS and BAS contain many similar definitions.

We collected all noun entries along with all definitions from the dictionaries and retained only items presented in the list of 60,000 most frequent words in the Russian National Corpus (RNC)[2]. This resulted in 21,957 unique nouns. Main statistics of the dictionary dataset can be found in Table 1.

Table 1. Main statistics of the data used in the study: **entries** – noun entries from six dictionaries; **defs** – their definitions; **entries@60K** and **defs@60K** – entries and their definitions after filtering out infrequent word entries; **WIKT** – words in **entries@60K**, for those hypernyms are presented in Wiktionary; **WIKT_pairs** – words in **WIKT**, whose definitions contain hypernyms from Wiktionary; unique definitions – unique strings after lowercasing and punctuation removal.

	entries	defs	entries@60K	defs@60K	WIKT	WIKT_pairs
EFR	55,499	84,072	18,812	37,918	5,085	3,031
BAS	29,877	42,229	14,888	24,812	4,579	2,604
MAS	33,664	51,182	17,922	32,627	4,794	2,899
BAB	1,810	2,011	1,510	1,689	569	217
USH	33,005	50,615	16,465	30,410	4,602	2,536
OZH	18,709	26,961	13,428	20,914	4,278	2,484
Unique	66,784	233,718	**20,312**	**134,460**	5,230	4,253

3.2 Reference Hyponym/Hypernym Pairs

Wiktionary[3] is a large online dictionary, whose content is curated by community members. At the time of writing, Russian Wiktionary contains more than 173,000 entries. A subset of dictionary entries contains semantic relations – synonymy, antonymy, hypernymy, and hyponymy. We collected 59,582 hyponym/hypernym noun pairs from Wiktionary. Then, we used the same 60K frequency list from RNC to filter out pairs containing rare words. Proper nouns, names, and narrow domain terms were removed, e.g. *Альза ≺ река* (*Alza ≺ river*), *Муцал ≺ имя* (*Muzzal ≺ name*), *алюмогидрид ≺ соединение* (*aluminumhydride ≺ compound*). This resulted in 10,826 pairs for 7,124 unique hyponyms.

We also matched these reference pairs with dictionary data (hyponym as a dictionary entry, one of whose definitions contains hypernym), see the last column in Table 1. It can be seen that about 39% of entries have hypernyms in one of their definitions.

---

[2] http://ruscorpora.ru.

[3] https://ru.wiktionary.org/.

3.3   Clustering of Definitions

In contrast to previous studies dealing with relation extraction from text, we deal with semi-structured data from several dictionaries. The multitude of data sources increases coverage and provides additional evidence in case of frequent items. At the same time, many definitions from different dictionaries are very similar [7]. To mitigate the redundancy problem we first cluster definitions from different sources corresponding to the same word entry. In our initial dataset, 93% of word entries have more than one definition.

We apply graph-based clustering that proved to be efficient for many NLP tasks [9,14,22]. All definitions corresponding to a word entry are vertices; there is an edge between them if cosine similarity is greater than 0.15. Table 2 shows clustering of definitions for a word *пиявка (leech)* as an example.

As a result, 134,460 definitions are clustered into 48,739 clusters, 21,199 of them contain a single definition.

Table 2. Example of clustered definitions for a word *пиявка (leech)*.

◇ червь класса кольчатых // *a worm that belongs to the class of annelid worms*
◇ пресноводный червь, питающийся кровью животных, к телу которых он присасывается // *a freshwater worm sucking blood from their host animal*
◇ о жадном и жестоком человеке, живущем за счет других // *a greedy and cruel person who extorts profit from others*
◇ тот, кто живёт за счёт чужого труда и ведёт паразитический образ жизни // *somebody who lives a parasitic life*
◇ щелчок, удар пальцами по телу человека, причиняющий острую жгучую боль // *a finger flip on a person's body causing acute burning pain*

4   Methods

4.1   Single Words as Hypernyms

As mentioned above, we use the method for hypernym extraction described in the paper [8] as a starting point for this study. The approach is very simple: the first noun occurring in definition is extracted, with a short list of exception words, for example *kind, species, section,* etc. The method uses data from a single dictionary. In our case, we use definitions from several dictionaries, previously combined into clusters that presumably correspond to different word senses. The modification of our method is that we extract a single most frequent candidate from a cluster. If there is a tie and we cannot make a choice based on the frequency, then the candidate with the highest RNC frequency is selected. The example below demonstrates how this strategy works in case of four definition clusters and extracted candidates in case of the word entry *искусство (art)*:

1. *система (system)* ×2, **отрасль (domain)** ×3 – most frequent entry;
2. **умение (skill)** ×3, *знание (knowledge)* ×1 – most frequent entry;
3. *отражение (reflection)* $freq_{RNC} = 5,789$, *воспроизведение (reproduction)* $freq_{RNC} = 1,323$, **деятельность (activity)** $freq_{RNC} = 46,298$ – RNC frequency;
4. **дело (occupation)** – single candidate.

Using this approach, we extracted 58,834 hypernym candidates from 48,739 definition clusters. In some cases identical hypernym candidates were extracted from different definition clusters. After eliminating duplicate pairs we ended up with 39,252 pairs in total. In case of 17% entries the method did not extract any candidate. 28% out of 39,252 extracted pairs are present in the list of hyponym/hypernym pairs extracted from the Wiktionary (see Sect. 3.2).

4.2   Multiwords as Hypernyms

If we look at single-word hypernym candidates extracted on the previous stage, we can notice that many of them are very abstract and though connected to their intended hyponyms, should probably reside not immediately above, but on a higher taxonomy level. For example: *адмирал ≺ чин (admiral ≺ rank)*, *ил ≺ масса (silt ≺ mass)*, *нездоровье ≺ состояние (ailment ≺ condition)*. When inspecting source definitions, we noticed that it is possible to extract more immediate hypernyms, when considering multiwords, for example *военно-морской чин (navy rank)*, *вязкая масса (viscous mass)*, and *болезненное состояние (painful condition)*.

Multiword expressions (MWE) have been underrepresented in traditional dictionaries [7], but became an essential part of modern electronic linguistic resources. For example, about 41% of Princeton WordNet entries are multiword [18]; MWE constitute about 47% of total 115,000 RuWordNet entries [11].

To extract verbose hypernym candidates from definitions, we used several morphosyntactic patterns, see Table 3.

These patterns have been used in many previous MWE extraction studies. Table 4 illustrates the productivity of patterns by dictionary. According to these figures, **EFR** seems to have a somewhat different definition structure.

All the patterns correspond to nominal phrases, thus the extracted expressions potentially refine the single-word candidates obtained by the previous method. Note that patterns are nested: we can extract several multiwords from the same definition. A possible downside of these more complicated structures is that they can lead to extraction of non-lexicalized expressions or extraction of almost entire definitions. This is especially obvious in case of 4-grams, for example: *рецептура – способ изготовления лекарственных веществ (formulation – a method of manufacturing pharmaceutical substances)*; *энциклопедист – представитель группы передовых мыслителей (encyclopaedist – a representative of a group of advanced thinkers)*; *стоянка – место поселения первобытного человека (settlement site – a settlement of the primitives)*. Nevertheless, even extracted items of this kind can be useful as glosses – many thesauri allow "empty" synsets containing a definition only that serve as a means for a more balanced hierarchy.[4]

Table 3. Multiword extraction patterns and their productivity (A – adjective; N – noun; N_G – noun, genitive case; Pr – preposition; in AN patterns adjective and noun must be coordinated).

Pattern	Example	#candidates
AN	аул ∷ горное селение (aul) (a mountain village)	41 891
NN_G	вихор ∷ прядь волос (cowlick) (a strand of hair)	23 145
AAN	баккара ∷ азартная карточная игра (baccara) (a gambling card game)	3 328
NAN_G	хронология ∷ последовательность исторических событий (chronology) (a sequence of historical events)	7 935
ANN_G	пядь ∷ русская мера длины (span) (a Russian unit of length)	2 469
NN_GA	арахис ∷ растение семейства бобовых (peanut) (a plant of the family of bean)	1 268
$NPrN$	ванна ∷ сосуд для купания (bathtub) (a vessel for bathing)	9 629
NN_GAN_G	ватт ∷ единица измерения активной мощности (watt) (a unit of active power)	1 123

Table 4. Percentage of definitions by dictionary, where the pattern has fired.

Pattern	BAS(%)	EFR(%)	OZH(%)	BAB(%)	MAS(%)	USH(%)
N	92	77	95	88	80	84
AN	37	28	35	31	31	30
NN_G	19	20	17	23	15	14
AAN	3	2	3	3	2	2
NAN_G	6	4	4	5	4	4
ANN_G	2	2	2	2	1	1
NN_GA	1	1	0	1	0	0
NN_GAN_G	1	1	1	1	1	0
$NPrN$	5	4	5	3	4	5

As with the single-word hypernyms, our goal was to extract one expression from each cluster of definitions. In order to filter out less reliable bigrams, we used two additional criteria. We required that the candidate is (1) selected from at least two clusters of our dataset and (2) present in the RNC frequency list of

---

[4] See for example GermaNet, http://www.sfs.uni-tuebingen.de/GermaNet/.

bigrams.[5] The former criterion can be seen as a simple sanity check: a hypernym is expected to have several 'children'; while the latter one is a light lexicalization test. Extracted candidates that meet both criteria are championed. Candidates that fail to meet both conditions, are discarded. Since the number of extracted 3- and 4-gram is much lower, we kept all of them for subsequent evaluation.

For example, for the entry *сыч (athene)* 11 multiword hypernym candidates are extracted, including *сумеречная птица (twilight bird)* and *хищная птица (a predatory bird)*. The latter one is championed, since it is potentially very productive as a hypernym. It is extracted from definitions of 17 entries, including *сип (sip)*, *стервятник (vulture)*, *кречет (gyrfalcon)*, etc.

In total, 61,134 bigrams and 17,628 3- and 4-grams were extracted. After applying our two criteria we ended up with 19,880 (18,360 bigrams and 1,520 3- + 4-grams) multiword hypernym candidates for 11,623 word entries.

4.3   Hypernym Extraction Based on Word Embeddings

An interesting property of word embeddings is that they capture not only semantic similarity of words, but also other relations between them. A famous example cited by Mikolov et al. [12] is that the closest vector to $vector(king) - vector(man) + vector(woman)$ is $vector(queen)$. There were several attempts to use pre-trained word embeddings for hypernymy detection, as well as learning dedicated embeddings for the task (see Sect. 2).

In this study we use pre-trained word2vec models from the Rus Vectōrēs project [10]. In particular, we employ 300-dimensional word vectors for Russian trained on Wikipedia and RNC (600 million tokens in total) using continuous skip-gram model.[6] We use `gensim` library[7] to find closest vectors. First, we trained a classifier using reference hyponym/hypernym pairs (see Sect. 3.2) as positive examples and pairs from a dataset originating from a semantic similarity shared task as negative examples. Then, using the classifier we extracted best candidate from the words closest to each item in our initial list of 21,957 words. The main difference of our approach from the similar experiments reported in the literature is that we aim at searching hypernyms not within a closed set of annotated word pairs, but 'in the wild'.

As positive examples we took 10,826 reference pairs, for 8,496 out of which pre-trained vectors are available. Out of these 8.5K reference pairs only in case of fewer than 3K a hypernym can be found as a similar entry in 1K words of word2vec outputs close to a given hyponym; 1.9K (23%) of them can be found in top-200 words. We sampled the same amount of negative pairs from 'unrelated examples' of the RuThes dataset[8]. We experimented with vector concatenations

---

[5] The list http://ruscorpora.ru/corpora-freq.html contains 6.8 million bigrams with frequency above 3. We also matched the extracted multiwords with Wikipedia titles, but there were fewer than 9% matches, and we did not use this as a selection criterion.

[6] The embeddings can be downloaded from http://rusvectores.org/models/, model `ruwikiruscorpora_upos_skipgram_300_2_2018`.

[7] https://radimrehurek.com/gensim/models/word2vec.html.

[8] http://russe.nlpub.org/downloads/.

(ConVec) and vector differences (DiffVec). We trained a classifier using SVM with RBF kernel implemented in `scikit-learn`[9]. Three-fold cross-validation resulted in 0.83 and 0.85 accuracy for ConVec and DiffVec, respectively, which is comparable with scores for similar experiment on closed-set hypernymy classification reported in the literature. To obtain hypernym candidates for initial word list not participating in training, we retain nouns from the top-200 words for each list item and feed them to the classifier. In addition, we require that the candidate words are present in RNC60K frequency list. The positively classified pair farthest from the separating hyperplane is considered the best match.

5   Evaluation

For evaluation we randomly sampled 300 words from the 14,833 entries from the initial list of 21,957 words excepting 7,124 unique entries from the Wiktionary hyponym/hypernym pairs. For these 300 words, we collected all hypernym candidates generated by all approaches. These pairs were evaluated by three assessors. The assessors judged each pair as correct (1) or incorrect (0). Cohen's kappa is 0.51, which indicates a moderate agreement. In addition, assessors could provide more fine-grained annotations:

- senses are *too distant* (**dist_sen**);
- the words is semantically similar, but reflect a *different relation* – synonymy, antonymy, co-hyponyms, association (**diff_rel**);
- expression is *grammatically incorrect*, expression is *incomplete, cognate* words, *inverse* relation, *definition* instead of hypernym (collapsed to **others** below).

Individual judgments were converted to common scores by applying an agreement rule – at least two assessors marked the pair as correct. Table 5 summarizes the evaluation results.

Table 5. Evaluation results.

Method	# candidates	precision	diff_rel	dist_sen	Others
1-word	485	0.57	0.04	0.04	0.28
2-word	236	**0.64**	0.03	0.00	0.23
3-word	14	0.57	0.00	0.00	0.43
ConVec	219	0.30	0.12	0.02	0.46
DiffVec	284	0.46	0.29	0.01	0.27
1w+DiffVec	399	**0.62**			
3|2w+(1w+DiffVec)	323	**0.61**			

It can be seen from the Table that candidate bigrams outperform unigrams in terms of precision, but have a lower coverage. Trigrams are represented scarcely

---

[9] http://scikit-learn.org/stable/modules/svm.html.

in the evaluated sample, though these few candidates demonstrate decent precision. DiffVec outperforms ConVec in terms of coverage and precision. Expectedly, embeddings-based methods often return semantically similar words that manifest a relation different from hypernymy. We can slightly improve the methods by their combination. For example, intersection of single-word extraction (1w) and DiffVec (i.e. the extracted noun is assigned to the positive class by DiffVec) increases precision, though decreases coverage. A simple heuristics 'take a longer candidate if present, a DiffVec-filtered single-word candidate otherwise' delivers about the same precision. The combined approach allows to extract candidates of different word length without losing accuracy.

Figure 1 illustrates relation between the positions of the extracted single-word candidates in definitions and their scores. By design, the majority of the extracted candidates are located towards the beginning of the definitions. It can be seen that the precision of the extracted hypernyms drops after the fifth position, though there are still correct hypernyms quite distant from the word being defined. Table 6 shows that two two-word patterns have approximately equal productivity; the examples of three-word candidates are too few to make reliable conclusions.

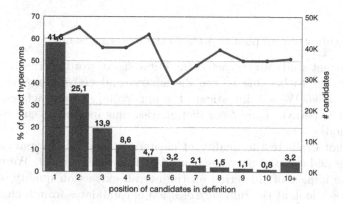

Fig. 1. Positions of extracted single-words candidates in original definitions (bars, right axis) and precision of each position (line, left axis). If the candidate is present in several definitions for the same entry, all positions are accounted for.

Table 6. Productivity of patterns.

	AN	NN_G	AAN	NAN_G	$NPrN$
#pairs	156	80	2	5	3
Precision	0.62	0.69	1.0	0.60	0.67

6   Conclusion and Future Work

In this paper, we presented several methods of extraction of hyponym/hypernym relations based on morphosyntactic templates, simple heuristics, and word embeddings. Manual evaluation demonstrated feasibility of the methods and their combinations. Table 7 demonstrate that our methods are able to produce at least a single candidate for 97% of the RNC frequently list.

Our methods are efficient for practical use and we believe that our findings will be applied to different natural language processing tasks and obtained data will extend existing lexicographic resources. The obtained candidate hyponym/hypernym pairs are freely available for research purposes[10].

Table 7. Coverage of methods: percentage of the most frequent nouns, for which at least one candidate hypernym is extracted.

	1-word	2-word	3-word	ConVec	DiffVec	Total
# entries	18,111	12,304	1,496	19,307	20,428	**21,330**
RNC60K, %	82	56	7	88	93	**97**

In the future work we plan to address the following issues:

- In the current study we restricted the initial list of word entries by the RNC frequency list to be able to use dictionary- and embeddings-based methods in parallel. We will investigate the opportunity to extract less frequent hyponym/hypernym pairs from dictionaries thus increasing the coverage of the method.
- We did not evaluate the quality of clustering, used ad-hoc parameters for clustering and selected only one candidate from each cluster. We will investigate the impact of clustering on the overall extraction quality, as well as take a closer look at the ranking/selection of candidates from a cluster. Multiple candidates can be useful for matching the extracted pair with existing synsets, enriching existing synsets, and increasing the overall quality of the method.
- To train the classifier we applied pairs of unrelated words as negative examples. The study [19] suggests that using other relations (synonymy, part/whole, etc.) as negative examples can increase the specificity of the classifier. We plan to investigate this hypothesis in our future work.
- We restricted evaluation to top-1 candidate of each classifier. For a better understanding of the methods and their potential we plan to evaluate top-5 candidates.

Acknowledgments. MK was supported by RFBR grant #15-37-50912, PB and YK were supported by RFH grant #16-04-12019.

---

[10] https://github.com/YARN-semantic-relations/hyponymic-relationship.

References

1. Baroni, M., Bernardi, R., Do, N.Q., Shan, C.: Entailment above the word level in distributional semantics. In: EACL, pp. 23–32 (2012)
2. Caraballo, S.A.: Automatic construction of a hypernym-labeled noun hierarchy from text. In: ACL (1999)
3. Cederberg, S., Widdows, D.: Using LSA and noun coordination information to improve the precision and recall of automatic hyponymy extraction. In: CoNLL, pp. 111–118 (2003)
4. Fu, R., Guo, J., Qin, B., Che, W., Wang, H., Liu, T.: Learning semantic hierarchies via word embeddings. In: ACL, pp. 1199–1209 (2014)
5. Hearst, M.A.: Automatic acquisition of hyponyms from large text corpora. In: COLING, pp. 539–545 (1992)
6. Hu, J., et al.: Enhancing text clustering by leveraging wikipedia semantics. In: SIGIR, pp. 179–186 (2008)
7. Kiselev, Y., et al.: Russian lexicographic landscape: a tale of 12 dictionaries. In: Dialog (2015)
8. Kiselev, Y., Porshnev, S., Mukhin, M.: Method of extracting hyponym-hypernym relationships for nouns from definitions of explanatory dictionaries. Softw. Eng. **10**, 38–48 (2015). (in Russian)
9. Kozareva, Z., Riloff, E., Hovy, E.: Semantic class learning from the web with hyponym pattern linkage graphs. In: ACL-HLT, pp. 1048–1056 (2008)
10. Kutuzov, A., Kuzmenko, E.: WebVectors: a toolkit for building web interfaces for vector semantic models. In: AIST, pp. 155–161 (2017)
11. Loukachevitch, N., Lashevich, G.: Multiword expressions in Russian thesauri RuThes and RuWordnet. In: AINL, pp. 1–6 (2016)
12. Mikolov, T., Yih, W.t., Zweig, G.: Linguistic regularities in continuous space word representations. In: NAACL, pp. 746–751 (2013)
13. Navigli, R., Ponzetto, S.P.: BabelNet: The automatic construction, evaluation and application of a wide-coverage multilingual semantic network. Artif. Intell. **193**, 217–250 (2012)
14. Navigli, R., Velardi, P., Faralli, S.: A graph-based algorithm for inducing lexical taxonomies from scratch. In: IJCAI, vol. 11, pp. 1872–1877 (2011)
15. Ponzetto, S.P., Strube, M.: Deriving a large scale taxonomy from Wikipedia. In: AAAI, vol. 7, pp. 1440–1445 (2007)
16. Roller, S., Erk, K., Boleda, G.: Inclusive yet selective: supervised distributional hypernymy detection. In: COLING, pp. 1025–1036 (2014)
17. Sabirova, K., Lukanin, A.: Automatic extraction of hypernyms and hyponyms from russian texts. In: AIST, pp. 35–40 (2014)
18. Shudo, K., Kurahone, A., Tanabe, T.: A comprehensive dictionary of multiword expressions. In: ACL-HLT, pp. 161–170 (2011)
19. Shwartz, V., Goldberg, Y., Dagan, I.: Improving hypernymy detection with an integrated path-based and distributional method. In: ACL, pp. 2389–2398 (2016)
20. Snow, R., Jurafsky, D., Ng, A.Y.: Learning syntactic patterns for automatic hypernym discovery. In: NIPS, pp. 1297–1304 (2005)
21. Vylomova, E., Rimell, L., Cohn, T., Baldwin, T.: Take and took, gaggle and goose, book and read: evaluating the utility of vector differences for lexical relation learning. In: ACL, pp. 1671–1682 (2016)
22. Widdows, D., Dorow, B.: A graph model for unsupervised lexical acquisition. In: COLING, pp. 1–7 (2002)

Sentiment Analysis of Telephone Conversations Using Multimodal Data

Alexander Gafuanovich Logumanov[✉], Julius Dmitrievich Klenin,
and Dmitry Sergeevich Botov

Chelyabinsk, Russia
logumanov@gmail.com

Abstract. Sentiment analysis of conversations is a widely studied topic, but the proposed solutions are mostly based only on text analysis, which in the real conditions of telephone conversations is not ideal and contains a lot of mistakes and inaccuracies arising at the stage of speech recognition. Today, there are almost no papers about the sentiment analysis of conversations using multimodal datasets for the Russian language. In this paper, we suggest the use of multimodal sentiment analysis of conversations, with both the recognized text and the audio signal used as the training data. To do this, we assemble our own dataset consisting of records of telephone conversations, labelled by sentiment intensity. The texts are obtained with the help of ready-made tools for automatic speech recognition. We carry out a number of experiments to find the best way to extract features from audio and texts and we also build models for determining the sentiment intensity for individual modalities and a combination of them. Different classification algorithms are compared: linear, neural networks and ensembles of decision trees, where XGBoost works best for audio, Logistic Regression - for text and LightGBM - for multimodal data. The results show that combining several modalities allows to achieve the best quality of classification.

Keywords: Sentiment analysis · Natural language processing
Audio processing · Multimodal data

1 Introduction

The information, which a person perceives is multimodal by nature, and our brain, through the course of evolution, has learned to work with such data. Most business processes are still based on phone calls, and the callers' mood can tell a lot about the success of the negotiations. Unfortunately, in telephone conversations we are limited to only two modalities - acoustic and language, but even this information is enough for people to sense someone else's mood. In this paper, we explore the possibilities of artificial intelligence in the task of multimodal sentiment detection in telephone conversations of a single telecommunications company.

© Springer Nature Switzerland AG 2018
W. M. P. van der Aalst et al. (Eds.): AIST 2018, LNCS 11179, pp. 88–98, 2018.
https://doi.org/10.1007/978-3-030-11027-7_9

Phone conversations have their own specifics, which are imposed by the GSM signal encoding, as well as the conditions of recording. The frequency range of the speech signal is limited to 300–3.400 Hz, while the power of the speech signal varies between 35 to 40 dB. At the time of the call the person might be anywhere, which adds a large amount of noise to the recording signal, which is difficult to get rid of without losing the information necessary for analysis.

2 Related Work

Sentiment analysis (SA) as a separate discipline has formed in the field of NLP (Natural Language Processing), and can be defined as the interpretation of feelings, opinions and subjectivity in texts. SA caught the eye of the scientists in 2002, after the publication of [23], where this task was also presented as an unsupervised learning problem, which used only statistical methods for calculating PMI (Pointwise Mutual Information) between phrases and words "excellent" and "poor". The first use of machine learning for text tonality determination was noted in [16], with the main idea of revealing the possibility to determine the tonality of a text with the same accuracy as its topic.

Almost all subsequent works include the advantages of both unsupervised and supervised learning [12]. [11] presents the combination of vector representation of documents and sentiment lexicons, with the SVM (Support Vector Machine) model reaching the best performance. Artificial neural networks in recent years have also become very popular and started to surpass the classic machine learning models, which can be seen in the [1], where a neural network based on GRU and trained on Word2Vec has shown the best classification performance metrics in the task of sentiment analysis for Russian language. [27] describes modern approaches to sentiment analysis based on deep learning.

Multimodal machine learning began in the 1970s [2], with the discovery of the McGurk effect, showing the interaction between hearing and sight in the speech perception: when a person hears the syllables "ba-ba" from a person who says "ga-ga", they identify it as "da-da". The study of this effect pushed researchers to develop new approaches, such as audio-visual speech recognition (AVSR) [24]. In that paper, audio was presented in the form of a spectrogram, an acoustic representation was extracted from the video using a neural network, and when combined, the recognition accuracy has increased significantly with noisy audio if compared to using only audio data.

The multimodal sentiment labelling is closely related to the recognition of emotions, which began to spread from the preprocessed for further analysis AMI Meeting Corpus dataset [3] published in 2005.

In [22] it was shown that even people, who tag the sentiment of the data, find it much easier to navigate when they simultaneously hear and see the conversation, rather than when using text alone. The agreement between experts when utilizing multimodal sentiment tagging grows significantly.

Sentiment intensity detection also did not stand still, and in 2016 a multimodal sentiment intensity corpus (MOSI) was introduced [25]. Several multimodal models were compared, the key idea of which is the alignment of information from each modality at the level of words and phonemes. A tool called P2FA[1] was used for this, the main disadvantage of which for our study is the lack of support for the Russian language. The most recent works on multimodal sentiment analysis [18,26] also suggest using utterance-level fusion of modalities.

The main idea of all these works is that combining several modalities improves performance. Despite the fact that the multimodal sentiment detection has been studied extensively for English language, there is practically no research for the Russian language. Also, the phone conversations, which are the subject of our research, have their own specificity, making the quality of text recognition one of our biggest issues, therefore it is desirable to use the information from both modalities (acoustic and language) together. We use the best, in our opinion, tool for Russian speech recognition, but it is quite difficult to recognize speech in telephone conversations. This tool substitutes words or simply skips them, which makes it useless to use tools for force alignment of modalities at the utterance level such as aeneas[2] which supports Russian language.

3 Methods

This section describes ways to represent each modality (audio and text), as well as a description of the models built on the features extracted from audio, text features and their combination. Hyperparameters of models were selected by 3-fold cross-validation (to speed up the calculations), then the model was trained on the entire training set and checked on the hold-out. In this paper we used scikit-learn [17], keras [5] with Tensorflow backend and gensim [20] implementations of machine learning algorithms. When choosing machine learning models, we were guided by the following considerations: for dense data we place great hopes on classifiers such as XGBoost, LightGBM, SVC, and in the case of sparse data we rely more on Logistic Regression and Neural Networks, as the most suitable algorithms for such data. Source code, hyperparameters and instructions are available at https://github.com/aminought/multimodal-telephone-sa.

3.1 Audio Representations

Audio for our task is stored in WAV format with a sampling frequency of 8 kHz and 16 bits per sample. To use audio in machine learning models, it could be presented in the following ways:

- Time sequence of signal amplitudes;
- Spectrogram;
- Acoustic features and VGG16

---

[1] https://github.com/srubin/p2fa-vislab.
[2] https://www.readbeyond.it/aeneas/.

Sequence of Samples. First of all, audio is a time sequence of signal amplitude values, taken with a certain sampling rate, so it can be used to train LSTM [8], which is the state-of-the-art for time sequences at the present [6]. The length of such sequence can reach 700000 samples, which can negatively affect the memory consumption and learning time of the model.

Spectrogram is a 2D-representation of audio signal presenting the signal power as a function of time and frequency. In this paper, a mel-spectrogram is used. On a mel-spectrogram, a linear frequency axis is replaced by a logarithmic one, thereby increasing the resolving power of the spectrogram at low frequencies and reflecting the perception of sound information by human hearing organs. The mel-spectrogram can be used as an input of the LSTM as a time sequence, or for CNN [10] as an image. In both cases, because of the distribution of the number of time windows for all audio fragments, we took an input sequence with the length of 100 time windows. Another way is to extract features from an image of a mel-spectrogram using the pre-trained VGG16 [21] and feed it to different classification algorithms.

Acoustic features extracted manually using specialized audio analysis tools for further use in the models for sentiment analysis, are described below. Librosa4 [13] and VGGish [7] was used for this purpose.

Acoustic features are represented as a matrix with a size $m \times n$, where m is a number of windows and n is a number of features. Next, different models such as Logistic Regression, LightGBM, XGBoost and Random Forest were built, based on the aggregation of acoustic features, concatenated in one vector: mean, standard deviation, median, minimum, maximum, excess, skewness.

VGGish is a Tensorflow definition of 128-dimensional embeddings of audio segments produced from a VGG-like audio classification model that was trained on a large YouTube dataset. Then, embeddings was fed to Logistic Regression, XGBoost and Recurrent Neural Network.

3.2 Text Representations

In general, text information can be represented in two forms: Bag-of-words and a vector representation (TF-IDF, word2vec and so on). Before converting the text into different representations, the following operations were performed:

- Tokenization. Words were extracted from texts with the help of a tokenizer based on regular expressions, each document was presented as a list of extracted words.
- Normalization. For each word, a lemmatization was carried out using pymorphy2 [9] library.
- Removal of stop words. From each document, Russian and English stop words from the nltk[3] set were deleted, except words expressing the person's attitude

---

[3] https://www.nltk.org/.

to the subject of conversation: "хорошо" ("good"), "да" ("yes"), "нет" ("no"), "можно" ("sure").

After that, models for detection of sentiment intensity of the text were constructed for each of the representations.

- Bag-of-Words, using CountVectorizer from scikit-learn;
- TF-IDF [19] or term frequency - inverse document frequency;
- word2vec [14] is a group of two-layer neural network models, of which we used:
  - trained skip-gram word2vec model with averaging of the vectors of all the words in the document;
  - word2vec models pretrained on, first, russian Wikipedia and, second, Russian Distributional Thesaurus [15].

For those representations the best model and hyperparameters were chosen using cross-validation. The model was chosen from the following space of models: Logistic Regression, Random Forest, SVC.

3.3   Multimodal Representations

We can combine modalities in at least two ways:

- Train the models for each modality and combine them with stacking. We used a simple scheme for averaging the results on a test dataset;
- Concatenate features of several modalities into one vector and build models that take information from two modalities as inputs: both audio and text features.

For the second case, we trained and compared the following models: XGBoost, Logistic Regression, Random Forest.

4   Dataset

In this section, we describe the process of building our own set of training data for the creation of a model for sentiment analysis of telephone conversations using multimodal data.

4.1   Data Acquisition

Work in a telecommunications company involves access to some confidential information, such as records of telephone conversations with customers. We extracted 1084 random telephone conversations from the database of such company and separated them by periods of silence into 10000 fragments with the help of pydub[4] library. Pauses were considered to be intervals of 2 s or more and a signal power of less than -16 dB. In each segment, speech was recognized with the help of Yandex SpeechKit - an automatic tool for recognizing Russian speech. Dataset statistics are shown in the Table 1.

---

[4] https://github.com/jiaaro/pydub.

Table 1. Characteristics of the dataset before preprocessing

Parameter	Value
Initial number of conversations	1084
Initial number of audio segments	10000
Number of segments with std of experts' labels > 1	1640
Number of "noisy" segments	559
Number of "clean" segments	8144
Number of recognized segments	7433

4.2 Dataset Labelling

As a measure of sentiment intensity, we took a linear scale from -3 to 3, which denotes labels of sentiment classes: from highly negative to highly positive. The experts who labeled dataset were 5 employees of the company where the research was conducted. Each expert was given access to a database of audio fragments that had been divided into eight classes of sentiment from highly negative to highly positive and "noise" class. The measure of agreement between experts, expressed by the Krippendorff's alpha, was 0.55. To increase this metric, we performed filtering of the segments, removing those for which the standard deviation was more than 1. After this procedure, the Krippendorff's alpha became equal to 0.72. It might seem that this operation made the corpus less realistic, but, in fact, this way we just got rid of segments that were noisy or too short and should be removed at the stage of preparation of the dataset. To assign segments to one of the classes, the values of the labels for each segment were averaged over all experts and rounded to integers. However, looking at the resulting distribution of classes, we found a strong class imbalance, since most audio segments were closer to a neutral mood with only a small deviation in one of the polarities, so it was decided to redistribute the labels to 3 classes (Table 2).

Table 2. Characteristics of the dataset before preprocessing

Label	Count
Negative	1099
Neutral	2490
Positive	2926

To compare models, the number of segments was aligned in both audio and text data (Table 3).

Table 3. Characteristics of the dataset after preprocessing

Parameter	Value
Number of conversations after alignment	1022
Number of segments after alignment	6515
Total duration (in hours)	8.99
Mean duration (in seconds)	4.97
Median duration (in seconds)	3.105
Total count of words in recognized segments	41867
Mean count of words in recognized segments	6.43
Median count of words in recognized segments	3.0

5 Evaluation

For an objective evaluation of the classifiers and the proof that trained models have a high generalizing ability, we divided our dataset into train and test speaker-independent subsets in a ratio 80:20.

To evaluate the classifiers, we used the usual classification metrics: accuracy and macro-averaged F1-measure (to account for class imbalance), as well as MAE (mean absolute error) as the most common criterion for evaluating the quality of sentiment analysis [4].

$$Accuracy = \frac{tp + tn}{tp + tn + fp + fn} \tag{1}$$

$$F1 = 2 \cdot \frac{precision \cdot recall}{precision + recall} \tag{2}$$

$$MAE = \frac{\sum_{i=1}^{n} |y_i - x_i|}{n} = \frac{\sum_{i=1}^{n} |e_i|}{n} \tag{3}$$

We took 2 dummy classifiers from the sklearn[5] library as the baselines: Stratified Random and Most Frequent. The tables below show classification results on a test dataset for individual modalities, as well as the combination of audio and text modalities in comparison with the baselines.

From Table 4 we can see that for audio the ensembles of decision trees, trained on acoustic characteristics, aggregated along the time axis, worked best. These features are quite complete and at the same time laconically reflect the frequency spectrum of the signal and have a representation that is convenient for learning models based on the gradient boosting of decision trees. XGBoost has proved itself well in solving problems for competitions, and, as we see, it shows itself perfectly in real problems. The best hyperparameters of XGBoost for this problem are: n_estimators = 50, max_depth = 5.

---

[5] https://github.com/scikit-learn/scikit-learn.

Table 4. Evaluation of sentiment analysis models using only audio features

Representation	Model	Acc	F1 macro	MAE
Baseline	Dummy Classifier (stratified random)	0.3852	0.3459	0.7574
	Dummy Classifier (most frequent)	0.4489	0.2065	0.7198
Sequence of samples	LSTM	0.3774	0.2147	0.7919
Spectrogram	CNN	0.4931	0.3907	0.5894
	LSTM	0.5170	0.4206	0.5753
Spectrogram, VGG16	Logistic Regression	0.5257	0.4342	0.6108
	DNN	0.5471	0.4555	0.5878
Acoustic features	LSTM	0.5573	0.4593	0.5400
Acoustic features (aggregated)	Logistic Regression	0.5825	0.5145	0.5341
	LightGBM	0.6001	0.5296	**0.5057**
	XGBoost	**0.6039**	**0.5378**	0.5065
VGGish	Logistic Regression	0.5801	0.4813	0.5410
	XGBoost	0.5533	0.473	0.5648
	LSTM	0.5786	0.5201	0.5395

TF-IDF is a fairly old technology (mid-20th century), but it has been widely used in practice to this day, since for many tasks it is competitive with other, more modern vector representations of words. As we can see from Table 5, Logistic Regression trained on TF-IDF showed the best metrics for the language modality with the following hyperparameters: C = 0.01, solver = "sag", tol = 1e−4.

Table 5. Evaluation of sentiment analysis models using only text features

Representation	Model	Acc	F1 macro	MAE
Baseline	Dummy Classifier (stratified random)	0.3852	0.3459	0.7574
	Dummy Classifier (most frequent)	0.4489	0.2065	0.7198
Count Vectorizer	Logistic Regression	0.5587	0.4926	0.5587
	Random Forest	**0.5717**	0.4640	0.5564
	SVC	0.5341	0.4814	0.5855
TF-IDF	Logistic Regression	0.5671	**0.5157**	**0.5395**
	Random Forest	**0.5656**	0.5111	0.5471
	SVC	0.5134	0.4182	0.5993
Word2Vec	Logistic Regression	0.5525	0.4245	0.5871
	Random Forest	0.5548	0.4134	0.5871
	SVC	0.5295	0.5117	0.6438
Word2Vec (Wiki)	LSTM	0.5556	0.5110	0.5402
Word2Vec (RDT)	LSTM	0.5640	0.5030	0.5425

The simple averaging of the probabilities of the two best models for each modality might seem like a good solution, however, as shown in Table 6, Light-GBM trained on concatenated acoustic audio features and text features based on TF-IDF, achieves the best results in comparison with other tested models because of its ability to work well with both dense and sparse data. Following hyperparameters performed best: n_estimators = 500, max_depth = −1.

Table 6. Evaluation of sentiment analysis models using multimodal features

Representation	Model	Acc	F1 macro	MAE
Baseline	Dummy Classifier (stratified random)	0.3852	0.3459	0.7574
	Dummy Classifier (most frequent)	0.4489	0.2065	0.7198
Mean	XGBoost (Audio) + Random Forest (Text)	0.6014	0.5218	0.5014
Concatenation	Logistic Regression	0.5694	0.5174	0.5372
	LightGBM	**0.6331**	**0.5941**	**0.4589**

6   Error Analysis

In the Fig. 1, we present the final confusion matrix, which lets us see when the classifier is most often mistaken. The most confusion arises between a negative and a positive class. Listening to records, on which the classifier has made a mistake, has led us to the following conclusions:

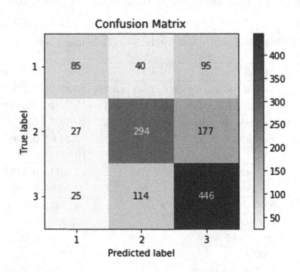

Fig. 1. Confusion matrix of the best multimodal model

– In most cases, the sentiment is transmitted through intonation, but not through the speech.

– Apparently, the classifier can not distinguish the negative sentiment from the positive because of a similar excitation of the voice, which is easy to separate from the neutral class, but is difficult to determine which direction the tone is shifted towards.

However, in some cases, it is sufficient to separate the positive or negative sentiment from neutral, since the emotional response is important to the companies, and no matter what it was, the main thing is that the client did not remain indifferent.

7 Conclusion

We assembled our own dataset, tested various ways of extracting features from audio and text, and also provided the results of experiments to build a model to recognize a sentiment intensity of a telephone conversation both on information about individual modalities and using multimodal data (audio and text). In future research, we are considering increasing the dataset for training, using audio preprocessing algorithms (noise reduction and voice activity detection) and using ready-made neural networks to extract features from audio and text, including pretrained of large datasets Word2Vec.

References

1. Arkhipenko, K., et al.: Comparison of neural network architectures for sentiment analysis of Russian tweets. In: Proceedings of the International Conference Dialogue (2016)
2. Baltrusaitis, T., Ahuja, C., Morency, L.P.: Multimodal machine learning: a survey and taxonomy. CoRR abs/1705.09406. arXiv:1705.09406 (2017)
3. Carletta, J., et al.: The AMI meeting corpus: a pre-announcement. In: Renals, S., Bengio, S. (eds.) MLMI 2005. LNCS, vol. 3869, pp. 28–39. Springer, Heidelberg (2006). https://doi.org/10.1007/11677482_3. ISBN 3-540-32549-2
4. Chen, M., et al.: Multimodal sentiment analysis with word-level fusion and reinforcement learning. In: Proceedings of the 19th ACM International Conference on Multimodal Interaction - ICMI 2017. ACM Press (2017). https://doi.org/10.1145/3136755.3136801
5. Chollet, F., et al.: Keras (2015). https://keras.io
6. Greff, K., et al.: LSTM: a search space Odyssey. CoRR abs/1503.04069. arXiv:1503.04069 (2015)
7. Hershey, S., et al.: CNN architectures for large-scale audio classification. CoRR abs/1609.09430. arXiv:1609.09430 (2016)
8. Hochreiter, S., Schmidhuber, J.: Long short-term memory. Neural Comput. **9**, 1735–1780 (1997)
9. Korobov, M.: Morphological analyzer and generator for Russian and Ukrainian languages. In: Khachay, M.Y., Konstantinova, N., Panchenko, A., Ignatov, D.I., Labunets, V.G. (eds.) AIST 2015. CCIS, vol. 542, pp. 320–332. Springer, Cham (2015). https://doi.org/10.1007/978-3-319-26123-2_31. ISBN 978-3-319-26122-5

10. LeCun, Y., Haffner, P., Bottou, L., Bengio, Y.: Object recognition with gradient-based learning. Shape, Contour and Grouping in Computer Vision. LNCS, vol. 1681, pp. 319–345. Springer, Heidelberg (1999). https://doi.org/10.1007/3-540-46805-6_19. ISBN 978-3-540-46805-9
11. Loukachevitch, N., et al.: Task on sentiment analysis of tweets about telecom and financial companies. In: Proceedings of International Conference Dialogue (2015)
12. Martínez-Cáamara, E., et al.: Sentiment analysis in Twitter. Nat. Lang. Eng. **20**, 1–28 (2014)
13. McFee, B., et al.: librosa/librosa: 0.6.1, May 2018. https://doi.org/10.5281/zenodo.1252297
14. Mikolov, T., et al.: Effcient estimation of word representations in vector space. CoRR abs/1301.3781. arXiv:1301.3781 (2013)
15. Panchenko, A., et al.: Human and machine judgements for Russian semantic relatedness. In: Ignatov, D.I., et al. (eds.) AIST 2016. CCIS, vol. 661, pp. 221–235. Springer, Cham (2017). https://doi.org/10.1007/978-3-319-52920-2_21. ISBN 978-3-319-52920-2
16. Pang, B., Lee, L., Vaithyanathan, S.: Thumbs up? In: Proceedings of the ACL 2002 Conference on Empirical Methods in Natural Language Processing - EMNLP 2002. Association for Computational Linguistics (2002). https://doi.org/10.3115/1118693.1118704
17. Pedregosa, F., et al.: Scikit-learn: machine learning in Python. J. Mach. Learn. Res. **12**, 2825–2830 (2011)
18. Poria, S., et al.: Multimodal sentiment analysis: addressing key issues and setting up baselines. CoRR abs/1803.07427 (2018)
19. Ramos, J.: Using TF-IDF to determine word relevance in document queries, January 2003
20. Řehůřek, R., Sojka, P.: Software framework for topic modelling with large corpora. In: Proceedings of the LREC 2010 Workshop on New Challenges for NLP Frameworks, pp. 45–50. ELRA, Valletta, Malta, May 2010. http://is.muni.cz/publication/884893/en
21. Simonyan, K., Zisserman, A.: Very deep convolutional networks for large-scale image recognition. CoRR abs/1409.1556. arXiv:1409.1556 (2014)
22. Somasundaran, S., et al.: Manual annotation of opinion categories in meetings. In: Proceedings of the Workshop on Frontiers in Linguistically Annotated Corpora 2006 - LAC 2006. Association for Computational Linguistics (2006). https://doi.org/10.3115/1641991.1641998
23. Turney, P.D.: Thumbs up or thumbs down? In: Proceedings of the 40th Annual Meeting on Association for Computational Linguistics - ACL 2002. Association for Computational Linguistics (2001). https://doi.org/10.3115/1073083.1073153
24. Yuhas, B.P., Goldstein, M.H., Sejnowski, T.J.: Integration of acoustic and visual speech signals using neural networks. IEEE Commun. Mag. **27**(11), 65–71 (1989). https://doi.org/10.1109/35.41402
25. Zadeh, A., et al.: MOSI: multimodal corpus of sentiment intensity and subjectivity analysis in online opinion videos. CoRR abs/1606.06259. arXiv:1606.06259 (2016)
26. Zadeh, A., et al.: Multi-attention recurrent network for human communication comprehension. CoRR abs/1802.00923. arXiv:1802.00923 (2018)
27. Zhang, L., Wang, S., Liu, B.: Deep learning for sentiment analysis: a survey, January 2018

Evaluation of Approaches for Most Frequent Sense Identification in Russian

Natalia Loukachevitch[1,2(✉)] and Nikolai Mischenko[1]

[1] Lomonosov Moscow State University, Moscow, Russia
louk_nat@mail.ru, laynikme@gmail.com
[2] Tatarstan Academy of Sciences, Kazan, Russia

Abstract. In this paper, we compare several approaches for determining the most frequent senses of ambiguous words for Russian. We compare several approaches (frequency-based, topic models, information-retrieval and embedding-based) and consider two representation forms of information about multiword expressions described in RuThes. We found that the information-retrieval approach is better than the method based on probabilistic topic models. The best results are obtained with the application of distributional vector representations with thesaurus path weighing.

Keywords: Lexical ambiguity · Most frequent sense · Thesaurus

1 Introduction

Word sense disambiguation is an important step within semantic analysis of natural language texts. This procedure includes automatic selection of an appropriate sense for ambiguous word usage from the pre-defined set of senses described in a dictionary or thesaurus [23]. Several evaluations devoted to lexical disambiguation have been organized [1,24]. The evaluations showed that there is a baseline that is difficult to overcome for many disambiguation systems [23]. This is so-called most frequent sense of the word (MFS). This also means that the most frequent sense is a useful heuristic in lexical sense disambiguation when the training or context data are insufficient.

MFS can be determined on the basis of a large sense-tagged corpus such as SemCor, which is labeled with WordNet senses [5,8]. However, the semantic annotation of a large corpus is a very labor and time-consuming task. In addition, the most frequent sense of a given word may vary according to the domain, therefore the automatic processing of documents in a specific domain can require re-estimation of MFS on the domain-specific text collection. The distributions of lexical senses can also depend on time [22]. Therefore, approaches to automatic calculation of the most frequent sense on a text collection without any semantic annotation should be studied [20].

This work was partially supported by Russian Science Foundation, grant N16-18-02074.

W. M. P. van der Aalst et al. (Eds.): AIST 2018, LNCS 11179, pp. 99–110, 2018.
https://doi.org/10.1007/978-3-030-11027-7_10

In this paper, we compare several approaches for determining the most frequent sense for Russian on the basis of the Thesaurus of the Russian language RuThes [17]. The approaches include frequency estimation of sense-related words and expressions, probabilistic topic modeling, information-retrieval, and embedding-based methods. We also compare single-word and phrase-based representations of documents. The methods are evaluated on a manually created gold standard.

The structure of the paper is as follows. In the second section we consider related work. The third section briefly describes the RuThes thesaurus. The fourth section presents the gold standard data used for the evaluation. In the fifth section we describe methods used for estimating MFS. The six section presents the experiments and the achieved results.

2   Related Work

It was found in various studies that the most frequent sense is a strong baseline for word sense disambiguation systems [23]. Because of difficulty to create sense-labeled corpora to determine MFS, techniques for automatic MFS revealing have been proposed.

McCarthy et al. [19,20] describe an automatic technique for ranking word senses on the basis of comparison of a given word with distributionally similar words. The distributional similarity is calculated using syntactic (or linear) contexts and the automatic thesaurus construction method described in [11]. WordNet similarity measures are used to compare the word senses and distributional neighbors. McCarthy et al. [20] report that 56.3% of noun SemCor MFS (random baseline – 32%), 45.6% verb MFS (random baseline – 27.1%) were correctly identified with the proposed technique.

In [7] the problem of domain specific sense distributions is studied. The authors form samples of ambiguous words having a sense in one of two domains: SPORTS and FINANCE. To obtain the distribution of senses for chosen words, the random sentences mentioning the target words in domain-specific text collections are extracted and annotated.

Lau et al. [10] proposed to use topic models for identification of the predominant sense. They train a single topic model per target lemma. To compute the similarity between a sense and a topic, glosses are converted into a multinomial distribution over words, and then the Jensen–Shannon divergence between the multinomial distribution of the gloss and the topic is calculated. In [2] the authors use embeddings obtained with word2vec for determining most frequent senses of nouns. They compare the word embedding of a word with all its sense embeddings and obtain the predominant sense with the highest similarity.

For Russian, Loukachevitch and Chetviorkin [15] studied the approach of determining the most frequent sense of ambiguous words using unambiguous related words and phrases described in the RuThes thesaurus. Lopukhina et al. [13] estimated sense frequency distributions for Russian nouns taken from the Active Dictionary of Russian.

Other works related to Russian word-sense disambiguation tasks include the following studies. In the all-word disambiguation task, Loukachevitch and Chuiko [16] studied parameters of word sense selection (types of the thesaurus paths, window size, etc.) on the basis of the RuThes thesaurus. Kobritsov et al. [6] developed disambiguation filters to provide semantic annotation for the Russian National Corpus. The semantic annotation was based on the taxonomy of lexical and semantic facets. In [9], statistical word sense disambiguation methods for several Russian nouns were described.

In the word sense induction task for Russian, Lopukhin et al. [12] evaluated four methods: Adaptive Skip-gram, Latent Dirichlet Allocation, clustering of contexts, and clustering of synonyms [14]. Ustalov et al. [26] performed unsupervised word sense disambiguation for Russian creating a word sense network from semi-structured resources such as synonym dictionaries. In 2018, the shared task on word sense induction in Russian has been organized [25].

3   RuThes Thesaurus of the Russian Language

RuThes Thesaurus of Russian language is a linguistic ontology for natural language processing, i.e. an ontology, where the majority of concepts are introduced on the basis of actual language expressions [17]. Each concept has a unique, unambiguous name. The concepts are linked with words and expressions conveying the corresponding concepts in texts (so-called text entries). As a resource for natural language processing, RuThes is similar to WordNet thesaurus [5], but has some distinctions. For the current study, the following differences are significant.

First, each concept in RuThes can have text entries of different parts of speech, multiword expressions (idiomatic or compositional expressions). For example, concept Кинофильм (*Movie*) has the following text entries: карти-на, кинокартина, кинолента, киноработа, кинофильм, киношка, кинемато-графическое произведение, кинематографическая продукция (*cinema film, cinematographic production, motion picture, moving picture, movie, etc.*).

Second, in RuThes the introduction of concepts based on multiword expressions is not restricted and even encouraged if this concept adds some new information to knowledge described in RuThes. This provides more monosemous (=unambiguous) entries than in WordNet-like resources. For example, such a phrase as дорожное движение (*road traffic*) has hyponyms: левостороннее движение (*left-hand traffic*) and правостороннее движение (*right-hand traffic*): the existence of such hyponyms cannot be inferred from its component words. Thus, it is useful to introduce the corresponding concept into the thesaurus.

Third, the thesaurus relations are concept-based. The main class-subclass relation roughly corresponds to the relation of hyponym-hypernym in WordNet. The scope of part-whole relations is different. Besides, one of the main RuThes relation types describes the dependence of the existence of one concept from another concept. For example, concept *Forest* cannot appear without the existence of concept *Tree*. This means that if, in our world, trees are absent then forests also cannot exist. The dependence relation is often used for describing relations between a phrase-based concept and the concept of its component word, which facilitates the introduction of phrase-based concepts [17].

Currently, publicly-available version of RuThes (RuThes-lite) contains 115,00 unique words and expressions[1].

4 Gold Standard Data

For evaluation of the MFS identification methods, we use the data set of the most frequent senses described in [15] and denoted further as MFS-RuThes. For each ambiguous word, MFS-RuThes contains name of the RuThes concept that denotes the most frequent sense of this word. The most frequent senses were labeled by experts after analysis of a sample of random word usages in the current news articles and reports provided the Yandex.news service[2]. Yandex.news service processes more than 100,000 news articles during a day. The news flow from different sources is automatically clustered into sets of similar news. When searching in the service, retrieval results are also clustered.

For a given ambiguous word, experts analyzed snippets on Yandex news service search engine pages. Considering several dozens of different usages of the word in news, the experts calculated the distribution of senses of the word, which allowed defining its most frequent sense. In news snippets, repetitions of the same sentences can be frequently met – such duplicates were dismissed from the analysis.

Because of insufficient amount of data under consideration, the experts could designate several senses as the most frequent ones if they saw that the difference in the frequencies does not allow them to decide what a sense is more frequent. For example, for word картина , two main senses (from six ones described in the thesaurus) were revealed: Фильм (*Moving picture*) and Произведение живописи (*Piece of painting*). Currently, the MFS-RuThes dataset contains 331 words for that two experts agreed in their most frequent senses.

5 Methods for MFS Estimation

We compare four methods for MFS estimation: a method based on frequencies of sense-related thesaurus entries, probabilistic topic model LDA, information-retrieval methods based on tf-idf, and embedding-based methods (word2vec, fastText). For all the methods, the following thesaurus conceptual contexts of an ambiguous word w in its specific sense C_i can be used:

– one-step context of word w with sense C ($ThesConw_1$) that comprises other words and expressions attached to the same concept C and concepts directly related to C as described in the thesaurus;

---

[1] http://www.labinform.ru/ruthes/index.htm.
[2] http://news.yandex.ru/.

- two (three, four)-step contexts of word w with sense C ($ThesConw_{2-4}$) comprising words and expressions from the concepts located at the distance of maximum 2–4 relations to the initial concept C (including C); the path between concepts can consist of relations of any types,
- one-step thesaurus context including only unambiguous words and expressions: $UniThesConw_1$.

Let us consider fragments of the sense-related lists of text-entries for three senses of word картина (two-step context):

- **piece of painting**: мозаичный, портретное изображение, пейзажное полотно, инсталляция, коллаж, стенная роспись, писание, произведение изобразительного искусства, произведение визуальный искусство, мозаика, импрессионизм, изобразительный, искусство живопись, масляный живопись, владетель галерея, стенопись, художественный мастерство, .. (*mosaic, portrait, visual art, oil painting, impressionism..*);
- **movie**: синефил, светочувствительная пленка, поставить фильм, кадр из кино, фильм, киновидеопродукция, кинонеделя, полнометражный, художественная картина, неигровой фильм, киномероприятие, мультипликационный кино, кинофотопленка, фестиваль, фильм идти, мультипликационный фильм, кинофорум, киноперсонаж, кинотеатр, кинозал, .. (*movie, film, animation, cinema, festival, fiction picture..*);
- **circumstances**: социальный обстановка, запущенный вид, санитарно-эпидемиологический состояние, тишь, безвыходный положение, положение, уютность, раздольность, состояние защищенность, обстановка, просуществовать, эксцесс, напряженный положение, молчать, нагой, незаполненность, демографический состав, невозбранный, .. (*circumstances, stressful situation, silence, economic state, existence..*);

It can be seen that the differences between senses are very clearly conveyed by the related words.

For statistical methods, it can be quite difficult to use multiword expressions in calculations. Therefore sets of sense-related text entries are divided to single words. The thesaurus context is represented as a vector of single words $VectorSing$ with frequencies of the word appearance in the thesaurus context. In such a way, it is possible to reveal repeated patterns in sense-related multiword expressions. The fragments of the word vector for картина are as follows:

- **piece of painting**: 'художественный': 13, 'произведение': 9, 'живопись': 7, 'холст': 7, 'скульптурный': 5, 'изображение': 5, 'эскиз': 4, 'галерея': 4,... (artistic, painting, canvas, gallery, sketch...);
- **movie**: 'фильм': 27, 'кино': 16, 'художественный': 5, 'экранизация': 4, 'демонстрация': 4, 'документальный': 4... (film, movie, artistic, adaptation, screening...)
- **circumstances**: 'состояние': 40, 'условие': 21, 'демографический': 11, 'экологический': 10, 'обстоятельство': 9, 'случай': 6, 'климат': 4, 'объективный': 4, 'окружающий': 4,... (state, conditions, demographic, ecological, circumstance...);

5.1   Frequencies of Related Text Entries

One of directions in WSD studies is the use of so-called monosemous relatives for ambiguous words, that are related words or expressions having only a unique sense. From these text and thesaurus contexts, the following features for ambiguous word w and its senses C_w were generated:

- the overall collection frequency of expressions from $UniThesConw_1$ – here we estimate how often unambiguous relatives of w were met in the collection – $Frcqdoc_1$ and logarithm of this value $logFreqdoc$. Table 1 depicts frequencies of monosemous relatives of word картина in the source collection,
- the frequency of expressions from $UniThesConw_1$ in texts where w was mentioned – $FreqdocW_1$,
- the overall frequency and the maximum frequency of words and expressions from $ThesConw_i$ co-occurred with w in the same sentences – $FreqSentWmax_i$ and $FreqSentWsum_i$ (i = 1, 2, 3),
- the overall frequency and the maximum frequency of words and expressions from $ThesConw_i$ occurred in the neighbor sentences with w – $FreqNearWmax_i$ and $FreqNearWsum_i$ (i = 1, 2, 3).

Table 1. Document frequencies of monosemous relatives of word picture in the source news collection of 2 mln. documents

Monosemous relatives	Senses	DocFreq
фильм (film)	moving picture	45,285
мультфильм (cartoon)	moving picture	4,097
документальный фильм (documentary film)	moving picture	3,516
живопись (painting)	piece of painting	3,200
съемка фильма (shooting a film)	moving picture	2,445
кинофильм (movie)	moving picture	1,955
произведение искусства (art work)	piece of painting	1,850
художественный фильм (fiction movie)	moving picture	1,391
изобразительное искусство (visual art)	piece of painting	1,102
режиссер картины (movie director)	moving picture	978
общая картина (general picture)	general circumstances	932

5.2   Use of Probabilistic Topic Models

Topic modeling approaches are statistical approaches for recognition of main topics in document collections, among which LDA (Latent Dirichlet Allocation) is most commonly used [3]. LDA is an unsupervised statistical algorithm that usually considers each document as a "bag of words". The LDA approach assumes that each document is a mixture of hidden topics and the LDA inference should discover topics' distribution over words and document distribution over topics. The algorithm generates a set of topics according to given parameters. Each topic looks as a set of words ordered by decreasing probability of the word in the topic (Table 2).

Table 2. Topics for Russian ambiguous word *kartina* (picture)

Topic 16th	Topic 19th
фильм (movie)	картина (picture)
роль (role)	миллион (million)
актер (actor)	год (year)
картина (picture)	аукцион (auction)
съемка (shooting)	работа (work)
режиссер (director)	продать (sell)
сыграть (play)	торг (bid)
год (year)	художник (artist)
проект (project)	полотно (piece)

For each word from the MFS-RuThes set, we extract documents mentioning this word, then we generate topic distributions for the extracted subcollection of documents. Using topic models, we suppose that words related to the most frequent sense of a word would be better represented in the created topic model, which means that the MFS-related words will have larger values of probabilities in created topics. For comparison of sense s_i and topic t_j, the following similarity measure is used:

$$sim(s_i, t_j) = 1 - JS(S||T)$$

where t_j – j-topic, s_i – i-sense of an ambiguous word, $JS(S||T)$ is the Jensen-Shannon divergence, where S is the probability distribution of words related to a specific sense and T is the probability distribution of words in the topic.

Jensen-Shannon divergence is a popular method of measuring similarities between two probability distributions:

$$JS(P||Q) = 1/2 \cdot D(P||M) + 1/2 \cdot D(Q||M) \tag{1}$$

where $M = 1/2(P+Q)$, the averaged sum of distributions P and Q, D(P||M) is Kullback-Leibler divergence, an asymmetric measure of differences between two distributions.

To learn the most frequent sense, the prevalence measure is calculated [10]:

$$Prevalence(s_i) = \sum_j^T sim(s_i, t_j) \cdot \frac{f(t_j)}{\sum_k^T f(t_k)} \tag{2}$$

where T is a set of topics, $f(t)$ is the topic's frequency, that is the sum of probabilities of words in the topic multiplied by the word frequencies. The prevalence score for sense s_i is higher when words related to this sense have relatively high similarity with high-probability topics.

5.3   Use of Information-Retrieval Features

Information-retrieval method for MFS of word w consists in the following steps:

- extracting of a subcollection of documents containing word w (w-subcollection),
- representation of the subcollection as a vector of words,
- calculation of similarity between the vector of the subcollection and vector of relatives for each sense of the word w, using cosine similarity (the more is the better),
- selection a sense obtaining maximal cosine similarity as the MFS sense of the word w.

A subcollection can be represented as a vector of word frequencies (tf), or $tf.idf$ vector, where $idf(w, D) = log(\frac{|D|}{|w \in d|})$. Factor idf gives more weight to rare words; tf is calculated on the w-subcollection, idf is calculated on the whole collection.

5.4   Use of Embedding Representations

Current approaches allow representing words using vectors of relatively low dimension (300–1000), so-called embeddings. In this paper we use gensim implementation of word2vec word embeddings [21][3] and fastText [4][4]. The embedding representation of words allows finding the most similar words to the given word in the vector space. It is supposed that the most frequent senses of ambiguous words are better represented in the created lists of similar words.

Table 3 shows the most similar words according to word2vec and fastText. It can be seen that both lists more support the movie-sense of word картина. But the lists are quite different. FastText reveals more word-derivations, some of them have low frequencies in the text collection, and are not represented in the thesaurus. Using embedddings, for each ambiguous word we calculate the list of the most similar words and compare the top of the list with sense-related words for each sense according to the thesaurus.

$$MFS(w_0) = \arg\max(score(C_i)) = \arg\max \sum_j (e_j \cdot t_{ij}))$$

where e_j is similarity of word w_j to initial ambiguous word w_0 according to an embedding model, $\overrightarrow{t}$ is a vector of words from the thesaurus context C_i.

---

[3] https://radimrehurek.com/gensim/.
[4] https://fasttext.cc/.

Table 3. The top of the most similar words to word *kartina*

Word2vec	fastText
фильм 0.810	фильм 0.872
кинолента 0.756	кинолента 0.864
кинокартина 0.727	кинокартина 0.856
короткометражка 0.690	фильм-экранизация 0.815
блокбастер 0.684	фильм-приквел 0.811
вестерн 0.624	фильм-притча 0.807
мультфильм 0.623	киношедевр 0.806
триллер 0.618	фильм-биография 0.804
байопик 0.616	фильм-вестерн 0.800
римейк 0.610	фильм-комедия 0.790
шедевр 0.608	киноперсонаж 0.786

6   Experiments

For calculation, we used a collection of 2 mln. news articles in Russian. Table 4 presents the most prominent results for all the above-mentioned methods according to accuracy. We underline the best accuracy for each group of the methods.

The first block of the Table 4 shows the results of several features based on frequencies of words and phrases related to each sense of ambiguous words. For example, feature $FreqDocW_1$, which calculates the number of documents where directly-related text entries and the initial word co-occur. The best result of single features is based on the number co-occurrences of a word and sense-related text-entries in the same sentences.

To combine the features, regression-oriented methods implemented in WEKA machine learning package[5] were utilized. The best quality of classification using labeled data was shown by the ensemble of three classifiers: Logistic Regression, LogitBoost and Random Forest. Each classifier ranged word senses according to the probability of this sense to be the most frequent one. The probabilities of MFS generated by these methods were averaged. The achieved result was 50.6% accuracy of MFS prediction (Table 4). The estimation is based on ten-fold cross validation.

The second block of the Table 4 shows the results obtained with the use of the LDA-based method. Two representation of multiword expressions were compared: divided into single words (probability-normalized vector *VectorSing*) and whole phrases. The thesaurus phrases found in the documents were added with their frequencies to the document single word representations and further the phrases were considered as single units. The best result is achieved with 1-step (synonyms and direct thesaurus relations) thesaurus context with division to single words.

The third block is based on comparison of vector representations for w-subcollection and vector of thesaurus context for specific word senses. Several

---

[5] https://www.cs.waikato.ac.nz/ml/weka/.

Table 4. Evaluation of methods for MFS prediction in Russian

Method	Accuracy
Sense-related word co-occurrences	
$FreqDoc_1$	42.4
$FreqDocW_1$	46.4
$FreqSentWsum_1$	41.2
$FreqSentWsum_2$	48.2
Supervised result	50.6
LDA-based feature	
1-step	52.6
2-step	48.3
LDA+phrases, 1-step	51.1
LDA+phrases, 2-step	52.2
Information-retrieval feature	
tf, 1-step	50.1
tf, 2-step	52.2
$tf.idf$, 1-step	50.1
$tf.idf$, 2-step	52.9
tf, 2-step,+phrases	49.8
$tf.idf$, 2-step,+phrases	55.6
Embedding-based features	
word2vec (dim = 400, win = 5, 200 words, 4-step, Lch)	**65.3**
word2vec (dim = 400, win = 5, 200 words, 4-step, no-Lch)	63.1
fastText (dim = 400, win = 5, 200 words, 3-step, Lch)	63.0
Random	23.5

variants are considered: different vectorizations and with/without phrases. The best result is better than for topic models; inclusion of the thesaurus phrases improved the results in this approach.

The fourth block of the Table 4 considers the best results achieved with embedding-based methods. We tested the following parameters of skip-gram method of word2vec and fastText: number of dimensions {100, 200, 300, 400, 500}, frequency threshold {5, 10}, the size of word window {5, 10}, the number of the most similar words for comparison with the thesaurus {100, 200, 300, 500}, the size of the thesarus context 1–4 steps (relations) from each sense of the word.

Besides initial representation of thesaurus context $VectorSing$, also the Leacock-Chodorow (Lch) measure of similarity between thesaurus nodes is used. The Lch measure estimates the similarity of two nodes by finding the path length between them in the hierarchy of thesaurus relations. It is computed as:

$$sim_{lch} = -log\frac{N_p}{2D}$$

where N_p is the distance between nodes and D is the maximum depth in the taxonomy. The distance is calculated in nodes, that is the distance between synonyms is equal 1, and the distance between a node and its hypernym is equal 2. The logarithm base is equal to 2D. In RuThes-lite, the maximum depth of the ontology accounting both types of relations is equal 14 [18].

The best result of MFS identification is obtained with word2vec skipgram using the Lch similarity measure.

7 Conclusion

Knowledge about the most frequent sense of an ambiguous word is very useful for semantic text analysis. In this paper, we compared four approaches for determining the most frequent sense for Russian and different approaches to phrase-representation. We found that a simple method based on frequencies of so-called monosemous relatives have quite high results, the information-retrieval approach is better than the method based on probabilistic topic models. The best results were obtained with the application of word embeddings (word2vec) with the Leacock-Chodorow weighing of the thesaurus paths.

References

1. Agirre, E., Soroa, A.: SemEval-2007 task 02: evaluating word sense induction and discrimination systems. In: Proceedings of the 4th International Workshop on Semantic Evaluations, pp. 7–12 (2007)
2. Bhingardive, S., Singh, D., Murthy, R.: Unsupervised most frequent sense detection using word embeddings. In: Proceedings of NAACL-2015 (2015)
3. Blei, D.M.: Probabilistic topic models. Commun. ACM **55**(4), 77–84 (2012)
4. Bojanowski, P., Grave, E., Joulin, A., Mikolov, T.: Enriching word vectors with subword information. arXiv preprint arXiv:1607.04606 (2016)
5. Fellbaum, C. (ed.): WordNet: An Electronic Lexical Database. MIT Press, Cambridge (1998)
6. Kobritsov, B., Lyashevskaya, O., Shemanayeva, O.: Surface filters for solving semantic homonymy in the textual case. In: Proceedings of International Conference on Dialogue-2005 (2005)
7. Koeling, R., McCarthy, D., Carroll, J.: Domain-specific sense distributions and predominant sense acquisition. In: Proceedings of EMNLP-2005, pp. 419–426 (2005)
8. Landes, S., Leacock, C., Tengi, R.I.: Building semantic concordances. WordNet: Electron. Lexical Database **199**(216), 199–216 (1998)
9. Lashevskaja, O., Mitrofanova, O.: Disambiguation of taxonomy markers in context: Russian nouns. In: Proceedings of the 17th Nordic Conference of Computational Linguistics (NODALIDA 2009), pp. 111–117 (2009)
10. Lau, J.H., Cook, P., McCarthy, D., Newman, D., Baldwin, T.: Word sense induction for novel sense detection. In: Proceedings of the EACL-2012, pp. 591–601. Association for Computational Linguistics (2012)

11. Lin, D.: Automatic retrieval and clustering of similar words. In: Proceedings of ACL-1998, pp. 768–774. Association for Computational Linguistics (1998)
12. Lopukhin, K., Iomdin, B., Lopukhina, A.: Word sense induction for Russian: deep study and comparison with dictionaries. In: Proceedings of International Conference on Dialogue-2017, vol. 1, pp. 121–134 (2017)
13. Lopukhin, K., Lopukhina, A.: Automated word sense frequency estimation for Russian nouns. In: Quantitative Approaches to the Russian Language, pp. 89–104. Routledge (2017)
14. Lopukhina, A., Lopukhin, K.: Word sense frequency estimation for Russian: verbs, adjectives, and different dictionaries. In: Proceedings of eLex 2017 Conference, pp. 267–280 (2017)
15. Loukachevitch, N., Chetviorkin, I.: Determining the most frequent senses using Russian linguistic ontology RuThes. In: Proceedings of Workshop on Semantic Resources and Semantic Annotation at NODALIDA 2015, pp. 21–27 (2015)
16. Loukachevitch, N., Chuiko, D.: Automatic resolution lexical ambiguity on basis of thesaurus knowledge, pp. 108–117 (2007)
17. Loukachevitch, N., Dobrov, B., Chetviorkin, I.: RuThes-Lite, a publicly available version of thesaurus of Russian language RuThes. In: Proceedings of International Conference on Dialogue-2014, vol. 2014 (2014)
18. Loukachevitch, N., Shevelev, A., Mozharova, V.: Testing features and methods in Russian paraphrasing task. In: Proceedings of International Conference on Dialog-2017, pp. 135–145 (2017)
19. McCarthy, D., Koeling, R., Weeds, J., Carroll, J.: Finding predominant word senses in untagged text (2004)
20. McCarthy, D., Koeling, R., Weeds, J., Carroll, J.: Unsupervised acquisition of predominant word senses. Comput. Linguist. **33**(4), 553–590 (2007)
21. Mikolov, T., Sutskever, I., Chen, K., Corrado, G.S., Dean, J.: Distributed representations of words and phrases and their compositionality. In: Advances in Neural Information Processing Systems, pp. 3111–3119 (2013)
22. Mitra, S., Mitra, R., Riedl, M., Biemann, C., Mukherjee, A., Goyal, P.: That's sick dude!: automatic identification of word sense change across different timescales. arXiv preprint arXiv:1405.4392 (2014)
23. Navigli, R.: Word sense disambiguation: a survey. ACM Comput. Surv. (CSUR) **41**(2), 10 (2009)
24. Navigli, R., Jurgens, D., Vannella, D.: SemEval-2013 task 12: multilingual word sense disambiguation. In: Second Joint Conference on Lexical and Computational Semantics SemEval 2013, vol. 2, pp. 222–231 (2013)
25. Panchenko, A., et al.: RUSSE'2018: a shared task on word sense induction for the Russian language. In: Proceedings of International Conference Dialogue-2018, pp. 547–564 (2018)
26. Ustalov, D., Panchenko, A., Biemann, C.: Watset: automatic induction of synsets from a graph of synonyms. In: Proceedings of the 55th Annual Meeting of the Association for Computational Linguistics, pp. 1579–1590 (2017)

RusNLP: Semantic Search Engine for Russian NLP Conference Papers

Irina Nikishina[1]([⊠])(iD), Amir Bakarov[1](iD), and Andrey Kutuzov[2](iD)

[1] National Research University Higher School of Economics, Moscow, Russia
ianikishina_1@edu.hse.ru, amirbakarov@gmail.com
[2] University of Oslo, Oslo, Norway
andreku@ifi.uio.no

Abstract. We present RusNLP, a web service implementing semantic search engine and recommendation system over proceedings of three major Russian NLP conferences (Dialogue, AIST and AINL). The collected corpus spans across 12 years and contains about 400 academic papers in English. The presented web service allows searching for publications semantically similar to arbitrary user queries or to any given paper. Search results can be filtered by authors and their affiliations, conferences or years. They are also interlinked with the NLPub.ru service, making it easier to quickly capture the general focus of each paper. The search engine source code and the publications metadata are freely available for all interested researchers.

In the course of preparing the web service, we evaluated several well-known techniques for representing and comparing documents: TF-IDF, LDA, and Paragraph Vector. On our comparatively small corpus, TF-IDF yielded the best results and thus was chosen as the primary algorithm working under the hood of RusNLP.

Keywords: Information retrieval · Semantic similarity
Scientific literature search · Document representations
Academic communities

1 Introduction

Russian natural language processing community is thriving and has a strong and interesting scholarly legacy: three leading venues indexed in the Scopus database, 20 years of active publishing, more than 1800 conference papers and dozens of smaller workshops, schools and conferences. Unfortunately, it is sometimes difficult to find NLP papers relevant for processing of Russian (for example, Russian named entity recognition) with mainstream scholarly search engines (like *Google Scholar*, *Semantic Scholar* or *Sci-Hub*). Russian NLP community lacks a unified database that could incorporate all knowledge accumulated by previous researchers and present it in a structured and interlinked way. Such a database (with the corresponding front-end user interface and a recommendation system)

W. M. P. van der Aalst et al. (Eds.): AIST 2018, LNCS 11179, pp. 111–120, 2018.
https://doi.org/10.1007/978-3-030-11027-7_11

can be of great help to those dealing with NLP problems related to Russian or those who study the structure of NLP academic and industrial communities.

In this paper, we describe the preparation and implementation of such a resource available online for everyone. The database and the corresponding semantic search engine are a part of the larger *RusNLP* project aimed at preserving and analyzing the structure and tendencies in the Russian computational linguistics community.

Note that it was not clear from scratch which document analysis technique is most efficient for recommending semantically similar texts on a relatively small collection of domain-restricted documents (in our case, papers from the Russian NLP conferences). Thus, we had to conduct a bunch of experiments with the well-known document representation algorithms and human annotators before choosing the best one for our task. These experiments are described below.

The rest of the paper is organized as follows. In Sect. 2, we briefly present the collected dataset. Section 3 delves into the evaluation experiments comparing different approaches to document representation regarding our task. Section 4 describes the features of the *RusNLP* web service and typical use cases. In Sect. 5, we put our work in the context of the previous research, while in Sect. 6 we conclude and outline future plans for the *RusNLP* project.

2   Dataset

The structure of our dataset (and thus, our publications corpus) is thoroughly described in [1]. That is why here we outline it very briefly.

In general, the *RusNLP* dataset contains texts and metadata for 1,794 academic papers (and 900 authors) related to computational linguistics and natural language processing, similar to the ACL Corpus [3]. Globally, NLP as a scientific field tends to value conferences at least as high as journals; the same is true for the Russian NLP scene. Thus, the papers in our collection come from the main program proceedings of 3 Russian conferences related to computational linguistics: **Dialogue**[1], **AIST**[2] and **AINL**[3]. We do not claim that this covers the whole Russian NLP landscape, but at least the most important venues are included, with the hope that in the future more will come.

The papers themselves were crawled from the **Dialogue** website and from the Springer digital library (for **AIST** and **AINL**). Each paper was parsed to extract metadata (title, abstract, authors' names, affiliations and emails). The author names and affiliations were normalized manually (for example, we merged '*Bauman Moscow State Technical University*' and '*МГТУ имени Н.Э. Баумана*' into one entity, etc.).

The database contains all the papers, but the *RusNLP* web service (see below) for the time being deals only with those written in English: 392 in total. This decision was made to streamline text processing workflow, and considering

---

[1] http://www.dialog-21.ru/en/.
[2] https://aistconf.org/.
[3] http://ainlconf.ru/.

the current tendency to publish most important research in English (**AINL** and **AIST** proceedings contain only papers in English, so this distinction is relevant only for **Dialogue**). In the future, we plan to integrate papers written in Russian into the search engine as well.

3   Experimental Setup

The purpose of our system is to find and recommend NLP papers (from the dataset described above) most similar by their content to a given paper or a given user query. Thus, some way to represent and compare texts had to be chosen. In this section, we describe the experiments we conducted to this end. Note that first all the 392 texts were pre-processed: that is, cleared from non-alphanumeric characters, and then lemmatized and PoS-tagged with *UDPipe* [15].

We tested three statistical approaches to document representation:

1. *TF-IDF* (term frequency - inverted document frequency), a term weighting scheme ubiquitous in information retrieval [14];
2. *LDA* (Latent Dirichlet Allocation), a widespread distributional topic modeling technique [4]; we tried variants with 10 and 20 topics;
3. *Paragraph Vector* (also known as *doc2vec*), a newer distributional parametric algorithm based on shallow feed-forward neural networks [9]; we tried variants with vector size 40 and 100.

These techniques are conceptually very different, with *TF-IDF* being the simplest and taking into account only word frequencies in the documents, *LDA* trying to model the hidden distribution of topics in the text collections, and *Paragraph Vector* extending the well-known *SGNS* and *CBOW* word embedding algorithms [11] to learn embeddings for sentences, paragraphs, or documents. We aimed to find out which of them would be the best in ranking documents by their similarity in our corpus[4].

Our evaluation workflow was quite simple and measured only the precision of the algorithms. First, 20 papers were randomly sampled from the collection (8 from **Dialogue**, 6 from **AINL**, and 6 from **AIST**). A set of 10 most similar documents (nearest neighbors) was produced for each of these papers, using each of the 5 models described above (*TF-IDF*, two variants of *LDA* and two variants of *Paragraph Vector*). Then, three human assessors independently annotated the sets with the number of non-relevant neighbors t, reflecting the amount of noise in the output of the model. The precision of each set is thus $1 - \frac{t}{10}$, and the precision of the model is the averaged precision for all 20 sets: $\frac{\sum_{i=1}^{20} t_i}{20}$. Note that there is no recall measurement in this setup, as this would require assessors to read through all the 400 papers to find relevant publications not selected by the models. This was not feasible with our time and human resources, and thus

---

[4] The models were trained on our English sub-corpus, using the algorithm implementations in the *Gensim* library [13].

these experiments are limited to measuring only the models' precision (their ability to avoid absolutely non-relevant papers in the 10 nearest neighbors). We leave finding the recall scores for future work. Additionally, we calculated the inter-rater agreement using Krippendorff's alpha [8].

We also considered using author-provided keywords as a 'gold standard' for evaluation (for example, with Jaccard similarity for keyword lists as the ground truth semantic distance between documents), but we found that in most of the documents the keywords were not actually representative since keyword choice for a paper is a highly subjective process.

Table 1. Precision scores and inter-rater agreement for the tested models.

Model	Assessor 1	Assessor 2	Assessor 3	Average	Agreement
TF-IDF	0.6	0.65	0.68	**0.64**	0.73
Paragraph Vector - 40	0.28	0.33	0.4	0.33	0.66
Paragraph Vector - 100	0.36	0.36	0.45	0.39	0.5
LDA - 10	0.17	0.2	0.21	0.2	0.66
LDA - 20	0.23	0.19	0.39	0.27	0.52

Table 1 reports the results of the comparison. First, for all models, the inter-rater agreement is at quite decent level, meaning the results are trustworthy[5]. The second outcome is somewhat unexpected: the simplistic *TF-IDF* model produced way better results than the sophisticated *LDA* and *Paragraph Vector* distributional algorithms.

We hypothesize that the reason for this is the size of our corpus, which is comparatively small. It seems that 400 documents with total word count of 1,340,957 are not enough to train a distributional model able to outperform the frequency-based approach. Note though that *TF-IDF* is in fact often used in production systems instead of modern algorithms, providing a good trade-off between performance, speed and model size: for example, [5] employ it as the main ranking model for their life science publications search engine. So, following these experiments, we used TF-IDF to implement the search engine and recommendation system described in the next section.

---

[5] Initially, the agreement levels for the *TF-IDF* and *LDA-10* models were below 0.5. We performed a reconciliation round with the assessors discussing their choices for these models, which resulted in changing some of the scores for particular documents. This increased the inter-rater agreement, but did not influence the final ranking.

4   RusNLP Web Service: Key Features

The *RusNLP* search engine provides web access to the collection of Russian NLP papers described above. Under the hood, it employs an *SQLite* database to store documents metadata and a *TF-IDF* model to find similar papers. For simplicity, and because of comparatively small absolute size of the corpus, at this moment we do not use the existing production-level text search libraries like *Sphinx*, *Solr* or *ElasticSearch*. However, in the future we plan to test whether using these libraries will improve user experience in our case.

Right now, the service already allows its users to:

- Find and rank papers most similar to an arbitrary user query (list of keywords); see Fig. 1 in Appendix A for an example.
- Find and rank papers most similar to any given paper; see Fig. 2 in Appendix A for an example.
- Filter search results by any combination of user-defined criteria:
  - publication venue,
  - publication year,
  - author names,
  - author affiliations,
  - paper title.

We envisage that *RusNLP* can be used either for the search of relevant previous work ('*I know this paper, what other similar papers are there in Russian NLP?*') or for studies in the history and structure of academic communities ('*What was published in 2008 by NLP scholars from Moscow State University?*' or '*Were there any papers about paraphrases detection at the **AINL** conference in 2015?*'). Figure 3 in Appendix A shows an example of a query for all papers presented at **Dialogue** or **AIST** by any author whose name starts with 'Chernj'.

For many papers, we also provide the so called 'task tags'. They link to the corresponding resource sections of the *NLPub* web service[6] [16]. For example, a hypothetical paper using a thesaurus for emotion detection would be interlinked with the *NLPub* pages listing existing tools for tasks related to thesauri and sentiment analysis. To this end, we manually selected sets of English keywords for each *NLPub* resource section. At query time, our service calculates the *TF-IDF* similarity of all papers found to each of these sets. If the similarity exceeds a predefined threshold (after some experimenting, we set it to 0.03), the paper is assigned the corresponding 'task tag'. This feature allows the users to find out at a glance what the paper is about and to quickly get an idea about the existing tools or resources relevant to this task.

The *RusNLP* web service we describe in this paper is publicly available at http://nlp.rusvectores.org.

---

[6] https://nlpub.ru.

5   Related Work

The importance of compiling reference in-domain datasets of texts related to NLP was expressed more than 10 years ago in [3]. This paper was the first towards describing and releasing a corpus of computational linguistics academic papers and their metadata: this dataset is now known as the *Association for Computational Linguistics Anthology*[7].

However, the *ACL Anthology* is not enriched with semantic search or a recommendation system. There exist more functional search engines and databases that are less related to NLP and computational linguistics: see, for instance, *Pulp* that includes semantic search engine, visualization of search topics (with topic modeling), and document-ranking algorithm [10]. Another system, *Bib2Vec* [17], embeds bibliographic information in a vector space and then uses these vectors to show relationships among entities in the ACL Reference Corpus. There are more and more similar service appearing each year, like *Semedico*, a life science semantic search engine [5], or *Citeomatic* which 'recommends papers you might want to cite' [2]. As for the Russian academic community, we are unaware of any domain-specific Russian scholarly search engines: there exist only global bibliographic services for all fields of Russian science like *CyberLeninka* or *Russian Science Citation Index* (RSCI).

Note that our research is highly related to studies in information retrieval (IR) on scientific publications. For instance, proceedings of International Workshop on Mining Scientific Publications (WOSP) provide a substantial amount of papers devoted to 'Automatic categorization and clustering of scholarly data' and 'Academic recommendation systems'. One can mention [6] and [12], that implement rather sophisticated approaches for metadata extraction which we plan to test and possibly use in the *RusNLP* project.

Considering previous research in the field of Russian NLP, the analysis of Russian NLP landscape was arguably started by Khoroshevsky [7]. Their research is similar to our project: they described academic communities, conference activity and links between scholar clusters for Russian computational linguistics community. We improve on this work in 3 aspects:

1. [7] is published in 2012, and uses data from 2009; our project mines conference proceedings up to 2917, and includes novel publishing venues (AIST and AINL). We believe that since 2009, the Russian NLP community significantly changed with regards to its primary trends and topics of interest; this is supported by our analysis of diachronic topical drift in Russian NLP academic papers [1].

---

[7] http://aclanthology.info/.

2. We implement the semantic search engine available online as a web service.
3. In [7], no datasets were released; we publish all our source code[8] and data[9]. Note that our contribution is not only crawling papers from the venues web sites, but also in consistent manual extraction and normalization of papers' metadata).

Thus, we argue that our project is novel, and is a step towards unification and structuring the research within the Russian NLP community.

6 Conclusion

In this paper, we described *RusNLP*, a web service for scholarly search in the collection of natural language processing papers. This is the first publicly available and manually curated database of publications in the most important Russian computational linguistics venues. Of course, contemporary science is global and as a rule one shouldn't limit herself to only papers published in this or that country. However, sometimes (especially in research dealing with particular language) it can be useful to focus more on the local academic landscape. This can be also interesting for those studying academic communities and the trends in NLP publishing activities.

Additionally, we described a set of experiments conducted to find out the most efficient algorithm for finding similar papers in our dataset. The outcome was that the simple *TF-IDF* model outperformed the sophisticated distributional algorithms of *LDA* and *Paragraph Vector* by a large margin, most probably because of comparatively small training corpus size. This model powers the nearest neighbors search under the hood of our web service (http://nlp.rusvectores.org).

The efforts described in this paper are part of a larger project aiming to analyze Russian NLP landscape (thus, the dataset is not complete and will be updated yearly). In the future, we plan to make it possible to use *RusNLP* to search for papers written in Russian, not only in English, thus making the resource multilingual. Another important direction in our nearest plans is the construction of the citation graph from all the papers in our database, which will allow to explore connections between scientific communities. Finally, we are considering the possibility of improving our keywords search by expanding user queries with the help of pre-trained word embedding models.

Acknowledgments. We thank numerous VPNs and *Tor Project*. At the time of finalizing this paper, they were the only ways for Russian-based scholars to collaborate with the colleagues abroad, because of Internet censorship carried by the Russian governmental agency called *Roskomnadzor*. It accidentally managed to temporarily block a whole bunch of academic resources, including *Softconf, Overleaf*, etc.

---

[8] https://github.com/bakarov/rusnlp/tree/master/code/web.
[9] http://nlp.rusvectores.org/about/.

Appendix A

Fig. 1. Searching scholarly papers in the database by user-provided query words (in this case, 'syntax neural classification').

Fig. 2. RusNLP recommending papers similar to the query paper.

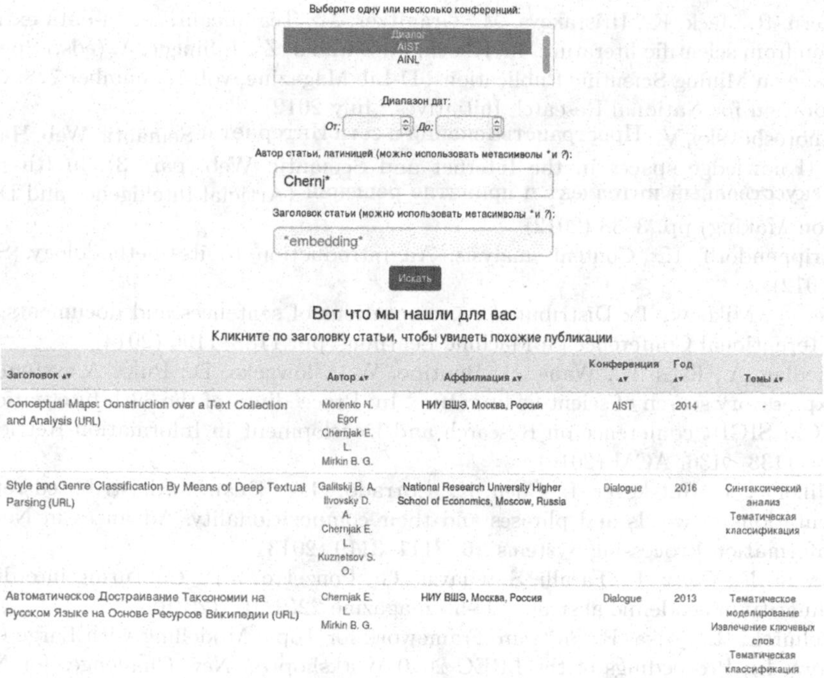

Fig. 3. RusNLP searching for all papers by a certain author in the **Dialogue** and **AIST** conferences.

References

1. Bakarov, A., Kutuzov, A., Nikishina, I.: Russian computational linguistics: topical structure in 2007–2017 conference papers. In: Proceedings of Dialogue-2018, online papers. ABBYY (2018), http://www.dialog-21.ru/media/4249/bakarov_kutuzov.pdf

2. Bhagavatula, C., Feldman, S., Power, R., Ammar, W.: Content-based citation recommendation. In: Proceedings of the 2018 Conference of the North American Chapter of the Association for Computational Linguistics: Human Language Technologies, Volume 1 (Long Papers). pp. 238–251. Association for Computational Linguistics (2018), http://aclweb.org/anthology/N18-1022

3. Bird, S., Dale, R., Dorr, B., Gibson, B., Joseph, M., Kan, M.Y., Lee, D., Powley, B., Radev, D., Tan, Y.F.: The ACL Anthology reference corpus: A reference dataset for bibliographic research in computational linguistics. In: LREC 2008 (2008), http://www.aclweb.org/anthology/L08-1005

4. Blei, D.M., Ng, A.Y., Jordan, M.I.: Latent Dirichlet allocation. Journal of Machine Learning Research 3(Jan), 993–1022 (2003)

5. Faessler, E., Hahn, U.: Semedico: a comprehensive semantic search engine for the life sciences. Proceedings of ACL 2017, System Demonstrations pp. 91–96 (2017)

6. Kern, R., Jack, K., Hristakeva, M., Granitzer, M.: Teambeam - meta-data extraction from scientific literature. In: Knoth, P., Zdrahal, Z., Juffinger, A. (eds.) Special Issue on Mining Scientific Publications, D-Lib Magazine, vol. 18, number 7/8. Corporation for National Research Initiatives (July 2012)

7. Khoroshevsky, V.: Пространства знаний в сети Интернет и Semantic Web, Часть 3 (Knowledge spaces in the Internet and Semantic Web, part 3); in Russian. Искусственный интеллект и принятие решений (Artical Intelligence and Decision Making) pp. 3–38 (2012)

8. Krippendorff, K.: Content analysis: An introduction to its methodology. Sage (2012)

9. Le, Q., Mikolov, T.: Distributed representations of sentences and documents. In: International Conference on Machine Learning. pp. 1188–1196 (2014)

10. Medlar, A., Ilves, K., Wang, P., Buntine, W., Glowacka, D.: Pulp: A system for exploratory search of scientific literature. In: Proceedings of the 39th International ACM SIGIR conference on Research and Development in Information Retrieval. pp. 1133–1136. ACM (2016)

11. Mikolov, T., Sutskever, I., Chen, K., Corrado, G.S., Dean, J.: Distributed representations of words and phrases and their compositionality. Advances in Neural Information Processing Systems **26**, 3111–3119 (2013)

12. Nanni, F., Dietz, L., Faralli, S., Glavaš, G., Ponzetto, S.P.: Capturing interdisciplinarity in academic abstracts. D-lib magazine 22(9/10) (2016)

13. Řehůřek, R., Sojka, P.: Software Framework for Topic Modelling with Large Corpora. In: Proceedings of the LREC 2010 Workshop on New Challenges for NLP Frameworks. pp. 45–50. Valletta, Malta (May 2010)

14. Sparck Jones, K.: A statistical interpretation of term specificity and its application in retrieval. Journal of documentation **28**(1), 11–21 (1972)

15. Straka, M., Straková, J.: Tokenizing, POS Tagging, Lemmatizing and Parsing UD 2.0 with UDPipe. In: Proceedings of the CoNLL 2017 Shared Task: Multilingual Parsing from Raw Text to Universal Dependencies. pp. 88–99 (2017)

16. Ustalov, D.: NLPub: a catalogue and a community for Russian linguistic resources. In: Selected Papers of XVI All-Russian Scientific Conference "Digital libraries: Advanced Methods and Technologies, Digital Collections". vol. 1297, pp. 56–60. RWTH (2014)

17. Yoneda, T., Mori, K., Miwa, M., Sasaki, Y.: Bib2vec: Embedding-based search system for bibliographic information. In: Proceedings of the Software Demonstrations of the 15th Conference of the European Chapter of the Association for Computational Linguistics. pp. 112–115. Association for Computational Linguistics (2017), http://aclweb.org/anthology/E17-3028

Russian Q&A Method Study: From Naive Bayes to Convolutional Neural Networks

Kirill Nikolaev[ID] and Alexey Malafeev[(✉)][ID]

National Research University Higher School of Economics,
Nizhny Novgorod, Russia
kinikolaev@edu.hse.ru, aumalafeev@hse.ru

Abstract. This paper deals with automatic classification of questions in the Russian language. In contrast to previously used methods, we introduce a convolutional neural network for question classification. We took advantage of an existing corpus of 2008 questions, manually annotated in accordance with a pragmatic 14-class typology. We modified the data by reducing the typology to 13 classes, expanding the dataset and improving the representativeness of some of the question types. The training data in a combined representation of word embeddings and binary regular expression-based features was used for supervised learning to approach the task of question tagging. We tested a convolutional neural network against a state-of-the-art Russian language question classification algorithm, an SVM classifier with a linear kernel and questions represented as word trigram counts, as the baseline model (60.22% accuracy on the new dataset). We also tested several widely-used machine learning methods (logistic regression, Bernoulli Naïve Bayes) trained on the new question representation. The best result of 72.38% accuracy (micro) was achieved with the CNN model. We also ran experiments on pertinent feature selection with a simple Multinomial Naïve Bayes classifier, using word features only, Add-1 smoothing and no strategy for out-of-vocabulary words. Surprisingly, the setting with top-1200 informative word features (by PPMI) and equal priors achieved only slightly lower accuracy, 70.72%, which also beats the baseline by a large margin.

Keywords: Natural language processing question answering
Machine learning · Deep learning · Convolutional neural networks
Relevant feature selection

1 Introduction

The coverage of the question answering domain for Russian language in recent years has been severely limited. Although a monograph by Sosnin [15] and a few research papers, such as [16] and [17], have been published, they generally contribute to the theory on the problem, rather than propose or evaluate efficient practical solutions.

Our recent work [14] differed from previously published papers as a semi-successful attempt to tackle the task of question tagging. The linear SVM algorithm we used had previously yielded state-of-the-art results for the English question

© Springer Nature Switzerland AG 2018
W. M. P. van der Aalst et al. (Eds.): AIST 2018, LNCS 11179, pp. 121–126, 2018.
https://doi.org/10.1007/978-3-030-11027-7_12

classification, as shown in a survey by Loni [12], reaching 95% accuracy on a coarse-grained dataset.

Our model showed 68.7% accuracy, which is significantly lower, but the results are not directly comparable, because the language is different, the dataset is much smaller and fewer NLP tools are available for Russian. In this paper, we attempt to surpass these results set as the baseline. We use an improved version of the dataset presented in [14], based on the original question typology (Table 1), which is the only publicly available one for the Russian-language question tagging task at the moment.

Table 1. Question typology.

Tag	Numerical tag	Examples
General	1	Что происходит в..? What is happening in..?
Verification	2	Правда ли, что? Is it true, that..?
Definition	3	Что означает/такое?.. What is..?
Example	4	Приведи пример..? Give an example of..?
Comparison	5	Чем похожи/отличаются..? What are the similarities/differences between..?
Choice	6	X или Y? X or Y?
Concept completion	7	Кто? Что? Где? Когда? Куда? Откуда? Во сколько? Who? What? Where? When? What time?
Quality	8	Какой? What kind of?
Quantity	9	Сколько? Как много? How many/much?
Action	10	Что делать, чтобы..? What do I do for..?
Instrument	11	С помощью чего/каким методом..? With what? How?
Goal	12	К чему? Зачем? For what?
Reason	13	Почему? Why?
Consequence	14	Каковы последствия? What are the consequences..?

In order to improve the results previously attained, we followed a different, deep learning approach. In the domain of text classification, recurrent convolutional neural networks (RCNNs) have recently been used [11]. The authors contrasted traditional learning methods based on human-designed features with an unsupervised text classifier. We decided to adopt a somewhat similar approach, but with a labeled set of questions and convolutional neural networks (CNNs) instead of RCNNs. More details on this follow in the next section.

2 Methodology

Questions can be treated as structured data. Problems where structure is important can be attempted via CNNs, which has proved useful in such tasks as image recognition [10] and video [18] representation. However, existing studies on CNNs tend to use simple convolutional kernels such as a fixed window [5]. The problem is that it is difficult to determine the window size: small window sizes result in information loss, whereas large sizes result in a large parameter space that is difficult to train. Recurrent neural networks (RNNs), in their turn, store the semantics of previously seen texts in a fixed-sized hidden layer [7], but are thus biased, with later inputs dominating over previous ones. RNNs adopted to text classification allow storing features, whereas the CNN-embedding of a max-pooling layer makes it possible to evaluate the features and their roles in question classification.

We experimented with RCNNs, somewhat of a hybrid of CNNs and RNNs, but it became clear that we do not have enough data to achieve satisfactory results. Thus, it was decided to use CNNs instead of RCNNs, due to the small dataset size. This decision proved fruitful; but before we discuss the results attained, it is necessary to give additional information about the data representation and neural network architecture used.

2.1 Dataset Modification and Question Representation

Drawing upon conclusions made in our previous work [14] and the specifics of the chosen methods, we decided to modify the dataset. The existing dataset of 2008 (training) and 150 (test) questions, manually annotated in accordance with a pragmatic 14-class typology, was modified. Classes 10 and 11 (Action and Instrument) are semantically very close, and there is little practical use in keeping the distinction, so we merged these two types, narrowing the typology to 13 classes. Additionally, we boosted the underrepresented classes (primarily 4 and 5, Example and Comparison) in both training and test sets, increasing the number of samples to 2086 and 181, respectively. Using the new dataset, we replicated the experiments carried out in [14] and reached the top accuracy of 60.22% (micro accuracy is reported here and further in the paper). This was significantly lower than the result of 68.7% reported in [14] due to a now larger representation of some question types that are very difficult to classify automatically. Thus, we made the task even more challenging. This result, 60.22% macro accuracy, was set as the baseline for further work.

It is widely known that a key problem in text classification and machine learning is text representation. In text classification, bag-of-word models are very common. For selecting more discriminative features, other feature selection methods, such as MI [6] and PLSA [3], can be applied. The word embedding approach that has become very popular recently is based on distributed semantic representations of words in any language. As is well known in the field, embeddings are a very efficient way of solving the data sparsity problem [2]. A number of tools have been developed for effectively working with, acquiring and processing word embeddings. One of the seminal ones is Word2Vec [8].

There is a pre-trained Word2Vec model available for Russian based on the Russian National Corpus of 250 million words, with 300-dimensional vectors [9]. However, the questions in the dataset are comprised of a limited vocabulary, of which many so-called "stop-words" or "noise-words" are quite meaningful and important. Yet there are no vectors for these words in the pre-trained model, as they were removed as stop-words. Thus, we had to use zero vectors for such words.

To obtain a matrix for each question, we took the vectors of first 8 words in the sentence (the mean number of words in all sentences in the training corpus is 7). If the sentence is shorter, it is filled with stop-words that have zero vectors. Longer sentences were simply truncated.

In order to compensate for the stop-words that are important for question tagging, we also introduced 40 binary word features. That is, each word in the sentence is matched against a number of hand-picked words that are often indicators of a particular question type. For example, the function word *или* (or) does not have a Word2Vec representation, yet it is an important word for detecting question type 6, Choice.

These binary feature values were appended to the semantic vectors for each word. After that the 8 vectors of 340 values for each question were converted into 340×8 matrices. The resulting matrices are input in the 2D convolutional layer. The labels are similarly transformed into a one-hot categorical matrix of size 2086×13, where each row is the one-hot encoding of the question type [4].

2.2 Architecture

We used three "blocks": a convolutional layer, a max pooling layer and a fully connected, or "dense" layer. The "input" layer accepts a structure of size $340 \times 8 \times 1$, receiving the following hyperparameters, tuned with 3-fold cross-validation on the training set:

Number of filters: 26. We attempted building multiple layers, 2 (26–52) and 3 (26–52–104), also using a different number of filters (13 or 52 for the first layer), the best result was shown by a single layer of 26 filters. This is probably due to the fact that our data is small and because 26 is divisible by the number of classes (13);

The spatial parameter (kernel size): 20×3;

Function "padding = same" provides for automatic calculation of the stride and the amount of zero padding, with the only condition being that the output shape should be identical to the input.

As an algorithm of learning from non-linear decision boundaries, we also relied on a Leaky version of the Rectified Linear Unit [13], which allows for a small gradient when the unit is not active. In order to avoid overfitting, we added a dropout (0.2) function after the maxpooling layer. Dropout randomly turns off a fraction of neurons during the training process, reducing the dependency on the training set by some amount.

3   CNN Evaluation and Feature Selection Experiments

Our CNN model [19] achieved 72.38% overall test accuracy (0.67 F-score), which is significantly higher than the baseline accuracy of 60.2%. Of other models trained on the same data representation (SVM 1 vs All, Logistic Regression, Bernoulli Naïve Bayes, Linear SVC), the highest result was shown by the LR model, with F-score and accuracy of 0.59 and 57%, respectively, which is lower than the SVM trigram baseline.

Given the small training set, we felt it was possible to achieve comparable classification accuracy by relying on a simple multinomial Naïve Bayes Classifier and automatically selecting pertinent word features in the training set. We used our own implementation [20] of the multinomial Naïve Bayes Classifier with Add-1 smoothing, written in Python. Only word features were used.

For feature selection, we used a simple procedure in which all words seen in the training set were ranked in accordance to top PPMI value in row, descending. Consequently, it became possible to select top n informative word features. During testing, only the known words were extracted from the test questions, i.e. no strategy for dealing with previously unseen words was used; these were simply ignored. The results of using different n values are shown in Table 2. The best result was achieved with top-1200 informative words: 70.72% accuracy. As can be seen, this does not merely beat the baseline, but is also very close to our CNN model result (72.38% accuracy).

Table 2. Pertinent word feature selection.

Words	All	2000	1500	1200	800	400	200
Acc.	59.7%	62.4%	65.7%	70.7%	69.1%	68%	61.9%

4 Conclusion and Future Work

The main contribution of our paper is the first implementation of Convolutional Neural Networks for the Russian-language question classification task. We have tested the algorithm on a hybrid question representation (word2vec + binary word features), transformed into the tensor form. The classifier achieved classification accuracy of 72.38%, in contrast to the Linear SVM 60.22% baseline result. To the best of our knowledge, this result can be considered state-of-the-art for Russian-language question tagging. Given the relatively small training set size, this may be close to the upper bound of accuracy attainable on this data. The most problematic question types (F1-score < 0.5), according to the typology shown in Sect. 1, are 1 (General) and 10&11 (Action/Instrument). Interestingly, comparable performance was shown by a classic method, multinomial Naïve Bayes Classifier with Add-1 smoothing using top 1200 informative word features on the lemmatized dataset.

We believe that it might be possible to improve question tagging results by further expanding the training and test data sets, using more complex (3-dimensional) question representations, and, possibly, calibrating the deep learning model parameters and using more advanced deep learning algorithms, namely RCNN (recurrent convolutional neural networks), which, as well as 3D-representations, require more significant volumes of data.

References

1. Abadi, M., et al.: TensorFlow: a system for large-scale machine learning. In: OSDI, vol. 16, pp. 265–283 (2016)
2. Bengio, Y., et al.: A neural probabilistic language model. J. Mach. Learn. Res. **3**(Feb.), 1137–1155 (2003)
3. Cai, L., Hofmann, T.: Text categorization by boosting automatically extracted concepts. In: Proceedings of the 26th Annual International ACM SIGIR Conference on Research and Development in Information Retrieval, pp. 182–189. ACM (2003)
4. Chollet, F., et al.: Keras (2015)
5. Collobert, R., et al.: Natural language processing (almost) from scratch. J. Mach. Learn. Res. **12**(Aug.), 2493–2537 (2011)
6. Cover, T.M., Thomas, J.A.: Elements of Information Theory. Wiley, Hoboken (2012)
7. Elman, J.L.: Finding structure in time. Cogn. Sci. **14**(2), 179–211 (1990)
8. Goldberg, Y., Levy, O.: Word2vec explained: deriving Mikolov et al.'s negative-sampling word-embedding method. arXiv preprint arXiv:1402.3722 (2014)
9. Kutuzov, A., Kuzmenko, E.: RusVectōrēs: distributional semantic models for the Russian (2017)
10. Krizhevsky, A., Sutskever, I., Hinton, G.E.: ImageNet classification with deep convolutional neural networks. In: Advances in Neural Information Processing Systems, pp. 1097–1105 (2012)
11. Lai, S., et al.: Recurrent convolutional neural networks for text classification. In: AAAI, vol. 333, pp. 2267–2273 (2015)
12. Loni, B.: A survey of state-of-the-art methods on question classification (2011)
13. Maas, A.L., Hannun, A.Y., Ng, A.Y.: Rectifier nonlinearities improve neural network acoustic models. In: Proceedings of ICML, vol. 30, no. 1, p. 3 (2013)
14. Nikolaev, K., Malafeev, A.: Russian-language question classification: a new typology and first results. In: van der Aalst, W.M.P., et al. (eds.) AIST 2017. LNCS, vol. 10716, pp. 72–81. Springer, Cham (2018). https://doi.org/10.1007/978-3-319-73013-4_7
15. Sosnin, P.I.: Question-answer modeling in the development of automated systems [Voprosno-otvetnoe modelirovanie v razrabotke avtomatizovannykh sistem], Ul'yanovsk, USTU (2007)
16. Suleymanov, D.Sh.: A study of the basic principles of building a semantic interpreter for questions and answers in natural language in AOS [Issledovanie bazovykh printsipov postroeniya semanticheskogo interpretatora voprosno-otvetnykh tekstov na estestvennom yazyke v AOS], Educational technologies and society [Obrazovatel'nye tekhnologii i obshchestvo], no. 3, pp. 178–192 (2001)
17. Tikhomirov, I.A.: Question-answering search in the intelligent search system Exactus [Voprosno-otvetnyy poisk v intellektual'noy poiskovoy sisteme Exactus]. In: Proceedings of the Fourth Russian Seminar on Evaluation of Information Retrieval Methods ROMIP [Trudy chetvertogo rossiyskogo seminara po otsenke metodov informatsionnogo poiska ROMIP], pp. 80–85 (2006)
18. Xu, Z., Yang, Y., Hauptmann, A.G.: A discriminative CNN video representation for event detection. In: 2015 IEEE Conference on Computer Vision and Pattern Recognition (CVPR), pp. 1798–1807. IEEE (2015)
19. RCNN Model. https://github.com/Pythonimous/Q-A-System. Accessed 13 Apr 2018
20. Naïve Bayes Model. https://github.com/WonderingTachikoma/naive_bayes. Accessed 13 Apr 2018

Extraction of Explicit Consumer Intentions from Social Network Messages

Ivan Pimenov[1]([✉]) and Natalia Salomatina[2]

[1] Novosibirsk State University, Novosibirsk, Russia
pimenov.1330@yandex.ru
[2] Sobolev Institute of Mathematics, Novosibirsk, Russia
salomatina_nv@live.ru

Abstract. In this paper we address the problem of automatic extraction of facts from Russian texts. The facts under examination are the intentions of social network users to purchase certain goods or use certain services. The utilized approach is machine learning with annotation. A training set for expert annotation consists of messages from the "VKontakte" social network, selected through the LeadScanner API. The invented system of semantic tags allows distinguishing between various intentional blocks: objects, their different properties and emphatic constructions. Pre-processing of the training set includes lemmatization and grammatical tagging with PyMorphy2. Then, on the material of the training set, a directed graph is constructed. Each node in this graph corresponds to an intentional block, including information about its expertly-assigned intentional tag, grammatical and/or lexical properties of its main word. The edges of the graph connect the intentional blocks that can be found in adjacent positions across all the messages of the training set. Extraction of intention objects and their properties is achieved by test set analysis in accordance to the constructed graph. Test set includes both messages containing non-consumer intentions or no intentions at all. The precision and recall of intention extraction with macro average is 82% and 74% respectively.

Keywords: Intention · Intention marker · Machine learning with annotation Directed graph · Fact extraction

1 Introduction

Of all the platforms of interest to digital marketing, social networks hold special importance as of now. Among the digital marketing objectives is that of lead generation, or detection of potential clients who post messages in social networks that indicate their interest in purchasing a certain product or service. The problem of identifying these intentions in a message belongs to a field of intellectual (factual) analysis.

The structure of any fact includes named entities which might be characterized by certain properties and might form relations between themselves. The following types of facts are considered to be prevalent: (1) object and its property, for example, a product and its price; (2) object and action (such as appointing someone to a particular position)

© Springer Nature Switzerland AG 2018
W. M. P. van der Aalst et al. (Eds.): AIST 2018, LNCS 11179, pp. 127–133, 2018.
https://doi.org/10.1007/978-3-030-11027-7_13

[1]. A fact is often preceded by special words that mark its presence in the text. The structure of a fact and its markers are determined by the nature of entities under analysis, which is the reason why almost every new fact demands creation of a new model that adjusts to diverse ways of its expression in the natural language.

Following formal models are used for the purposes of facts description: (1) regular expressions, such as those of Jape language [2], (2) context-free grammars [3], (3) templates and frames [4], etc.

Methods of building models of fact can be separated into three groups: knowledge-based (done by a human expert) [4], machine learning-based and hybrid. Methods of machine learning can be further divided into supervised learning [5] and unsupervised learning [6]. An example of a hybrid method is the iterative one, where expertly-created patterns are expanded with names of entities and relations between those by means of an iterative search [7]. In the present work, the method used is that of machine learning with annotation [8], which, while similar to supervised learning, relies not on the machine finding hidden correlations between the positive examples, but creating a generalized structure of the ways of fact expression that is accessible to an observer and allows identification of different semantic components of the fact.

Some of the already existing services for lead generation, such as Qualia or Leadsift, do not perform textual analysis at all, while some of the others (Socedo) operate only with English texts and only through keywords detection. There do exist services that work with texts in Russian (LeadScanner, Leaderator, It's great), yet these only identify a thematic group of a located intention-expressing message without actually extracting an exact intention object nor its properties.

Goals of the research are defined as follows: (1) to automatically construct rules for the extraction of intention object and its relevant properties through the use of expertly-annotated (in accordance to the intentional structure of its messages) training set; (2) to evaluate the constructed rules in terms of precision and recall on the test set.

2   The Problem of Intention Recognition

This paper addresses the problem of identifying and extracting intentions that are expressed in a text explicitly – in the meaning that a special marker indicates their presence within. Intentions of this type can be represented as sequences of intentional blocks, where each block corresponds to a single word or a group of words in the original message. For the purposes of the research we distinguish between four different types of intentional blocks, with the basis being what we call their intentional semantic. These types are as follows: intention marker (Mr), which indicates that a psychological category of intentionality is expressed in the message; intention object (Ob), or the object in the direct focus of an intention; properties of the intention object (Pp) and emphatic elements (Em), such as irrelevant properties of the object or expressions of politeness. The former two types are obligatory for any intention expression, while the latter two are optional.

The following steps are required for constructing rules of detecting intention and extracting its content: (1) creation of the training set of texts that contain consumer intention; (2) creation of the markers dictionary; (3) determination of the degree of intentional blocks variability on grammatical and lexical levels; (4) introduction of adjustments to G_k structure on account of possible transposition of the intentional blocks; (5) aggregation of G_k on the basis of their structural similarity.

3   Methods

3.1   Training Stage

Annotation of the training set is conducted by an expert, who transforms each of the messages into a sequence of intentional blocks and assigns to each of the blocks an intentional semantic tag. These tags are determined through an analysis of the intention structure of the message, with each of the intentional block types specified further in accordance to its actual role. For instance, markers might consist both of a single word and multiple words. Among the latter there exists a distinction between markers corresponding to a part of the intention object (mrob) and markers completely detached from it (mrpl). These markers are characterized by different variability on the lexical level, which is reflected in their corresponding intentional blocks in G_k. All of the intentional semantic tags are listed in Table 1, the last column of which offers examples from the training set messages with corresponding words in italics.

Table 1. Semantics tags for intentional blocks annotation.

Mr:	mrsg	standalone marker	*hochu, podskazhite, nuzhen*
	mrpl	marker with abstract meaning	*kto znaet, gde kupit'*
	mrob	marker that is a part of the object	gde *sdelat'* UZI
Ob:	obmn	main object	posovetujte *dietu*
	obnd	additional elements of the object	podskazhite repetitora *po himii*
	obin	object indicator	dajte *koordinaty* massazhista
Pr:	pppl	location	ishchu ortopeda *v Rostove*
	ppad	addressee of the desired service	nuzhen logoped *dlya shkol'nika*
	pptm	time or date	ishchu fotografa *na 8 marta*
	pppr	price	hochu vzyat' kredit *okolo 30000*
	ppgl	goal	ishchu repetitora *dlya sdachi EGEH*
	ppqn	quantity	nado *1000* druzej dobavit'
Em:	emph	expression of politeness	podskazhite *pozhalujsta* vracha
	emmd	extension of marker	kuda *luchshe\mozhno* skhodit'
	emql	irrelevant object property	podskazhite *horoshego* vracha

The further processing of texts is conducted automatically and includes:

(1) lemmatization of each word and determination of its grammatical tags (through the use of PyMorphy2 module, version 3.5.2 [9]);

(2) construction of convolution rules for intentional blocks that consist of two or more words; these rules replace grammatical tags of all words with grammatical tags of the main one among them (determined by an expert during the annotation stage);

(3) deletion of grammatical properties that are highly variable while also irrelevant to differentiation of intentional blocks with different intentional meaning;

(4) creation of a dictionary of markers that appear in the initial intentional blocks;

(5) formation of $G^j = \cup_{k=1}^{K_j} G_k$ as a directed graph ($j = 1,..., M$, with M being number of markers in the dictionary and K_j – the number of messages marked j-th marker in the training set). Nodes from G^j correspond to intentional blocks and contain information about intentional, grammatical and/or lexical tags, while edges connect them in accordance to their sequence in the original message.

Intention graphs always have an intention marker as their initial node, while terminal nodes might correspond both to intention objects and their properties. In Fig. 1 we provide an example of an intentional graph, one for initial marker = "gde", simplified for the purpose of an image clarity: only intentional tags are taken into consideration, without including any information about grammatical tags and lemmas of their corresponding intentional blocks. Nodes that can potentially be terminal are drawn in a rectangular shape, while all others are elliptic. The absence of "obin" and "emph" blocks serves as an example to indicate that graphs for different markers might differ in compositions of their blocks.

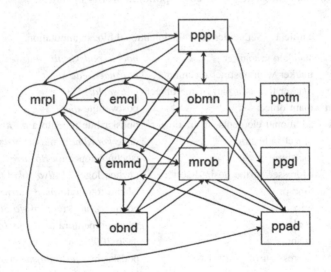

Fig. 1. Intention graph for initial marker = "gde".

3.2   Identifying and Extracting Intentions

The procedure of intention identification includes the following steps:

(1) text preprocessing: lemmatization, morphological analysis of its words and their separation into sequential blocks in accordance to the convolution rules, assignation to each of the blocks its lexical and grammatical tags;

(2) detection of potential markers: for each of the markers in the corresponding dictionary, its lexical and grammatical tags are compared to those of each of the blocks in the transformed message. If a perfect match occurs, the procedure described in (3) is started for the corresponding graph. If (3) has not been completed successfully for any of the markers, a decision about the absence of intentions in the message is made.

(3) comparison of a message block sequence to the paths in the chosen graph: for the current block in the sequence, it is compared on both lexical and grammatical level to each of the nodes directly reachable from the node corresponding to the previous block. If a perfect match occurs or no more than two blocks have been skipped yet, the procedure is repeated for the next block in the sequence. For the longest completed path from the initial node to any of the potentially terminal, blocks corresponding to nodes of intentional object and its properties are displayed.

4 Results

The training set contained 1000 annotated messages from the "VKontakte" social network, selected by an expert through the use of the "Leadscanner" API [10], where each expressed exactly one consumer intention. Annotation of these messages, belonging to 9 different commercial categories, was conducted by another expert. 483 convolution rules were formulated on the material of the training set, and the created dictionary of initial markers contains 35 elements, of which 15 appeared just once.

Two test sets were created for evaluating constructed rules of intention recognition:

1. Set S_500, containing 500 messages from the "Vkontakte" social network.
2. Set S_386, containing 387 texts from the Russian National Corpus [11], where half of the messages, expressing intention, expressed one of a non-consumer type.

Table 2 demonstrates the percentage scores of precision, recall and F-measure, calculated upon the test sets, along with their macro average. Errors of recognition encountered can be attributed to two types, false positives and false negatives. Practically the only cause for type I errors is that messages meant for advertising are sometimes deliberately written to resemble consumer messages. Proper distinguishing between messages of these two kinds requires analysis of the context.

Type II errors are more varied, their possible causes being incompleteness of the constructed graphs (48%), mistakes made by the morphological analyzer (32%), improper application of the convolution rules (12%) and mistakes made by the message author (8%). Among these, errors of the first and third kind can be solved through the use of a larger training set. Successful recognition of non-consumer intentions is also to be noted, as it indicates the possibility of extending this method area of application without significantly changing the recognition rules.

Table 2. Intention recognition scores for the test sets.

	S_500	S_386	Macro average
P	80	83	82
R	69	78	74
F-measure	74	80	77

5 Conclusion

In this paper we address the problem of automatic extraction of a new fact – consumer intention in its expression in a social network message. Unlike other facts of the "object – property" type, intention is characterized by a more complex structure, able to express a wide variety of properties.

The distinctive features of the approach tested are: (1) intention analysis in accordance to its semantic structure; (2) representation of the recognition rules through a directed graph model with nodes accumulating multi-level information about intentional blocks.

The results of the testing stage show that the approach used allows ascertaining if a particular message expresses intention, and, if it does, extracting the intention object along with its relevant properties – while reaching reasonable scores in both precision and recall for consumer and non-consumer intentions alike. Afterwards, the data extracted can be used for further refinement of message classification.

Acknowledgements. The study was supported by the Russian Academy of Science (the Program of Basic Research, project 0314-2016-0015).

References

1. Pivovarova, L.M.: Faktograficheskij analiz teksta v sisteme podderzhki prinyatiya reshenij. Vestnik SPbGU, Ser. 9, Vyp. 4, pp. 190–197 (2010)
2. Cunningham, H., Maynard, D., Tablan V.: JAPE: a Java Annotation Patterns. Technical report CS–00–10, University of Sheffield, Department of Computer Science (2000)
3. Tomita-parser. http://tech.yandex.ru/tomita. Accessed 26 Mar 2018
4. Bol'shakova, E., Baeva, N., Bordachenkova, E.: Leksiko-sintaksicheskie shablony v zadachah avtomaticheskoj obrabotki tekstov. In: Komp'yuternaya lingvistika i intellektual'nye tekhnologii (Dialog 2007), vol. 2, pp. 70–75. RGGU, Moskva (2007)
5. Tang, J., Hong, M., Zhang, D., Liang, B., Li, J.: Information extraction: methodologies and applications. In: Emerging Technologies of Text Mining: Techniques and Applications, chap. 1, pp. 1–33 (2008). http://keg.cs.tsinghua.edu.cn/jietang/publications/Tang-et-al-Information_Extraction.pdf. Accessed 21 Mar 2018
6. Nguyen, T., Grishman, R.: Event detection and domain adaptation with convolutional neural networks. In: ACL-IJCNLP 2015 - 53rd Annual Meeting of the Association for Computational Linguistics and the 7th International Joint Conference on Natural Language Processing of the Asian Federation of Natural Language Processing, Proceedings, vol. 2, pp. 365–371 (2015)

7. Lukashevich, N.: Iteracionnoe izvlechenie shablonov opisaniya sobytij po novost-nym klasteram. In: RCDL 2012, pp. 353–359. Pereslavl'-Zalesskij (2012)
8. Navigli, R., Velardi, P.: Learning Word-Class Lattices for Definition and Hypernym Extraction. http://www.aclweb.org/anthology/P10-1134. Accessed 23 Mar 2017
9. Pymorphy2. http://pymorphy2.readthedocs.io/en/latest. Accessed 26 Mar 2018
10. Leadscanner. https://leadscanner.ru. Accessed 24 Feb 2018
11. NCRL. http://www.ruscorpora.ru/search-main.html. Accessed 11 Feb 2018

Probabilistic Approach for Embedding Arbitrary Features of Text

Anna Potapenko[✉]

National Research University Higher School of Economics, Moscow, Russia
anna.a.potapenko@gmail.com

Abstract. Topic modeling is usually used to model words in documents by probabilistic mixtures of topics. We generalize this setup and consider arbitrary features of the positions in a corpus, e.g. "contains a word", "belongs to a sentence", "has a word in the local context", "is labeled with a POS-tag", etc. We build sparse probabilistic embeddings for positions and derive embeddings for the features by averaging of those. Importantly, we interpret the EM-algorithm as an iterative process of intersection and averaging steps that reestimate position and feature embeddings respectively. With this approach, we get several insights. First, we argue that a sentence should not be represented as an average of its words. While each word is a mixture of multiple senses, each word occurrence refers typically to just one specific sense. So in our approach, we obtain sentence embeddings by averaging position embeddings from the E-step. Second, we show that Biterm Topic Model (Yan et al. [11]) and Word Network Topic Model (Zuo et al. [12]) are equivalent with the only difference of tying word and context embeddings. We further extend these models by adjusting representation of each sliding window with a few iterations of EM-algorithm. Finally, we aim at consistent embeddings for hierarchical entities, e.g. for word-sentence-document structure. We discuss two alternative schemes of training and generalize to the case where the middle level of the hierarchy is unknown. It provides a unified formulation for topic segmentation and word sense disambiguation tasks.

Keywords: Topic models · Word embeddings · EM-algorithm

1 Introduction

Recently there have been proposed a lot of frameworks for embedding short pieces of text: [2,6,8] to name a few. Many of them suggest that a sentence embedding is represented as an average of its word embeddings. It can be done on top of pre-trained word embeddings [2] or enforced during training [8]. However, averaging implies bringing all uncertainty of individual word senses into a fairly smooth combination. For instance, consider the clear concept conveyed by the phrase "support vector machine". Ideally, this embedding should be much more specific (e.g. sparse) then the embeddings of its constituent words.

© Springer Nature Switzerland AG 2018
W. M. P. van der Aalst et al. (Eds.): AIST 2018, LNCS 11179, pp. 134–140, 2018.
https://doi.org/10.1007/978-3-030-11027-7_14

To resolve this inconsistency, let us start with a question: "What are the smallest units of language, such that all other units could be expressed using them?". Arora [1] argues that those units are so called *atoms of discourse*, that represent *what is talked about in the current moment*. They show that each word can be expressed as a linear combination of several (e.g. $k = 5$) atoms, and that every atom can be interpreted as a sense of a word, thus providing promising results for polysemy. They have 2000 atoms in their experiments, which makes the atoms rather generic, thus leaving it unclear whether the model has enough granularity.

We are interested in investigating those smallest units for several reasons: (1) it would shed some light on the structure of the embedding space, (2) it could be used for inducing more sparsity and interpretability of the embeddings, (3) it would give the right building blocks to embed more complicated unites, such as sentences.

Our claim is that these units should be *individual word occurrences* in the corpus. Then the embeddings for other entities will naturally come out as some aggregates. E.g. *word embedding* is an average of all occurrences of this word, while *sentence embedding* is an average of all word-occurrences within this sentence. With this approach, both words and sentences are made of the *word-occurrence* embeddings, and thus neither of them is a "sub-embedding" of the other.

In this work we develop a framework for consistent embedding of arbitrary text features. The framework is based on topic modeling and provides probabilistic embeddings with interpretable components.

2   EM-algorithm as Intersection-Averaging in Embeddings Space

Probabilistic Latent Semantic Analysis (PLSA, [4]) explains observed words in documents with a mixture of latent topics. Training is performed with EM-algorithm, where E-step computes posterior probabilities of topics for each word in each document, and M-step performs maximum likelihood updates for the parameters, namely for probabilities of words in topics and for probabilities of topics in documents. In this section we provide a new interpretation of this algorithm as an iterative intersection-averaging procedure in the embedding space of arbitrary text features.

Let us denote word embeddings with ϕ_w, document embeddings with θ_d, and embeddings for positions in the corpus with ψ_i (i runs from 1 to the length of the corpus N). We define the embedding space as a probability simplex $\{x \in \mathbb{R}^k : \sum_{j=1}^{k} x_j = 1; \forall j \, x_j \geq 0\}$. Thus we can interpret each dimension as a *topic*, and elements of the vectors as topic probabilities. The following two lemmas represent formulas for the EM-algorithm in our notation, providing a new intuition behind the updates.

Lemma 1. *E-step performs* a soft intersection of topics *in word and document embeddings related to position i:*

$$\psi_i = \frac{1}{Z_i} \, \tau \circ \phi_{w_i} \circ \theta_{d_i}, \tag{1}$$

where $\tau = \left(\frac{1}{p(t_1)}, \ldots, \frac{1}{p(t_k)}\right)$ *is a vector of inverted topic probabilities in the corpus,* $\circ$ *denotes an element-wise product, and* Z_i *is a partition function.*

As a corollary, it's seen that position embeddings ψ_i are more sparse (specific) then the corresponding feature embeddings ϕ_{w_i} and θ_{d_i}. This also follows an intuition, since positions are more specific objects then words or documents. In future, we will denote the weighted element-wise product (1) as a *mult* operation.

Lemma 2. *M-step performs* averaging *of position embeddings:*

$$\phi_w = \frac{1}{\sum_i [w_i = w]} \sum_{i:w_i=w} \psi_i; \quad \theta_d = \frac{1}{\sum_i [d_i = d]} \sum_{i:d_i=d} \psi_i, \tag{2}$$

where squared brackets denote the indicator function.

As a corollary, it's seen that feature embeddings ϕ_{w_i} and θ_{d_i} are more smooth (generic) then the corresponding position embeddings. Let us denote element-wise averaging (2) as a *mean* operation.

Note About Equivalence Relations. Let us consider two *equivalence relations* for the set $S = \{1, 2, \ldots, N\}$ of all positions in the corpus. The first relation $\mathcal{W}$ considers two positions equivalent if they hold the same words. The second relation $\mathcal{D}$ considers two positions equivalent if they are contained in the same document:

$$i \sim_{\mathcal{W}} j \text{ if } w_i = w_j; \quad i \sim_{\mathcal{D}} j \text{ if } d_i = d_j.$$

Then $S/\sim_{\mathcal{W}}$ is the vocabulary and $S/\sim_{\mathcal{D}}$ is the set of all documents. The M-step of the algorithm computes embeddings for every equivalence class. The E-step computes embeddings for every equivalence class with respect to the *intersection* of the defined relations $\sim_{\mathcal{W}} \cap \sim_{\mathcal{D}}$. It means that the choice of the equivalence relations $\mathcal{W}$ and $\mathcal{D}$ *defines* what are the smallest unites distinguished by the algorithm.

Arbitrary Text Features. Word-in-document unit might be not fine-grained enough, so one can consider *arbitrary features of position i* and straightforwardly incorporate them to the M-step instead of the document feature $d_i = d$. Here are some examples:

1. Local context: $c_i = c$, where $c_i = \{w_{i-h}, \ldots, w_{i+h}\}$ or some other local neighbourhood of the position i.
2. Sliding window: $j - h \leq i \leq j + h$, where h is the window size.
3. Arbitrary annotation: tag_i = tag by some system of tagging, e.g. POS-tags, grammar roles, types of named entities, WordNet nodes, etc.
4. Additional modality: $a_i = a$, e.g. authors of each token in call center dialogues.

Some of these features are commonly used in NLP tasks, e.g. as an input to CRFs. In this work we claim that they can also be freely used in topic modeling.

3   Incorporating Local Contexts and Hierarchies

A straightforward way of using local context features (1) in the EM-algorithm, would require to learn parameters for all unique $2h$-words contexts, which is not feasible. Thus one might suggest to reparametrize context embeddings as an average of individual contexts. This will amount to Word Network Topic Model (WNTM), proposed in [12]. It was initially considered as a modification of PLSA for modeling word co-occurrences in short text analysis.

One could go further and notice that in this case Φ and Θ matrices have the same dimension and represent *word and context embeddings* respectively. According to modern literature [5,10], it might be beneficial to tie input and output embeddings, thus reducing the parameter space.

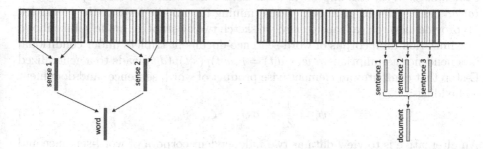

Fig. 1. Consistent embeddings for hierarchical structures.

Lemma 3. *Condition $\Theta = \Phi$ set in the initialization of Word Network Topic Model is preserved with EM iterations.*

We omit the proof due to space limitations and provide the final update rules:

$$\text{E-step: } \psi_i = mult(\phi_{w_i}, \underset{v \in c_i}{mean}\, \phi_v); \qquad \text{M-step: } \phi_u = \underset{i:w_i=u}{mean}\, \psi_i. \qquad (3)$$

Remarkably, these update rules represent EM-algorithm for Biterm Topic Model (BTM) [11], which was initially learned with Gibbs Sampling. BTM was proposed before WNTM and studied independently. Here we highlight the close relatedness of two models and confirm this by their equal performance in the experiments section.

Fitting Window Embeddings. Incorporating sliding window feature (2) from the list above, would amount to the following EM-updates:

$$\text{E-step: } \psi_i = mult(\phi_{w_i}, \theta_i); \qquad \text{M-step: } \theta_i = \underset{i-h \leq j \leq i+h}{mean}\, \psi_j, \ \phi_u = \underset{i:w_i=u}{mean}\, \psi_i.$$
$$(4)$$

With a naive approach, one would need to learn θ vectors for every position in the corpus. With a modification of *online EM-algorithm* [9], this can be done on the fly. For this, we propose to compute θ vectors as a moving average and never store them. Updates (4) suggest to embed local context by averaging specific word position embeddings ψ_i. BTM updates (3) are remarkably similar, but average ambiguous word embeddings ϕ_u instead. We argue that this is an important difference and prove this empirically for a more simple case. Namely, we compare sentence embeddings obtained by averaging of word embeddings vs fitted position embeddings.

Embedding Hierarchical Structures. To build consistent representations of hierarchical text features, let us consider a *word* embedding as an average of the *word sense* embeddings, which are the averages of the corresponding *word-occurrence* embeddings (see Fig. 1). Similarly, a document embedding is an average of its paragraph embeddings, which are the averages of the word-occurrence embeddings. With this approach, both *word sense induction* and *topic segmentation* can be viewed as a task of determining the middle level of the hierarchy. To provide hierarchical consistency, we sketch two alternative models. Likelihood maximization for a corpus of word-sentence-document triplets under conditional independence assumptions $p(w, s, d|t) = p(w|t)p(s|t)p(d|t)$ leads to a generalized E-step that computes an element-wise product of word, sentence, and document embeddings:

$$\psi_i = \frac{1}{Z_i}\,\tau^2 \circ \phi_{w_i} \circ \zeta_{s_i} \circ \theta_{d_i} \tag{5}$$

An alternative is to view data as two independent corpora of word-sentence and sentence-document pairs. This leads to two separate EM-procedures that share sentence embeddings. One can show that the update for position embeddings amounts exactly to (5), but the iteration scheme differs.

4   Experiments

On Equivalence of BTM and WNTM. Table 1 shows an equal performance of BTM and WNTM on a range of word similarity tasks. Both models were trained on Wikipedia 2016-01-13 dump with open-source topic modeling library BigARTM[1]. Word similarity was obtained as the dot product of ϕ vectors. Refer to [9] for any details, since we fully follow their setup. Fitting window embeddings with (4) is left for future work.

Sentence Embeddings Though Positions. We closely follow the setup from [3] to evaluate sentence embeddings for semantic textual similarity tasks, namely SICK-2014 and STS-2014. For SICK dataset we train a linear model on top of the embedding features with SentEval tool[2]. With the topic modeling framework [7], we prepare averaging of word embeddings (BOW) and averaging of position embeddings (fitted). Position embeddings are obtained with 10 passes of EM-algorithm for the target sentences given fixed ϕ vectors. In both cases,

---

[1] bigartm.org
[2] github.com/facebookresearch/SentEval

we use word vectors pre-trained on Wikipedia. The important take-away from Table 2 is that position fitting improves performance on all datasets, thus justifying our argument on consistent sentence representations. Our approach produces sentence embeddings that perform on par with word2vec baseline, being at the same time more interpretable and sparse as shown in [9].

Table 1. Spearman correlation of BTM and WNTM embeddings on word similarity tasks.

Model	WS-353 Sim	WS-353 Rel	WS-353 All	SimLex Hill et al.	MEN Bruni et al.	RareWords Luong et al.	Radinsky M. Turk
BTM	**0.68**	**0.59**	**0.61**	0.24	0.65	0.32	0.54
WNTM	0.67	0.58	0.60	0.24	**0.66**	**0.33**	**0.55**

Table 2. Sentence embeddings performance for unsupervised tasks. We report Person/Spearman correlations for SICK relatedness and STS-2014 datasets and accuracy for SICK entailment.

Model	STS-2014						SICK	
	Forum	News	Headlines	Images	Tweets	Average	Rel	Ent
BOW (ours)	0.41/0.42	0.70/0.62	0.60/0.53	0.76/0.71	0.68/0.63	0.64/0.60	0.77/0.70	76.27
Fitted (ours)	**0.45/0.46**	0.70/0.62	0.61/0.55	**0.76**/0.71	0.68/0.62	**0.65**/0.62	0.78/**0.71**	**76.96**
BOW (w2v)	0.39/0.46	0.67/**0.66**	**0.64/0.60**	0.76/**0.72**	**0.70/0.69**	0.65/**0.65**	**0.79**/0.69	75.62

5   Conclusions

In this work we interpreted EM-algorithm for learning a topic model as a process of intersection-averaging steps in the embedding space. This gave us a number of new insights: (1) how to use topic modeling for embedding arbitrary text features; (2) how to obtain consistency between representation of word occurrences, word senses, words, sentences, and documents; (3) how to generalize and improve local context topic models. In the experiments, we confirmed that sentence embeddings should be built from *word-in-document embeddings* rather than from more ambiguous *word embeddings*, and also showed the equivalent performance of WNTM and BTM. However, many other interesting outcomes of our theory, such as building hierarchies of embeddings or representing syntactic role features, were left for future work.

Acknowledgements. The research was supported by Russian Foundation for Basic Research (17-07-01536).

References

1. Arora, S., Li, Y., Liang, Y., Ma, T., Risteski, A.: Linear algebraic structure of word senses, with applications to polysemy. CoRR abs/1601.03764 (2016)
2. Arora, S., Liang, Y., Ma, T.: A simple but tough-to-beat baseline for sentence embeddings. In: International Conference on Learning Representations (2017)
3. Conneau, A., Kiela, D., Schwenk, H., Barrault, L., Bordes, A.: Supervised learning of universal sentence representations from natural language inference data. In: Proceedings of EMNLP, pp. 670–680. Association for Computational Linguistics (2017)
4. Hofmann, T.: Probabilistic latent semantic analysis. In: Proceedings of the Fifteenth Conference on Uncertainty in Artificial Intelligence, UAI 1999, pp. 289–296. Morgan Kaufmann Publishers Inc., San Francisco (1999)
5. Inan, H., Khosravi, K., Socher, R.: Tying word vectors and word classifiers: A loss framework for language modeling. CoRR abs/1611.01462 (2016)
6. Kiros, R., et al.: Skip-thought vectors. In: Proceedings of the 28th International Conference on Neural Information Processing Systems, NIPS 2015, pp. 3294–3302. MIT Press, Cambridge (2015)
7. Kochedykov, D., Apishev, M., Golitsyn, L., Vorontsov, K.: Fast and modular regularized topic modelling. In: Proceeding of the 21st Conference of FRUCT Association, ISMW, pp. 182–193 (2017)
8. Pagliardini, M., Gupta, P., Jaggi, M.: Unsupervised learning of sentence embeddings using compositional n-gram features. In: Proceedings of NAACL (2018)
9. Potapenko, A., Popov, A., Vorontsov, K.: Interpretable probabilistic embeddings: bridging the gap between topic models and neural networks. In: Filchenkov, A., Pivovarova, L., Žižka, J. (eds.) AINL 2017. CCIS, vol. 789, pp. 167–180. Springer, Cham (2018). https://doi.org/10.1007/978-3-319-71746-3_15
10. Press, O., Wolf, L.: Using the output embedding to improve language models. In: Proceedings of ACL: Volume 2, Short Papers, pp. 157–163. ACL (2017)
11. Yan, X., Guo, J., Lan, Y., Cheng, X.: A biterm topic model for short texts. In: Proceedings of WWW, pp. 1445–1456 (2013)
12. Zuo, Y., Zhao, J., Xu, K.: Word network topic model: a simple but general solution for short and imbalanced texts. Knowl. Inf. Syst. **48**(2), 379–398 (2016)

Learning Representations for Soft Skill Matching

Luiza Sayfullina$^{(\boxtimes)}$, Eric Malmi, and Juho Kannala

Aalto University, Espoo, Finland
sayfullina.luiza@gmail.com

Abstract. Employers actively look for talents having not only specific hard skills but also various soft skills. To analyze the soft skill demands on the job market, it is important to be able to detect soft skill phrases from job advertisements automatically. However, a naive matching of soft skill phrases can lead to false positive matches when a soft skill phrase, such as friendly, is used to describe a company, a team, or another entity, rather than a desired candidate.

In this paper, we propose a phrase-matching-based approach which differentiates between soft skill phrases referring to a candidate vs. something else. The disambiguation is formulated as a binary text classification problem where the prediction is made for the potential soft skill based on the context where it occurs. To inform the model about the soft skill for which the prediction is made, we develop several approaches, including soft skill masking and soft skill tagging.

We compare several neural network based approaches, including CNN, LSTM and Hierarchical Attention Model. The proposed tagging-based input representation using LSTM achieved the highest recall of 83.92% on the job dataset when fixing a precision to 95%.

1 Introduction

Since 1980, the number of jobs with high social skills requirements has increased by around 10% from the total US labor force [1]. When an industry becomes more competitive, job applicants need to possess additional skills that help them to start smoothly at work and to be distinguished from candidates with similar qualifications [18].

According to Collins dictionary, soft skills are desirable qualities for certain forms of employment that do not depend on acquired knowledge: they include common sense, the ability to deal with people, and a positive flexible attitude.

Detecting soft skills in the unstructured text corpora is not straightforward, since some matches can turn out to be false positives. For example, the word tolerant might be matched as a part of fault tolerant systems phrase. So, the task is to increase the precision of soft skill detection with an additional requirement: to match only those soft skills that would describe a candidate. For example, friendly team would be a false positive, despite friendly is a soft skill.

W. M. P. van der Aalst et al. (Eds.): AIST 2018, LNCS 11179, pp. 141–152, 2018.
https://doi.org/10.1007/978-3-030-11027-7_15

The task of detecting candidate soft skills is related to word sense disambiguation (WSD), where the sense of a word being used in a particular context has to be determined [6]. However, our task is different since we are not aiming at mapping a skill to a particular sense such as a one from WordNet [12]. Instead, the goal is to differentiate between the two cases of interest without requiring exact sense. Besides, we work both on the word and phrase levels.

Dependency parsing [13] could be helpful for disambiguation of the skill by detecting the entity to which the skill refers. However, the amount of possible entities referring to the candidate is quite high, since among such obvious entities as `candidate`, `individual`, a candidate might be referred by a profession name or a university degree level. Thus, entities referring to candidate and others can be hard to differentiate. Working with skill phrases, also makes it harder to use dependency parsing, that usually outputs a relation between a pair of words.

Massive skill extraction was done and actively employed by LinkedIn [2], although with a focus on hard skills. They observed that people provide a list of comma separated skills in the specialties section of their profile that can be used for mining potential skills. The authors disambiguated skills by using clustering based on the co-occurring phrases. Some skills were tagged with many senses belonging to different clusters with an industry label. Kivimäki et al. [9] suggested a system for automatic detection of new skills in free written text using spreading activation algorithm on the Wikipedia graph and skill lists from LinkedIn. However, these skill extraction approaches aim at mining and extracting new skills, rather than using a ready list of skills and detecting them.

We formulate the problem of soft skill matching as a binary classification, where the positive class refers to the skill describing the candidate and the negative class refers to all other cases. We do classification on the text snippets or the context sentences where the soft skill occurs.

One baseline approach for tackling the classification task is to use any existing text classifier to classify the context sentence as such. A limitation of this approach is that the classifier is not explicitly informed about the skill it is trying to disambiguate. To address this limitation, we investigate various alternative input representations where the context sentence is modified in order to indicate the soft skill to be classified.

The contributions of this work can be summarized as follows:

1. We are the first one to formalize the problem of soft skill matching and release a benchmark dataset collected via crowdsourcing for this problem.
2. We introduce various input representations for the classifier, including modifying the context with soft skill masking, soft skill tagging and augmenting it with soft skill embeddings.
3. We evaluate the performance of several strong text classification baselines, including LSTM, CNN and the Hierarchical Attention Model, using the proposed input representations. Our experiments show that the best performance is achieved using soft skill tagging as well as unmodified sentence representations.

The paper is organized as follows. First, we provide a motivation for our work by analysis of entities that soft skills relate to in Sect. 2. In Sect. 3, we briefly describe both CV (Curriculum Vitae) dataset and the job ads dataset. In Sect. 4, the process of obtaining the data with the help of crowdsourcing is explained. Then, in Sect. 5, we give a brief overview of applied neural network–based classifiers and describe the proposed approach for input representations. Experimental setup and the analysis of the obtained results with different approaches is presented is Sect. 6.

2   Motivation

When it comes to detection, some skills can refer to various entities and not only to a candidate. Consider the following two sentences found from job ads: "Produce **accurate**, legible documentation packs..." and "Fault **tolerant** systems...". In these cases, words **accurate** and **tolerant**—two potential soft skills—do not explicitly refer to the qualities of the desired candidate. In general, shorter soft skill phrases, especially one-word skills, are more probable to result in false positive matches. Some skills like **responsible** can be treated alone as soft skill, but with additional phrase, like **responsible for finding new clients**, these become job duties.

In order to understand the problem better, we use Stanford Dependency Parser [3] to find entities with adjectival modifier (amod), particularly **adjective + noun** phrases. Since we are interested in cases where certain adjectives or skills are used to describe other entities, this method allows us to mine easily some of such cases. Thus, we selected only **amod** relations with adjectives that are part of soft skill list from the first 10,0000 sentences from the job ad dataset (to be described in the next section) for the preliminary study.

The top nouns used with one-word soft skills are shown in Fig. 2. The majority of these nouns, like **personality, individual, person** indeed refer to the

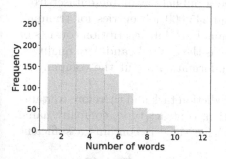

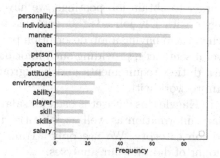

Fig. 1. The histogram of soft skill phrase lengths. Due to presence of stop words, the length reaches 12 words. The majority of the skills are shorter than 6 words.

Fig. 2. Frequencies of the most common nouns used with one-word adjective soft skills. The statistics was obtained using 10,000 sampled sentences from the job ads dataset.

candidate, but frequent non-candidate nouns like `team, environment, salary` can be found as well.

Despite the fact that dependency parsing helps to find some of the entities that soft skills describe, this method occasionally fails to find the objects to which a soft skill refers. This can happen when objects are simply absent in the sentence, e.g. `communication skills are required at work`. Secondly, dependency parsing usually finds the relations for a pair of words, that makes it hard to apply for soft skill phrases.

3   Datasets

In this work we use three datasets for the analysis: the first is a corpus of job advertisements (ads), the second is a corpus of CVs or resumes of people and the third is a resource with soft skills list. Both job ads and CVs contain soft skills and candidate descriptions. However, job ads provide more examples with cases where soft skills are used in different contexts and describe company, team, environment, etc. The purpose of using a CV dataset is to have additional data from a different domain to test our approach. Both datasets are used for soft skill annotation that provides data for using supervised approaches. Finally, the third dataset is the set of soft skills that we learn to disambiguate.

We got access to a comprehensive list of soft skills available online[1]. There are 1072 skills and 234 skill clusters respectively. Clusters group skills with a similar meaning but phrased differently. For example, the `polite` cluster consists of `polite manners, polite` and `ability to speak politely`. 155 skills out of 1072 are one-word, while the longest one consist of 12 words. The histogram of soft skill phrase lengths is shown on Fig. 1. We see that most of the soft skill phrases consist of less than 6 words.

3.1   Kaggle Job Ads and CV Corpora

In order to obtain job postings, we have used a publicly available dataset from Kaggle with UK job ads.[2] It contains around 245,000 job entries for training, each of them having a job description paragraph. Each job description consists of several sentences, providing such information as the desired candidate qualities, work duties, requirements and background information about the company or team to work with.

In Kaggle this dataset is used for salary prediction task and therefore contains salary information as well as job title, location, contract type, company name and job category. We use only "Train_rev1" file since it contains a sufficient amount of data for our analyses.

We collected manually 525 CV samples from *indeed.com*. Each CV entry was segmented either into a set of sentences or long phrases using a various set of

---

[1]   https://github.com/AnonymousWWW18/soft_skills.
[2]   The dataset is available at: https://www.kaggle.com/c/job-salary-prediction.

delimiters and special symbols used for list creation. In [17], you can find an example of CV from the same dataset. After segmenting each CV, we got 2517 text snippets in total.

We provided training and test data sizes in Table 1.

Table 1. Class distribution in training and test datasets. Positive class refers to candidate soft skills. CV corpus was used only for testing. Both test datasets have naturally imbalanced class distribution. At the same time training data was intentionally made to be nearly balanced.

	Positive class	Negative class
Job ads train	16,656	15,205
Job ads test	1984	222
Cv test	536	77

4 Crowdsourcing Skill Annotations

In order to make a supervised dataset for training, we have established a crowdsourcing experiment both for the job ads and for the CV corpora. We extracted all text snippets containing one soft skill from the list of 1 072 soft skills. Each text snippet consists of the matched soft skill and a maximum of 10 words around a soft skill from both sides. Each snippet was shown with the soft skill highlighted in bold and a worker had to answer whether the highlighted skill is a soft skill referring to the candidate or not.

A crowdsourcing experiment was done using CrowdFlower[3], a crowdsourcing platform where workers can contribute to data annotation process. Each sentence was assessed by at least 3 workers. During the test run, the platform has selected only workers who got an accuracy of more than 85% for the CV corpus (80% for job ad corpus) on a limited labeled set of 40 sentences. The test run is made to select trusted workers for the main task. The rest 597 sentences were used for the main crowdsourcing experiment. The annotated sentences contained 149 unique skill phrases in total. All of the tagged CV data was used for the test experiment only. Obviously, some of the soft skills in the CV dataset were not seen during the training.

4863 sentences from the job ad corpus were labeled in a similar way as the CV corpus. Crowdsourcing experiment resulted in 427 skills labeled as non-candidate ones and 3984 as candidate ones. So we had to create more training data due to class imbalance. For that we picked all 9 soft skills that described non-candidate in more than 70% of our annotations. Then we extracted 15,000 text snippets with those skills from the job ad corpus and marked them as non-candidate skills after discarding those containing "candidate", "individual" and "looking for" words near the soft skill. We also selected 123 skills that described only the

---

[3] https://www.crowdflower.com.

candidates and mined 15,000 text snippets from the job ad corpus containing them and labeled these skills as candidate ones. The job ads test data contains only hand-annotated samples with 1984 positive (about a candidate) and 222 negatives labels.

5   Overview of the Methods

In this section we give an overview of three neural network classifiers used for our predictions and describe in detail various ways to represent a soft skill and its context for these classifiers.

5.1   LSTM and CNN

We have selected CNN [8] and LSTM [5,16] models since they are both frequently used as text classification benchmark algorithms. In both models, words in the sentences are embedded with word2vec [11] vectors. LSTM employs sequential nature of the text, handles long-term dependencies and allows making predictions on a variable length input. To make a prediction on variable length input, one can take the output of the hidden layer from the last word to make a final prediction.

CNN model for text classification developed by Kim [8] takes as input only a fixed-size 2D tensor, where rows correspond to words and columns correspond to word embedding dimensions. By applying convolutions with a window of N words covering full embedding dimension, CNN allows to take into account the word order, limited by the size of the window. We refer the reader to [8] for more details about the model.

5.2   Hierarchical Attention Model

Hierarchical Attention Model (HAN) [21] was shown to outperform CNN [8], LSTM, Conv-GRNN, LSTM-GRNN [20] on six large classification benchmarks, including Yelp, IMDB, Amazon, Yahoo reviews.

The motivation of the model is the following: the output of classification depends more on certain words in certain sentences to which the most attention should be payed. Attention in the model is context-dependent, which means that similar words might be of different importance in different context. The model is called *hierarchical* since it models documents using sentences which in their turn are modeled sequentially using words. Gated Recurrent Units [4] with bidirectional structure are used to sequentially encode each sentence, as well as document as the sequence of sentences.

5.3   Proposed Approaches for Input Representation

In order to match candidate soft skills, it is important to consider the context in which the skill is detected. One baseline way to do this is to take the sentence where the skill is detected and train any text classification method to classify such sentences into those that refer to the candidate and those that refer to something else. However, a significant limitation of this approach is that the classifier is not informed about the location of the skill in the sentence.

We propose three alternative ways of input representation for soft skill disambiguation:

1. Mask the soft skill phrase with **xxx** tokens for each word, in other words, apply *soft skill masking*. E.g. the sentence "bar and kitchen business seek a **dedicated person** who want to be" will be transformed into "bar and kitchen business seek a xxx xxx who want to be".
2. Combine *soft skill masking* with a soft skill embedding.
3. Surround a soft skill phrase with <begin> and <end> tokens. We call this approach *soft skill tagging*.

While masking the soft skill phrase, we use the context around the skill, the position of the soft skill itself and the number of words in soft skill phrase. However, this approach does not employ for the prediction useful information about the skill itself, compared to two other approaches. In fact, each soft skill has a prior probability of describing a candidate and with soft skill masking this prior is not taken into account.

For the second alternative, we use both soft skill masking as well as soft skill embedding. Since the soft skill is hidden and not used while making the prediction, we came up with the idea of augmenting the classifier's input with a soft skill embedding. This way the prior information about the skill to make a prediction for is not lost, although the soft skill is still separated from its context. We suggest to augment the input to the last hidden layer of a neural network classifier with a soft skill embedding. Particularly, a neural network takes as a main input masked context, which goes through a forward pass. Then, an input to the last hidden layer is concatenated with a soft skill embedding and is used for a final prediction.The benefit of the this approach is that one can easily integrate it with any chosen neural network used in text classification. One approach to build a soft skill embedding is to use word2vec [11] model and present text as the mean of its word-embeddings.

The third input representation would be a sentence with added <begin> and <end> tags around the soft skill in question. This approach is simple to implement, since it does not require any model changes as with soft skill embedding. Besides, the soft skill phrase is used for the classification, unlike with soft skill masking, and tagged as being "important" compared to raw sentence input from baseline approach.

6 Experimental Evaluation

6.1 Experimental Settings

All models except Hierarchical Attention were implemented from scratch using Python and PyTorch [14]. The code and the datasets will be publicly released here[4].

---

[4] https://github.com/muzaluisa/soft-skill-matching.

Text preprocessing included case-folding, lemmatization with NLTK [10] WordNet lemmatizer as well as removing all syntax tokens, except commas.

We have set the width of CNN filters to be $\{2, 3, 4\}$. Since the network tends to overfit, we have used only 50 filters of each size. Maximum document length was set to 30 since we had a window of maximum 10 words around the soft skill from both sides. The dropout [19] rate was set to 0.5 and we found it to be useful for model regularization.

LSTM network was implemented using variable-length inputs. The hidden layer size is 100, the dropout rate of 0.2 is applied to the input to the last fully connected layer to prevent over-fitting. We tested different dropout rates between 0.2 and 0.5 and found 0.2 to be the most optimal one.

We have experimented with two types of word embeddings: Glove 6B pre-trained word vectors on Wikipedia 2014 and Gigaword 5 datasets[5] and our own embeddings on full job ads data prepared using gensim library [15]. Finally, we chose own embeddings of size 100 for both CNN and LSTM. Furthermore, both networks used Adam optimizer with a learning rate of 0.001 due to its fast convergence and good performance. We used a batch size of 16 since often smaller batch sizes lead to better generalization [7]. Cross-entropy loss is used for learning by backpropagation. Half of the training data was used for validation of the model.

Hierarchical Attention Model[6] has a GRU encoder with 50 hidden units, resulting in 100 units in both directions for sentence embedding. We left the majority of parameters of the implementation intact, including a batch size of 64, but used own pre-trained word vectors of size 100.

The training data consists of only job ads corpus, while for testing we used both job ads and CV corpora. The purpose of the CV experiment is to check how different approaches generalize on the new domain, where the distribution of the data is different. Since test data is imbalanced towards the positive class, we used F1-weighted measure. F1-weighted measure is a modification of F1-macro measure, but weighted with a class support. F1-macro measure computes precision and recall for each class separately and then takes the average with equal weight for each class.

The performance of compared approaches is evaluated based on the precision-recall balance. While using naive approach by selecting all soft skill phrases we get 89.50% precision on the job corpus and 87.5% precision on the CV corpus with 100% recall. Since the aim is to increase the precision while soft skill matching with still reasonable recall, we fixed a precision to be 95% and 90% for the job and CV corpora respectively by adjusting a decision boundary between two classes. A lower precision for the CV corpus was chosen, since it is not used during the training and therefore it is more difficult to achieve 95% of precision with a reasonable recall.

---

[5] https://nlp.stanford.edu/projects/glove/.

[6] The implementation of the model was adopted from: https://github.com/EdGENetworks/anuvada.

Table 2. F1-weighted score, precision and recall, (%) for compared classifiers across four different input representations. For comparison, we fixed the precision for job corpus to be 95% and 90% for CV corpus by adjusting the decision boundary between two classes. Naive approach means labeling all matched skills as candidate soft skills and F1-weighted score is not applicable. Soft skill tagging showed the highest recall on the test samples from the job corpus used for the main evaluation.

	Job test			CV test		
Method	F1	Pr	Re	F1	Pr	Re
Naive approach	N/A	89.50	100	N/A	87.4	100
Unmodified sentence						
CNN	80.07	95.5	76.69	**82.23**	**90.26**	**88.80**
LSTM	82.81	95.03	81.69	**82.20**	**90.24**	**88.80**
HAN	77.85	95.35	73.29	78.14	90.68	80.00
Soft skill masking						
CNN	79.44	95.19	76.02	79.92	90.36	84.00
LSTM	78.95	95.06	75.34	78.79	90.79	81.12
HAN	82.8	95.07	81.7	47.04	90.34	34.95
Soft skill masking and embedding						
CNN	78.17	95.30	73.80	75.08	90.14	75.18
LSTM	82.08	95.06	80.40	76.632	90.52	77.46
HAN	79.87	95.01	76.81	44.75	91.05	32.34
Soft skill tagging						
CNN	80.36	95.13	77.54	80.9	90.23	86.19
LSTM	**84.18**	**95.05**	**83.92**	81.33	90.11	87.31
HAN	81.28	95.21	78.94	90.07	89.91	82.56

6.2 Baseline Approach

For the baseline approach we used unmodified sentences without replacing a soft skill. This way, it is not clear from the representation for which soft skill the prediction is made. At the same time, many sentences contain only one skill that can be given higher weight for the prediction without additional tagging of the skill. Thus, the classifier could learn soft skills from the training data as well as features, indicating whether the sentence will contain candidate soft skills.

We combined the results of all our experiments in Table 2. Surprisingly, but from Table 2 we can see that *unmodified sentence* representation gives the best result on the CV corpus. CNN and LSTM using this representation result in almost the same performance on the CV corpus and we highlight both methods as the strongest ones.

6.3 Soft Skill Masking

Masking a skill (with and without soft skill embedding) from the sentence turned out to be overall the weakest from the proposed representations. At the same time the sentence a with hidden or masked skill have a reasonable performance, meaning that the context itself can be used for the prediction. For CNN and LSTM soft skill masking did not bring any improvement compared to the raw sentence representation.

The third input representation used additionally the mean of word2vec [11] embedding computed from the soft-skill phrase. Since the number of added dimensions for both CNN and RNN is almost (90 for CNN, 100 for LSTM) the size of the last hidden layer, the prior coming from the soft skill is strong. Surprisingly, but the embedding resulted in worse performance than with the soft skill masking. There can be several reasons for that including the lack of various soft skill samples for learning from their embeddings as well as overfitting due to a strong soft skill prior.

6.4 Soft Skill Tagging

Soft skill tagging similarly to an unmodified sentence representation keeps the soft skill inside its context. These two approaches turned out to show overall the highest performance. From Table 2 we can see that a soft skill tagging using LSTM gives the highest recall of 83.92% on the job corpus across all methods and representations. Besides, the performance of CNN on the test samples from the job corpus is also the highest with soft skill tagging representation.

In addition, HAN model did not outperform other models in our task across various representations. This could happen because our inputs consisted of only one sentence, while a HAN model assumes input samples to consist of several sentences. Besides, the recall on the CV corpus was pretty low while using soft skill masking.

6.5 Skill Disambiguation Use Case

We decided to show how drastically soft skill frequencies might change after we filter out words not related to the candidate. For that we took randomly 100,000 job ads with soft skills of maximum length equal to two since most ambiguous phrases tend to be short. Then we ran disambiguation classifier based on the LSTM and soft skill tagging and filtered out the skills with a high confidence score of not referring to the candidate. Skills with top difference are given in Fig. 3. Moreover, there are 58 skills that had more than 50 words filtered out. These results show the amount of possible false positive results for certain skills while using a naive matching giving raw counts.

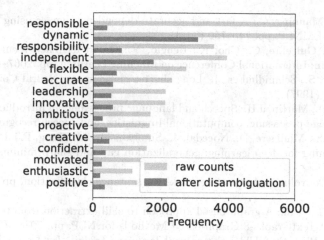

Fig. 3. Raw counts in pink correspond to raw frequencies, while the bars in gray correspond to counts after filtering out skills not referring to the candidate. We showed the skills for which the difference between these counts was the largest to show the amount of possible false positive matches for some skill phrases with a naive approach. (Color figure online)

7   Conclusion

Soft skill analysis and automatic detection of soft skills in unstructured text has not been systematically approached previously. A naive detection of soft skills in unstructured text leads to false positive matches, when a skill does not refer to the candidate. To tackle this problem, our work proposes a method for differentiating between soft skills referring to a candidate and other skills using binary classification. The prediction is made using the context where the soft skill occurs. We propose several approaches for making an input representation to the classifier, including raw context sentences, context sentences with soft skill masking, soft skill tagging and augmentation with soft skill embedding.

We compare several strong neural network baselines for text classification and show that LSTM yields the best disambiguation performance on sentences with tagged soft skills on the job corpus and with a raw sentence representation on the CV corpus. The developed soft skill disambiguation methods open up new possibilities for refined analyses of soft skill demands and supply on the job market.

References

1. Bakhshi, H., Downing, J.M., Osborne, M.A., Schneider, P.: The future of skills employment in 2030. Technical report, Pearson PLC (2017)
2. Bastian, M., et al.: LinkedIn skills: large-scale topic extraction and inference. In: Proceedings of the 8th ACM Conference on Recommender systems, pp. 1–8. ACM (2014)

3. Chen, D., Manning, C.: A fast and accurate dependency parser using neural networks. In: EMNLP, pp. 740–750 (2014)
4. Chung, J., Gulcehre, C., Cho, K., Bengio, Y.: Gated feedback recurrent neural networks. In: International Conference on Machine Learning, pp. 2067–2075 (2015)
5. Hochreiter, S., Schmidhuber, J.: Long short-term memory. Neural Comput. **9**(8), 1735–1780 (1997)
6. Jurafsky, D., Martin, J.H.: Speech and language processing: an introduction to natural language processing, computational linguistics, and speech recognition (2009)
7. Keskar, N.S., Mudigere, D., Nocedal, J., Smelyanskiy, M., Tang, P.T.P.: On large-batch training for deep learning: generalization gap and sharp minima. In: ICLR (2017)
8. Kim, Y.: Convolutional neural networks for sentence classification, pp. 1746–1751 (2014)
9. Kivimäki, I., et al.: A graph-based approach to skill extraction from text. In: Proceedings of TextGraphs-8 Graph-based Methods for NLP, pp. 79–87 (2013)
10. Loper, E., Bird, S.: NLTK: the natural language toolkit. In: Proceedings of the ACL-02 Workshop on Effective Tools and Methodologies for Teaching Natural Language Processing and Computational Linguistics, ETMTNLP 2002 , pp. 63–70 (2002)
11. Mikolov, T., Sutskever, I., Chen, K., Corrado, G.S., Dean, J.: Distributed representations of words and phrases and their compositionality. In: NIPS, pp. 3111–3119 (2013)
12. Miller, G.A.: WordNet: a lexical database for English. Commun. ACM **38**(11), 39–41 (1995)
13. Nivre, J.: An efficient algorithm for projective dependency parsing. In: Proceedings of the 8th International Workshop on Parsing Technologies (IWPT). Citeseer (2003)
14. Paszke, A., Gross, S., Chintala, S., Chanan, G., Yang, E., et al.: Automatic differentiation in PyTorch. In: NIPS-W (2017)
15. Řehůřek, R., Sojka, P.: Software framework for topic modelling with large Corpora. In: Proceedings of the LREC 2010 Workshop on New Challenges for NLP Frameworks, pp. 45–50 (2010)
16. Rumelhart, D.E., Hinton, G.E., Williams, R.J.: Learning representations by back-propagating errors. Nature **323**(6088), 533 (1986)
17. Sayfullina, L., Malmi, E., Liao, Y., Jung, A.: Domain adaptation for resume classification using convolutional neural networks. In: van der Aalst, W.M.P., et al. (eds.) AIST 2017. LNCS, vol. 10716, pp. 82–93. Springer, Cham (2018). https://doi.org/10.1007/978-3-319-73013-4_8
18. Schulz, B.: The importance of soft skills: education beyond academic knowledge (2008)
19. Srivastava, N., Hinton, G.E., Krizhevsky, A., Sutskever, I., Salakhutdinov, R.: Dropout: a simple way to prevent neural networks from overfitting. J. Mach. Learn. Res. **15**(1), 1929–1958 (2014)
20. Tang, D., Qin, B., Liu, T.: Document modeling with gated recurrent neural network for sentiment classification. In: EMLNP, pp. 1422–1432 (2015)
21. Yang, Z., Yang, D., Dyer, C., He, X., Smola, A., Hovy, E.: Hierarchical attention networks for document classification. In: Proceedings of the 2016 Conference of the North American Chapter of the ACL: Human Language Technologies, pp. 1480–1489 (2016)

Analysis of Images and Video

Application of Fully Convolutional Neural Networks to Mapping Industrial Oil Palm Plantations

Artem Baklanov[1,2](✉)[ID], Michael Khachay[3,4][ID], and Maxim Pasynkov[3]

[1] National Research University Higher School of Economics, St. Petersburg, Russia
[2] International Institute for Applied Systems Analysis, Laxenburg, Austria
baklanov@iiasa.ac.at
[3] Krasovsky Institute of Mathematics and Mechanics, Yekaterinburg, Russia
{mkhachay,pmk}@imm.uran.ru
[4] Ural Federal University, Yekaterinburg, Russia

Abstract. This research is motivated by sustainability problems of oil palm expansion. Fast-growing industrial Oil Palm Plantations (OPPs) in the tropical belt of Africa, Southeast Asia and parts of Brazil lead to significant loss of rainforest and contribute to the global warming by the corresponding decrease of carbon dioxide absorption. We propose a novel approach to monitoring of the expansion of OPPs based on an application of state-of-the-art Fully Convolutional Neural Networks (FCNs) to solve Semantic Segmentation Problem for Landsat imagery. The proposed approach significantly outperforms per-pixel classification methods based on Random Forest using texture features, NDVI, and all Landsat bands. Moreover, the trained FCN is robust to spatial and temporal shifts of input data. The paper provides a proof of concept that FCNs as semi-automated methods enable OPPs mapping of entire countries and may serve for yearly detection of oil palm expansion.

Keywords: Mapping tree plantations · Remote sensing
Semi-automated methods · FCN-8s

1 Introduction

Oil palm is a dominant source of vegetable oil, which causes degradation to the environment [2, 4, 6, 18] and exceedingly threatens forest conservation in the tropics. To support the sustainability of Oil Palm Plantations (OPPs), one should be able to monitor their development. Remote sensing is the most promising way to allow for independent monitoring in all oil producing regions. As suggested in [6], there are the following applications of remote sensing in OPP monitoring: land-cover classification; change detection; tree counting; estimation of yield, age, biomass, carbon production; pest and disease detection. To study complex

This research was supported by Russian Science Foundation, grant no. 14-11-00109.

W. M. P. van der Aalst et al. (Eds.): AIST 2018, LNCS 11179, pp. 155–167, 2018.
https://doi.org/10.1007/978-3-030-11027-7_16

issues, e.g., how patterns of oil palm driven deforestation change due to zero-deforestation commitments, a fusion of the above applications is required [2].

Two approaches are suggested to map OPPs: visual interpretation ('manually' performed by experts) and semi-automated methods based on supervised learning procedures. Visual interpretation of Landsat imagery was successfully used in [2,4,11,22] and relies on a rather easily detectable pattern: OPPs are mostly *'organized in rectilinear patterns and co-occur with context indicators such as road networks'* [2].

The conventional semi-automated methods rely on classification algorithms to assign every pixel in an image to a specific class; these are so-called per-pixel classifiers [19, Table 1]. For these methods, human expertise is needed to generate training examples (reference data) for automated models to learn; see [6,22] for a review of the most recent related results for OPPs mapping. The limitation of existing semi-automated methods for OPPs mapping becomes obvious if we consider to which extent they are used. Namely, for high-resolution imagery, these methods are applied only to small regions, e.g., $637.2\,\text{km}^2$ [21], $1020\,\text{km}^2$ [18]. We are not aware of any high-resolution OPPs map covering a country that was obtained by semi-automated methods. Basically, the only sources for high-resolution OPPs maps for entire countries are infrequent publications with maps obtained by visual interpretation. In contrast, the situation with deforestation mapping is absolutely different: Hansen's deforestation map [14] is available for the entire world and updated yearly thanks to a semi-automated implementation. The prominent study by Gaveau et al. [11] published in 2016 sheds light on the issue. They suggest that OPP patterns *'are easily detected by the human eye, but are difficult to capture with computer codes. Object-Oriented codes cannot separate between the great mixtures of land parcels arrangements. Spectral-Color algorithms cannot distinguish closed-canopy oil-palm plantations from other closed-canopy vegetation types, or open-canopy plantations from bare lands.'*[1] Therefore, they delineated OPPs by visual interpretation.

We propose the novel semi-automated technique for OPPs monitoring based on the reduction of the initial problem to the well-known Semantic Segmentation Problem and subsequent solution of the latter with Fully Convolutional Neural Networks (FCNs). In particular, we use the state-of-the-art FCN-8s network, showed excellent performance on the prominent PASCAL-VOC competition [23]. Our results provide a proof of concept that deep learning is at the stage that enables us to monitor OPPs without laborious visual interpretation; thus, a map product for OPPs similar to Hansen's deforestation map is indeed feasible.

To show the potential for replacement of experts by FCNs for OPPs mapping, we ensure that the following crucial criteria are fulfilled

- an FCN uses all sources of information (dataset) available for experts;
- the dataset is available globally in the same format and updated frequently;

---

[1] Object-oriented classifiers first perform image segmentation to merge pixels into objects and then apply classification based on the objects [19] (but not on an individual pixel as per-pixel classifiers do). Spectral-Color algorithms are trained using multi-spectral features of each pixel.

– the dataset is available for scientific community and well documented;
– the trained FCN is indeed spatially and temporally scalable.

The paper is structured as follows. In Sect. 2, we introduce the problem and describe our task. In Sect. 3, we present our assumptions and approach, provide details on data and performance metrics, introduce baseline classification methods and training procedure for the FCN. In Sect. 4, we report performance results and illustrate spatial and temporal scalability of the proposed approach. Section 5 concludes.

2   Problem Statement

Recall that expansion of OPPs is a major cause of deforestation and degradation of environment leading 'to deterioration of biodiversity, disruption of the carbon cycle and social issues' [6]. To enable OPPs monitoring, change detection-related studies focus on detection of oil palm expansion. A straightforward way to implement change detection is to perform multi-temporal classification. Commonly, it involves two images, 'before' and 'after', for which the OPPs' areas are compared. The accuracy of this approach is determined by the accuracy of the method by which the land cover maps (for different time instants) were obtained through classification. A thorough analysis of this type was performed by Carlson et al. (see [4, p. 6 of SI]) by visual interpretation of areas cleared for or planted with oil palm in 1990-, 2000-, and 2010s in Kalimantan.

Since the land-cover analysis that is done once every 10 years can hardly serve as OPPs monitoring tool, a challenging and urgent problem in the context of oil palms remains. Namely, there are no publicly available and scientifically tested tools for monthly or yearly monitoring of OPPs expansion based on actual remote sensing data. This paper is the first step to deal with this challenging problem since we accomplished the related task. Given publicly available Landsat imagery, provide a spatially robust method that enables yearly monitoring of industrial OPPs with characteristic road patterns. All presented experiments can be reproduced since our code and Landsat imagery are publicly available. This ensures that presented results can further serve as a benchmark.

3   Proposed Approach

The main idea of our approach is quite simple and is based on the following assumption. OPPs have the characteristic texture on Landsat imagery; the texture only weakly depends on geolocation and slowly varies in time. Namely, the major part of industrial OPPs in Indonesia can be easily recognized by a human in band 'B8' of Landsat 8 due to their characteristic road networks (Fig. 1).

Consequently, to monitor expansion of industrial OPPs, we propose

(i) using a historical Landsat image, to train on a classifier to perform high-quality segmentation of areas with specific textural, geometric, color features, i.e. to reduce the initial problem to the known semantic segmentation image analysis problem (see, e.g. [5])

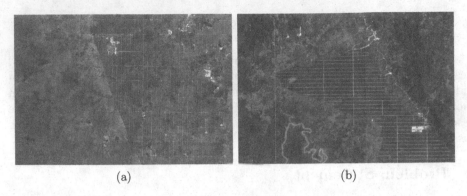

         (a)   (b)

Fig. 1. Industrial oil palm plantations in Kalimantan (a) and Sumatra (b), Landsat 8 images, band 'B8' at native 15 m/pixel resolution

(ii) to perform the training among FCN architectures since they have excellent applicability for solving similar problems (see, e.g. [5,9,10])
(iii) to use the best FCN for OPPs mapping at other locations or in the future.

3.1 Semantic Segmentation and FCNs

In semantic segmentation, we have gray-scale or multi-spectral images (photos, scenes) containing visual representations of objects belonging to given classes $C_1, \ldots, C_k$. For instance, in the famous PASCAL Visual Object Classes (VOC2012) Challenge, there were 20 classes divided into 4 groups: *people, animals, vehicles, and indoor things.* Every image contains a single or multiple objects from these classes presented over some natural background, which is regarded as an additional *void* class C_0. The goal is to train an algorithm to perform segmentation (pixel-wise classification) of the given images by assigning every pixel to a certain class (Fig. 2).

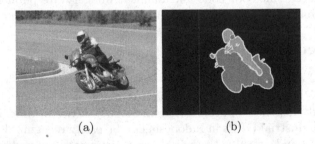

         (a)   (b)

Fig. 2. Example of semantic segmentation: (a) an initial scene; (b) a segmentation of three classes (including the background). Source: [1]

Generally speaking, we may assign a separate visual class to any type of industrial plantations. In this paper, we focus only on a single foreground class

C_1, *OPP* featuring an authentic visual pattern, and train an FCN to discriminate it against the background class C_0, *non-OPP* (Fig. 3).

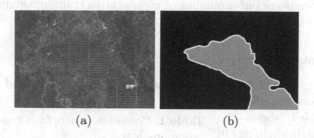

(a)   (b)

Fig. 3. The initial scene (a) and the corresponding true segmentation of the *OPP* (b)

Convolutional Neural Networks (CNN) is a widely known tool for solving problems of image analysis, in particular, the segmentation problem (see, e.g. [10]). The conventional approach is based on classification networks mapping a query image to the set of class labels. Although CNNs remain quite applicable for *detection* and *tracking* problem, their employment to segmentation (pixel-wise classification) problems appears to be substantially limited due to weak quality and high overall time-complexity [16, 23]. On the other hand, FCNs show promising results [9, 23] by assigning to an image a *heat map* of the same size and, therefore, classify every pixel separately.

To evaluate our approach, we introduce available data and performance metrics for OPPs mapping. Then we describe baseline classifiers and two FCNs.

3.2 Data

We used Google Earth Engine (GEE, http://earthengine.google.org) as the public source of remote sensing data. GEE is not only a storage that keeps almost all Landsat imagery and updating this collection near real time but it is also a large-scale computational cloud available for all sorts of geospatial analysis, e.g., Lee et al. [18] showed the potential of GEE as '*a low-cost, accessible and user-friendly oil palm detection tool*'.

The Region Of Interest (ROI) is the rectangle[2] at the map of South cost of Kalimantan with the total area equal to $61170\,km^2$. As a picture, the rectangle's dimensions are 24492×11143 pixels. We performed a manual segmentation (mapping of OPPs) of the satellite image of the ROI made in 2014 by visual interpretation. We consider this map as the ground truth for training and benchmarking segmentation algorithms. A part of the ROI was manually segmented for 2015, 2016, and 2017, to estimate accuracy of OPPs expansion

---

[2] In GEE, rectangles are represented by the minimum and maximum corners as lists $(xMin, yMin, xMax, yMax)$. The ROI is defined by $(110.3, -3, 113.6, -1.5)$.

detection. Additionally, we performed the manual segmentation of a smaller ROI in Sumatra for 2014 to explore generalization ability with respect to geolocation.

To choose the semantic segmentation algorithm with the best generalization performance, we carry out a number of comparative training-validation experiments for several well-known algorithms.

3.3 Performance Metrics

Performance of the algorithms was evaluated using the ground truth map. For a two-class classification problem, Table 1 shows the confusion matrix helping to define the following performance metrics related to a particular class i: precision (P_i), recall (R_i), producer's accuracy (PA_i), and user's accuracy (UA_i); overall accuracy (OA) is an aggregate metric. Note that the notions of producer's and user's accuracy correspond to common definitions in machine learning:

Table 1. Confusion matrix for a two-class classification problem

		Ground truth		Total
		Class C_1	Class C_0	
Prediction	Class C_1	$p_{1,1}$	$p_{1,0}$	$p_{1,\cdot}$
	Class C_0	$p_{0,1}$	$p_{0,0}$	$p_{0,\cdot}$
Total		$p_{\cdot,1}$	$p_{\cdot,0}$	$p_{\cdot,\cdot}$

$$P_i = UA_i = \frac{p_{i,i}}{p_{i,\cdot}}, \quad R_i = PA_i = \frac{p_{i,i}}{p_{\cdot,i}}, \quad OA = \frac{p_{1,1} + p_{0,0}}{p_{\cdot,\cdot}}, \quad i = 0, 1.$$

3.4 Baseline Algorithms

Our two classes are rather unbalanced: about 79.6% of all ROI pixels belong to *non-OPP* class. Thus, the first (naive) baseline classifier always predicting *non-OPP* has accuracy around 0.796.

We argue that Random Forest with different sets of features (predictors) is a suitable baseline classifier. Firstly, Random forest had numerous applications in remote sensing [3]. Secondly, in the recent study [18], a per-pixel classification via supervised learning in GEE environment was performed. Random Forest consistently outperformed other well-known classifiers available in GEE.

Since OPPs have spatial patterns, we ensure that neighborhood pixels are taken into account when performing per-pixel classification. As reported, texture metrics based on Gray Level Co-occurrence Matrix were helpful to increase accuracy of many tasks (see Table 3 and references in [19]) and, in particular, of OPPs mapping (see [8,17]). This motivates us to employ three sets of baselines:

(i) Random Forest trained on the 8th band ('B8') of Landsat 8 with native 15 m resolution and texture metrics from the GLCM around each pixel of this band. Note that GEE outputs 18 GLCM metrics (see [15, Appendix I] and [7, Sect. 2.3]). Thus, in total we have 19 numeric features. We consider three possible sizes of the neighborhood for computing each GLCM: 1, 5, and 10 pixels. We call these models 'G(1)', 'G(5)', and 'G(10)', respectively.

(ii) Random Forest trained on all available bands of Landsat 8 with 15 m resolution (after downscaling) and additional computed NDVI band[3]; in total, we have 12 numeric features. This model is called 'ALL'.

(iii) Random Forest trained on different combination of feature sets from the above baselines; the models are 'ALL+G(1)', 'ALL+G(5)', 'ALL+G(10)'.

We highlight that feature set 'ALL' contains important spectral information that was not provided for training of FCNs. Thus, some baseline algorithms utilise a richer set of information.

3.5   Proposed Algorithm

As a counterpart to classic machine learning algorithms, we use the slightly modified version[4] of FCN-8s fully convolution network proposed in [23]. The original FCN-8s was developed for semantic segmentation of PASCAL-VOC2012 images, which are RGB-color scenes containing objects from 21 visual classes. Therefore, to adapt this model to our problem, we reduce the number of input channels to 1 and the number of classes to 2 corresponding to OPPs and non-OPPs areas, respectively.

Actually, we consider two modifications of this FCN. The first one, we call it FCN-8s-TL, is obtained by conventional transfer learning approach. Following this approach, we exclude from the training machinery all unmodified layers taking all corresponding parameters from the trained FCN-8s network directly. Another network, FCN-8s-EL is trained end-to-end from the scratch. For the weights initialization, we use Caffe[5] implementation 'xavier' of the algorithm from [12] while all biases were filled with zeros. Both networks are trained by the classic SGD algorithm with momentum 0.99 and learning rates 10^{-11} and 10^{-10}, respectively.

Results of exploratory training for 2×10^5 epochs show the superiority of FCN-8s-EL over FCN-8s-TL. Although the latter network lead to a high initial accuracy (about 0.88 vs. 0.77 for FCN-8s-EL), its training progress was poor and unreliable (Fig. 4). Therefore, we exclude this network from subsequent experiments. Furthermore, we limit the experiments to 3×10^4 epochs since FCN-8s-EL starts to overfit around this stage.

---

[3] An image from Landsat 8 has 11 bands, 'B1' – 'B11'. NDVI stands for Normalized Difference Vegetation Index that varies between −1 and 1 and utilizes the relationship between vegetation brightness in the red and infrared bands. Namely, absorption of red light ('B4') and reflection of infrared radiation ('B5') by plants allows to separate non-vegetation (e.g., water, bare soil, man-made objects) from living vegetation. Basically, NDVI serves as an index of photosynthesis in a pixel.

[4] Implemented on the basis of https://github.com/shelhamer/fcn.berkeleyvision.org.

[5] Caffe deep learning framework http://caffe.berkeleyvision.organized.

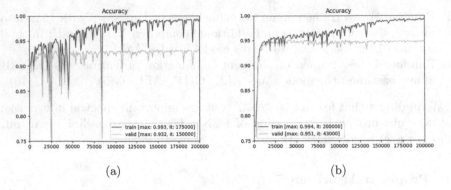

(a)   (b)

Fig. 4. Overall accuracy s.t. epoch number in exploratory learning for (a) FCN-8s-TL and (b) FCN-8s-EL

4   Experiments

This section presents numerical results and demonstrates that the best performing algorithm, FCN-8s-EL, has high generalization ability with respect to location and time by examining the predicted segmentation of two ROIs in Indonesia (in Kalimantan and Sumatra) for 2014–2017.

4.1   Baselines

We sampled 50000 pixels from the ROI that were split into a dataset for training with 5-fold cross-validation (80%) and a dataset for validation (20%).[6] The training was based on grid search of two parameters: m (number of variables randomly sampled as candidates at each split) and ns (minimum size of terminal nodes). We explored $m = 2, 5, 7, 10$ and $ns = 10, 30, 100$.[7]

Recall that the naive classifier's accuracy is 0.796. Table 2 shows results for 5-fold cross-validation for the best pairs (m, ns). For the best model 'ALL+G(10)', $OA = 0.864$ on the validation dataset. The contribution of texture features is hardly noticeable; the multi-band data does somewhat increase accuracy. Clearly, the traditional pixel-based methods can not significantly outperform naive baseline even if the texture and multi-spectral features are available.

---

[6] For models with features 'G(10)' included, we sampled 20000 random pixels due to computational constraints of GEE.

[7] The training was performed in R using library 'randomForest' with all other default parameters except number of trees to grow was set to 300.

Table 2. OA for the baseline algorithms and the best values for m and ns

	G(1)	G(5)	G(10)	ALL	ALL+G(1)	ALL+G(5)	ALL+G(10)
OA for 5-CV	0.8	0.815	0.824	0.853	0.852	0.857	**0.861**
Best (m, ns)	5, 100	7, 30	5, 30	10, 30	10, 30	10, 10	10, 10

4.2   Fully Convolutional Neural Network

Since the satellite image is prohibitively large for FCN-8s, we partition the image into disjoint tiles of 500×500 pixels. Using the set of all tiles, we employ the standard 5-fold cross-validation procedure by randomly partitioning this dataset into 5 disjoint subsets of equal size. Then, we perform 5 stages of training/testing taking; every stage 4 parts of the initial dataset were used for training and the remaining part was left for testing.

Table 3 combines the performance metrics for two baseline algorithms and the best FCN. We see that the FCN significantly outperforms the baseline algorithms, which were considered as state of the art just recently. Though both baselines access multi-spectral data, the FCN delivers better results using only the panchromatic band 'B8'.

Table 3. Results obtained by 5-fold cross-validation

	ALL	ALL+G(10)	FCN-8s-EL
Overall accuracy (OA)	0.82	0.86	**0.95**
OPP precision (UA_1)	0.71	0.73	**0.95**
OPP recall (PA_1)	0.48	0.5	**0.96**
non-OPP precision (UA_0)	0.87	0.89	**0.99**
non-OPP recall (PA_0)	0.95	0.95	**0.98**

4.3   Spatial and Temporal Scalability

Recall that the above training and testing was performed using exclusively the cloud-free composite obtained in 2014 for the ROI in Kalimantan. To illustrate generalization ability of FCN-8s-EL trained on Kalimantan data from 2014, we use this FCN for mapping of an ROI in Sumatra[8] (Fig. 5) in 2014 and for a part of Kalimantan ROI[9] (Fig. 6) in 2015–2017. As before, we used band 'B8' of

---

[8] The image size is 2048×1688 pixels; Lat.: 1.14868; Long.: 99.18078. The distance to the boundary of Kalimantan ROI is around 1300 km.
[9] The image size is 1336×1069 pixels; Lat.: -1.93356; Long.: 111.21357.

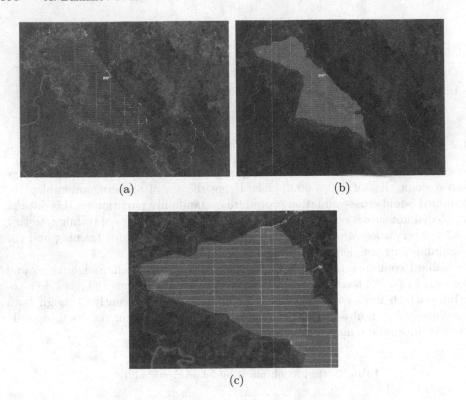

(a)   (b)

(c)

Fig. 5. Segmentation of Sumatra ROI in 2014: (a) Initial ROI image; (b) FCN-8s-EL segmentation; (c) Zoomed top-left part of the segmented area. We use the following color scheme: green - $p_{1,1}$, gray - $p_{0,0}$, red - $p_{0,1}$, and blue - $p_{1,0}$. For this ROI we have $OA = 0.99, UA_1 = 0.95, UA_0 = 0.99, PA_1 = 0.97, PA_0 = 0.99$. (Color figure online)

Landsat 8 cloud-free composite produced in GEE. We highlight that the data in these experiments is either from a completely different location or is taken after 2014; thus, the trained FCN has never seen this data.

Figures 5 and 6 show that the FCN-8s-EL is robust to spatial and temporal shifts of input data. This demonstrates that today deep learning indeed enables us to monitor OPPs within countries without laborious visual interpretation, i.e., it is possible to develop a semi-automated method for OPPs mapping of entire countries, but also a tool for yearly (monthly) detection of oil palm expansion.

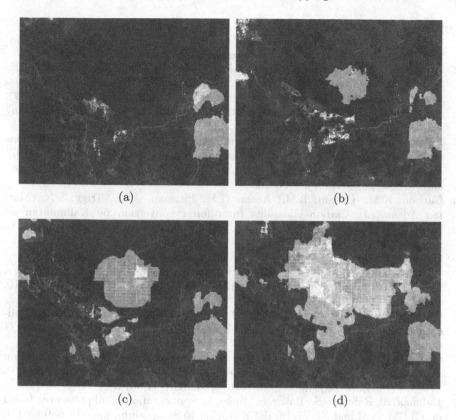

Fig. 6. Expansion of an OPP in Kalimantan. Highlighted regions correspond to OPPs detected by FCN-8s-EL in 2014 (a), 2015 (b), 2016 (c), and 2017 (d) with overall accuracy equal to 0.988, 0.982, 0.969, and 0.97 respectively.

5   Conclusion and Future Work

We proposed the novel semi-automated method for OPPs monitoring based on the state-of-the-art network FCN-8s. The achieved performance is on a par with producer's and user's accuracies reported in [2, 22] for the visual interpretation method. We plan to push the performance even higher by using all spectral bands available for Landsat imagery.

Further, we are interested to apply our methods to other regions. For example, in Peru, large-scale OPPs have the similar structure (see [13, Fig. 2]) to the one we observed in Kalimantan. Thus, one may expect that the trained FCN should perform well even without additional learning. In some other region, the complete retraining of the net or transfer learning may be needed. Based on the obtained results, we conjecture that the recent advances in FCNs are promising for successful detection of diverse types of industrial plantations (e.g., coffee, rubber, and eucalyptus) and call for new applications in a wider context of data mining in agriculture [20].

References

1. PASCAL VOC 2012. www.host.robots.ox.ac.uk/pascal/VOC/voc2012/
2. Austin, K., Mosnier, A., Pirker, J., McCallum, I., Fritz, S., Kasibhatla, P.: Shifting patterns of oil palm driven deforestation in Indonesia and implications for zero-deforestation commitments. Land Use Policy **69**, 41–48 (2017). https://doi.org/10.1016/j.landusepol.2017.08.036. http://www.sciencedirect.com/science/article/pii/S0264837717301552
3. Belgiu, M., Drăguţ, L.: Random forest in remote sensing: a review of applications and future directions. ISPRS J. Photogram. Remote Sens. **114**, 24–31 (2016). https://doi.org/10.1016/j.isprsjprs.2016.01.011. http://www.sciencedirect.com/science/article/pii/S0924271616000265
4. Carlson, K.M., Curran, L.M., Asner, G.P., Pittman, A.M., Trigg, S.N., Marion Adeney, J.: Carbon emissions from forest conversion by Kalimantan oil palm plantations. Nat. Clim. Change **3**, 283–287 (2013). https://doi.org/10.1038/nclimate1702
5. Chen, L., Kokkinos, I., Murphy, K., Yuille, A.: DeepLab: semantic image segmentation with deep convolutional nets, atrous convolution, and fully connected CRFs. IEEE Trans. Pattern Anal. Mach. Intell. **40**, 834–848 (2018). https://doi.org/10.1109/TPAMI.2017.2699184
6. Chong, K.L., Kanniah, K.D., Pohl, C., Tan, K.P.: A review of remote sensing applications for oil palm studies. Geo-spatial Inf. Sci. **20**(2), 184–200 (2017). https://doi.org/10.1080/10095020.2017.1337317
7. Conners, R.W., Trivedi, M.M., Harlow, C.A.: Segmentation of a high-resolution urban scene using texture operators. Comput. Vis. Graph. Image Process. **25**(3), 273–310 (1984). https://doi.org/10.1016/0734-189X(84)90197-X
8. Daliman, S., Rahman, S., Bakar, S., Busu, I.: Segmentation of oil palm area based on GLCM-SVM and NDVI. In: IEEE Region 10 Symposium, pp. 645–650 (2014). https://doi.org/10.1109/TENCONSpring.2014.6863113
9. Fu, G., Liu, C., Zhou, R., Sun, T., Zhang, Q.: Classification for high resolution remote sensing imagery using a fully convolutional network. Remote Sens. **9**(5), 498 (2017). https://doi.org/10.3390/rs9050498. http://www.mdpi.com/2072-4292/9/5/498
10. Garcia-Garcia, A., Orts-Escolano, S., Oprea, S., Villena-Martinez, V., Jose Garcia-Rodriguez, V.: A review on deep learning techniques applied to semantic segmentation. Manuscript **1** (2017)
11. Gaveau, D.L.A., et al.: Rapid conversions and avoided deforestation: examining four decades of industrial plantation expansion in Borneo. Sci. Rep. **6** (2016). https://doi.org/10.1038/srep32017
12. Glorot, X., Bengio, Y.: Understanding the difficulty of training deep feedforward neural networks. In: Teh, Y.W., Titterington, M. (eds.) Proceedings of the Thirteenth International Conference on Artificial Intelligence and Statistics, Proceedings of Machine Learning Research, vol. 9, pp. 249–256. PMLR, Sardinia, 13–15 May 2010. http://proceedings.mlr.press/v9/glorot10a.html
13. Gutiérrez-Vélez, V.H., DeFries, R.: Annual multi-resolution detection of land cover conversion to oil palm in the Peruvian Amazon. Remote Sens. Environ. **129**, 154–167 (2013). https://doi.org/10.1016/j.rse.2012.10.033. http://www.sciencedirect.com/science/article/pii/S003442571200421X
14. Hansen, M.C., et al.: High-resolution global maps of 21st-century forest cover change. Science **342**(6160), 850–853 (2013). https://doi.org/10.1126/science.1244693. http://science.sciencemag.org/content/342/6160/850

15. Haralick, R., Shanmugam, K., Dinstein, I.: Textural features for image classification. IEEE Trans. Syst. Man Cybern. **SMC–3**, 610–621 (1973). https://doi.org/10.1109/TSMC.1973.4309314

16. Huang, Z., Pan, Z., Lei, B.: Transfer learning with deep convolutional neural network for SAR target classification with limited labeled data. Remote Sens. **9**(9), 907 (2017). https://doi.org/10.3390/rs9090907. http://www.mdpi.com/2072-4292/9/9/907

17. Kamiran, N., Sarker, M.L.R.: Exploring the potential of high resolution remote sensing data for mapping vegetation and the age groups of oil palm plantation. IOP Conf. Ser.: Earth Environ. Sci. **18**(1), 012181 (2014). http://stacks.iop.org/1755-1315/18/i=1/a=012181

18. Lee, J.S.H., Wich, S., Widayati, A., Koh, L.P.: Detecting industrial oil palm plantations on Landsat images with Google Earth Engine. Remote Sens. Appl. Soc. Environ. **4**, 219–224 (2016). https://doi.org/10.1016/j.rsase.2016.11.003. https://www.sciencedirect.com/science/article/pii/S235293851630129X

19. Lu, D., Weng, Q.: A survey of image classification methods and techniques for improving classification performance. Int. J. Remote Sens. **28**(5), 823–870 (2007). https://doi.org/10.1080/01431160600746456

20. Mucherino, A., Papajorgji, P.J., Pardalos, P.M.: Data Mining in Agriculture, 1st edn. Springer, New York (2009). https://doi.org/10.1007/978-0-387-88615-2

21. Nooni, I., Duker, A., Van Duren, I., Addae-Wireko, L., Osei Jnr, E.: Support vector machine to map oil palm in a heterogeneous environment. Int. J. Remote Sens. **35**(13), 4778–4794 (2014). https://doi.org/10.1080/01431161.2014.930201

22. Petersen, R., et al.: Mapping tree plantations with multispectral imagery: preliminary results for seven tropical countries. Technical report, World Resources Institute (2016). www.wri.org/publication/mapping-tree-plantations

23. Shelhamer, E., Long, J., Darrell, T.: Fully convolutional networks for semantic segmentation. IEEE Trans. Pattern Anal. Mach. Intell. **39**, 640–651 (2017). https://doi.org/10.1109/TPAMI.2016.2572683

CalciumCV: Computer Vision Software for Calcium Signaling in Astrocytes

Valentina Kustikova[1], Mikhail Krivonosov[1], Alexey Pimashkin[1],
Pavel Denisov[1], Alexey Zaikin[1,2], Mikhail Ivanchenko[1], Iosif Meyerov[1(✉)],
and Alexey Semyanov[1,3(✉)]

[1] Lobachevsky State University of Nizhni Novgorod,
Nizhni Novgorod 603950, Russian Federation
valentina.kustikova@itmm.unn.ru, mike_live@mail.ru,
alexey.zaikin@ucl.ac.uk, ivanchenko.mv@gmail.com, meerov@vmk.unn.ru,
{pimashkin,denisov,semyanov}@neuro.nnov.ru
[2] Department of Mathematics and Institute for Women's Health,
University College London, London WC1E 6AU, UK
[3] Shemyakin-Ovchinnikov Institute of Bioorganic Chemistry,
Moscow 117997, Russian Federation

Abstract. Developing computational analysis of time-lapse imaging of calcium events in astrocytes is a challenging task in neuroscience. Here we report the implementation of an algorithm that solves this task. After noise reduction with the block-matching and 3D filtering (BM3D) algorithm, calcium activity is identified as fluorescence elevation above the baseline level. Individual events are detected by sliding window approach applied to the variation of pixel intensity relative to the baseline level. The maximal projection and duration of astrocytic calcium events are then assessed. The novelty of the proposed method is an adaptive construction of the baseline level. The statistical results generated by our program are consistent with the previous algorithm reported and used by us for the reference. The software is publicly available.

Keywords: Bioinformatics · Calcium signaling in astrocytes
Computer vision · Image analysis · Statistical analysis

1 Introduction

Astrocytes are electrically inactive brain cells, which are connected through gap-junctions and form a syncytium-like functional network [1,2]. Although the mechanisms of astrocytic Ca^{2+} dynamics have been studied for more than quarter of a century, assessment of spatial-temporal properties of individual Ca^{2+} events in the entire astrocyte remains a challenging task.

This research financially supported by Russian Science Foundation (AZ to 16-12-00077, algorithm; AS to 16-14-00201, data analysis).

W. M. P. van der Aalst et al. (Eds.): AIST 2018, LNCS 11179, pp. 168–179, 2018.
https://doi.org/10.1007/978-3-030-11027-7_17

Imaging of calcium activity is a basic method for calcium activity analysis. The simplest approach to this problem is an analysis of Ca^{2+} dynamics in adjustment regions of interest (ROI) [3–5] in two-dimensional plane. For example, if an event starts on one ROI and propagates to the neighboring ROIs, one can estimate its spread by counting the number of involved ROIs. However, this method is laborious and provides less accurate information about size and duration of Ca^{2+} events than measuring their actual contours in each imaging frame. Recently, several approaches to measure automatically the parameters of individual Ca^{2+} events in time-lapse imaging data have been suggested [6–8], however, none of these methods has become a standard. Nowadays, advanced methods of three-dimensional Ca^{2+} imaging are developed [9], but application of these methods requires the appropriate hardware for the experiments that is not widespread.

The state-of-the-art approaches for image preprocessing and solving computer vision problems are based on deep learning, especially, on convolutional neural networks [10–15]. Application of deep learning supposes the existence of a large train dataset. During astrocyte activity analysis, to get new data means to carry out new biological experiment which requires corresponding materials and hardware. The labeling data is performed by experts, but there are astrocyte activities which may be invisible for some of them that is why the labeling is a subjective process. Therefore, the application of deep learning to the astrocyte activity analysis is complicated by the high-level complexity of data preparation.

We improved the previously developed method [16] to analyse time lapse imaging records for the calcium activity of an astrocyte. To do this, we made two main modifications in the event detection flow. First, we implemented a new precise and adaptive algorithm for a baseline level approximation to detect the inactive state of the astrocyte. Second, we proposed using the sliding window approach to detect calcium events. We show that the algorithm detects Ca^{2+} events in imaging data and extracts their duration and maximal projection. The new implementation improves performance by 16–40%. The results are in good agreement with the previous report [16].

2   Calcium Event Detection

2.1   Problem Statement

The proposed method analyses time lapse imaging records for the calcium activity of an astrocyte. Each frame displays a spatial distribution of fluorescence intensity in several planes parallel to the substrate, yielding a set of planar images that reflect calcium activity. Further on, we introduce and consider an instantaneous maximal projection, defined as the maximal intensity at each point over this set (Fig. 1):

$$V = \{V_s, 0 \leq s \leq k - 1\}, V_s = (V_{ij}^{(s)} : 0 \leq i \leq h - 1, 0 \leq j \leq w - 1), \quad (1)$$

where k is a number of frames, and V_s is a frame with a resolution $w \times h$.

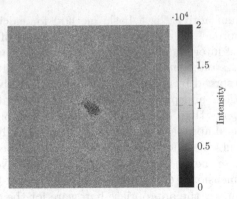

Fig. 1. The instantaneous maximal projection of the fluorescence intensity over a set of planar sections for an astrocyte. Shifting from the blue to red end of the color map codes increasing brightness, units are arbitrary (Color figure online).

Moreover, we have a movie

$$N = \{N_r, 0 \leq r \leq n-1\}, N_r = (N_{ij}^{(r)} : 0 \leq i \leq h-1, 0 \leq j \leq w-1), \quad (2)$$

that corresponds to the camera noise (Fig. 2) and represents a video without astrocyte. Here, n is a number of noise video frames, and N_r is a noise frame. The principle of noise generation on videos is unknown; the noise model is Gaussian noise.

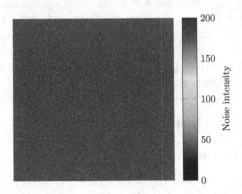

Fig. 2. The noise frame sample. Shifting from the blue to red end of the color map codes increasing brightness, units are arbitrary (Color figure online).

The proposed method processes time lapse imaging records and noise records, and extracts a set of calcium events in the astrocyte. A single *event* is assigned with start and end times (frame identifiers), and described by a set of points (frame pixels) that belong to it over the whole time span (a set of frames).

For the extracted events the statistical analysis is carried out. We analyse the statistical properties of event time spans and maximal projections.

2.2   Method

General Pipeline. The developed method includes the following steps (Fig. 3):

1. Video preprocessing:
   - Aligning an image (eliminating jitter).
   - Calculating noise parameters and subtracting the camera noise.
   - Filtering video frames.
   - Evaluating a noise level on the filtered video.
2. Calculating a baseline of the fluorescence intensity and its relative variation.
3. Detecting calcium events based on the relative intensity variation.

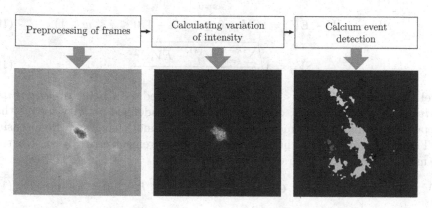

Fig. 3. Pipeline of the proposed method (Color figure online).

Video Preprocessing. This step consists of 4 operations. Let us consider each operation in detail.

1. Aligning an image. To align each video frame V_s we compute normalized cross-correlation $C(i, j)$ of the first frame and the current frame shifted by a vector (i, j), where $i = \overline{-w, w}$, $j = \overline{-h, h}$. The optimal shift should be computed as a decision of the following optimization problem:

$$(i_{opt}, j_{opt}) = \arg\max_{(i,j)} C(i, j). \tag{3}$$

Next, we denote aligned video as V.

2. Calculating noise parameters and subtracting the camera noise. To identify noise parameters noise video N should be considered. We calculate mean noise frame by time

$$EN = \langle N \rangle_r = (EN_{ij} : 0 \leq i \leq h - 1, 0 \leq j \leq w - 1), \tag{4}$$

$$EN_{ij} = \frac{1}{n} \sum_{r=0}^{n-1} N_{ij}^{(r)} \tag{5}$$

and standard deviation of the noise

$$SN = \sigma_r(N - EN) = (SN_{ij} : 0 \le i \le h - 1, 0 \le j \le w - 1), \tag{6}$$

$$SN_{ij} = \sqrt{\frac{\sum_{r=0}^{n-1} (N_{ij}^{(r)} - EN_{ij})^2}{n - 1}}. \tag{7}$$

Similarly we compute mean frame and standard deviation of the input video:

$$EV = \langle V \rangle_s = (EV_{ij} : 0 \le i \le h - 1, 0 \le j \le w - 1), \tag{8}$$

$$EV_{ij} = \frac{1}{k} \sum_{s=0}^{k-1} V_{ij}^{(s)}, \tag{9}$$

$$SV = \sigma_s(V - EV) = (SV_{ij} : 0 \le i \le h - 1, 0 \le j \le w - 1). \tag{10}$$

$$SV_{ij} = \sqrt{\frac{\sum_{s=0}^{k-1} (V_{ij}^{(s)} - EV_{ij})^2}{k - 1}}. \tag{11}$$

Pixel (i, j) for which $SV_{ij} < \alpha \cdot SN_{ij}$ is considered as a point of **undefined activity** (α is a parameter, default value is 3). Undefined activity means that the point constains camera measurement error or minor changes of the intensity.

Thereafter we need to eliminate camera measurement error by subtracting the mean noise frame of the video frames

$$OV = \{OV_s, s = 0...k - 1\}, OV_s = (OV_{ij}^{(s)} : 0 \le i \le h - 1, 0 \le j \le w - 1), \tag{12}$$

$$OV_{ij}^{(s)} = V_{ij}^{(s)} - EN_{ij}. \tag{13}$$

3. *Filtering video frames.* Filtering video OV using block-matching and 3D filtering (BM3D) [17] and Gaussian filters allows to smooth video and to prepare it for post-processing. BM3D is based on the idea that real-world images contain similar patches. The similarity of these patches is calculated by L_2-norm. Patches are grouped, and 3D-transform is applied to the group. The transform output is spectral representation of the block group. Filtering spectral coefficients by internal filter and applying inverse transform allow to restore filtered image. Let us denote the filtered video as FV:

$$FV = \{FV_s, 0 \le s \le k - 1\}, FV_s = (FV_{ij}^{(s)} : 0 \le i \le h - 1, 0 \le j \le w - 1), \tag{14}$$

where FV_s is a filtered video frame. The application of BM3D is explained by the successful experience of its using in [16]. One of the directions for the future research related with the investigation of applying more advanced denoising methods [18–21].

4. *Evaluating a noise level on the filtered video.* Further we consider pixels with defined activity to evaluate a noise level on the filtered video FV. The

task is to estimate standard deviation $\sigma_s(FV - RV)$ the intensity of the filtered video FV relative to the unknown real signal RV, where

$$RV = \{RV_s, 0 \leq s \leq k-1\}, RV_s = (RV_{ij}^{(s)} : 0 \leq i \leq h-1, 0 \leq j \leq w-1). \quad (15)$$

We assume that if $SN = \sigma_r(N - EN)$ is a standard deviation of the noise video then $SFN = \sigma_r(FN - EN)$ is a standard deviation of the filtered noise video FN that described as follows:

$$FN = \{FN_r, 0 \leq r \leq n-1\}, \quad (16)$$

$$FN_r = (FN_{ij}^{(r)} : 0 \leq i \leq h-1, 0 \leq j \leq w-1). \quad (17)$$

Therefore, a coefficient of variation of the standard deviation after filtering the noise video can be computed as follows:

$$R = \left\langle \frac{SN}{SFN} \right\rangle_{ij} = \frac{1}{w \cdot h} \sum_{i=0}^{h-1} \sum_{j=0}^{w-1} \frac{SN_{ij}}{SFN_{ij}}. \quad (18)$$

Hence, $\sigma_r(N - EN) \approx R \cdot \sigma_r(FN - EN)$, and we can suppose that

$$\sigma_s(V - RV) \approx R \cdot \sigma_s(FV - RV). \quad (19)$$

Since real signal RV is unknown and applied filter is good enough (i.e. $FV \approx RV$), it can be assumed $\sigma_s(V - RV) \approx \sigma_s(V - FV)$. Consequently, we can estimate standard deviation $\sigma_s(FV - RV)$ of the intensity of the filtered video FV relative to the unknown real signal RV as follows:

$$\sigma = SR = \sigma_s(FV - RV) \approx \frac{\sigma_s(V - FV)}{R}. \quad (20)$$

Using this, we classify pixels as noise if intensity variation relative to the baseline level is less than $3 \cdot \sigma$ (Fig. 4).

Calculating a Baseline of the Fluorescence Intensity and Its Relative Variation.

A baseline of the fluorescence intensity corresponds to the inactive state of the astrocyte. Let us introduce some auxiliary notation: $F = \{F^t, 0 \leq t \leq k-1\}$ is a pixel intensity realization in time, $F_0 = \{F_0^t, 0 \leq t \leq k-1\}$ is a baseline pixel intensity that corresponds to the inactive astrocyte state in the pixel. Calculating the baseline intensity F_0 for each pixel assumes an iterative approximation by applying moving average for the current estimate of a baseline (Fig. 5). At the first iteration we suppose that the baseline is equal to the pixel intensity $F_0 = F$. The fixed-size moving average is applied to the intensity realization, and we compute first approximation of the baseline F_0. The window size is a parameter and its value is choosen based on the video capture parameters. Further, we compute the difference between pixel intensity realization F and current baseline approximation F_0. If the difference is less than zero then we set it to zero since it is an impossible situation from the point of view of astrocyte activity. Then the standard deviation of such difference is

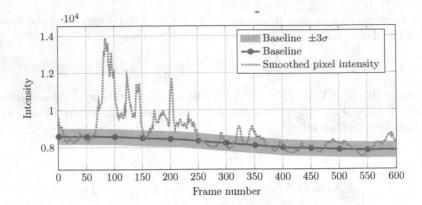

Fig. 4. Baseline pixel intensity and three standard deviations from it (Color figure online).

computed. The iterative process stops when this standard deviation is less than the noise level represented earlier. It should be noted that the novelty of the method consists in the iterative procedure of the fluorescence intensity baseline.

The relative intensity variation for each pixel is calculated as follows:

$$\frac{dF}{F_0} = \frac{F - F_0}{F_0}. \tag{21}$$

This equation is used to compute calcium fluorescence [22].

We classify the pixel as active at the frame number t if $\frac{F^t - F_0^t}{F_0^t} > 0$ otherwise the pixel is considered to be inactive.

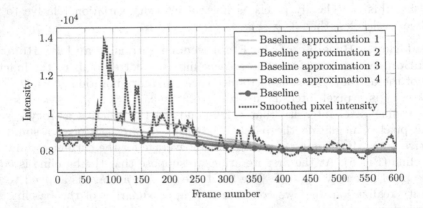

Fig. 5. Iterative approximation of a baseline of the fluorescence intensity (Color figure online).

Detecting Calcium Events Based on the Relative Intensity Variation.
First of all we need to identify time gaps where each pixel is active. This step is
implemented by clustering the activity moments using DBSCAN [23] method.
Next, these time gaps are combined over space and time using sliding window
approach. The constructed time gaps for the set of all pixels are represented
then as a set of graph vertices. Two vertices are connected by an edge if the
corresponding pixels belong to the same window location and the corresponding
time gaps intersect. We construct connectivity components for this graph, where
each connectivity component represents an astrocyte event. Consequently all
event points may be reconstructed based on the data stored in the graph vertices.

3   Implementation

The proposed method is implemented in C++ programming language, tools
for statistical events analysis are implemented in MATLAB. The code is distributed free and open source. The source code can be downloaded from GitHub:
https://github.com/UNN-VMK-Software/astro-analysis. We also submit a short
test movie and step-by-step tutorial for building and executing developed program, and for fine-turning method parameters.

4   Experiments

We tested the developed method on the available experimental data and verified
its validity. We used 10 records with a duration from 100 to 3000 frames, with
frame resolution is not exceeding 512×512 pixels (Table 1).

Table 1. Test movie parameters.

#	Resolution	Duration	Number of detected events
1	512×512	1500	465
2	451×441	3000	1204
3	421×512	3000	2048
4	512×512	100	65
5	512×512	1000	458
6	512×512	1000	591
7	500×390	1000	260
8	512×512	200	61
9	512×512	500	97
10	512×512	600	364

The correctness of event identification in a given subject area is determined
today by the method of expert assessments. Unfortunately, we do not know the
accurate metrics.

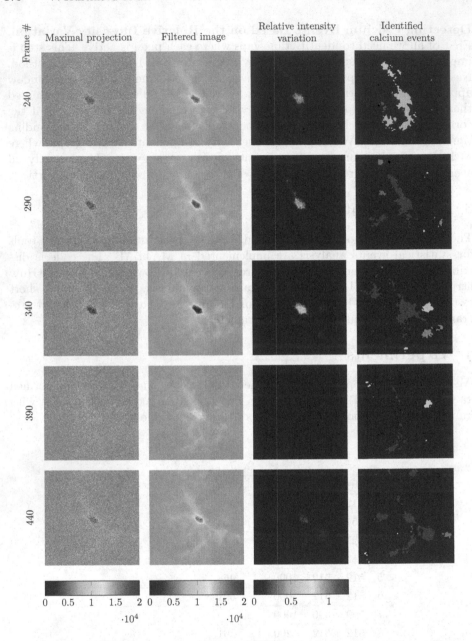

Fig. 6. Intermediate and final results of the astrocyte calcium events detection method for a test movie for the frames from 240 to 440 with the 50 frame step (Color figure online).

Firstly, identification of calcium events was confirmed by visual inspection. Figure 6 exemplifies the snapshots after filtering, the intensity variation patterns, and the set of detected events. A detailed verification can be done, for example, with the prepared movie: https://cloud.mail.ru/public/4Zhh/BCjo7KN86. Clearly, the method adequately yields the regions of calcium activity.

Secondly, we implemented a statistical analysis of calcium events. The complementary cumulative distribution functions for durations and maximal projections of events are approximated well by power laws, as confirmed by Kolmogorov-Smirnov test. The typical values of the exponents are consistent with the previously reported data (Figs. 7 and 8). A comparison with the results of the approach described in [16] demonstrates that the novel method allows more accurate baseline approximation, crucial in certain cases. While in [16] it is taken that fluorescence at a pixel belongs to background if its intensity is below a certain threshold, our method employs the adaptive threshold both in time and space. Moreover, parameters of the baseline level are different for different pixels. It tackles the typical pitfalls of global thresholding, such as picking noise for its overly low value and missing actual events when it is set overly high, with inevitable mistakes if the amplitude ranges of noise and events overlap.

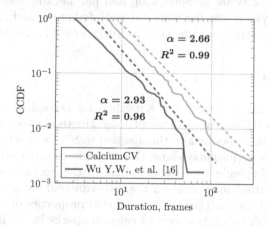

Fig. 7. The complementary cumulative distribution functions (CCDF) for event durations and the power law fits in logarithmic scale: α is a value of the exponent, R^2 is an approximation accuracy (Color figure online).

Performance of the software has been also tested as following. The total execution time for the sample movie that contains 600 frames with 512×512 resolution approximately equals 10.39 min, that compares to the method from [16] (about 12.38 min), given the test computational environment with Intel® Core[TM] i7 870, 2.93 GHz, 4 cores, 8 GB RAM, OS Windows (64-bit). The new implementation improves performance by 16%. This measurement value grows to 40% when processing more complicated movies which contain large number of events. At the same time the implementation of the proposed method was serial,

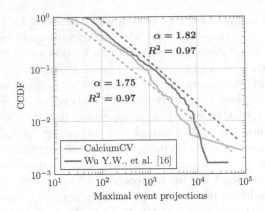

Fig. 8. The complementary cumulative distribution functions (CCDF) for event maximal projections and the power law fits in logarithmic scale: α is a value of the exponent, R^2 is an approximation accuracy (Color figure online).

and multi-threaded in case of [16]. Finally, we notice the potential for accelerating the developed software by optimizing and parallelizing some method steps. Performance analysis is not the main subject of this research, here we demonstrate that the developed implementation works in a reasonable time.

5   Conclusion

In summary, we have demonstrated a method for detecting calcium signalling events in astrocytes. The novelty of the proposed method consists in constructing the baseline level corresponding to the inactive state of astrocyte. Moreover, the analysis of software performance shows that the time of astrocyte event detection for the developed implementation is less than for the existing one. The software is available at GitHub[1] for testing on raw experimental data. Our program can be used by researchers analyzing spatiotemporal properties of calcium events in astrocytes and other cell types such as calcium sparks in cardiomyocytes. The method is open for future development.

References

1. Bennett, M.V., et al.: New roles for astrocytes: gap junction hemichannels have something to communicate. Trends Neurosci. **26**(11), 610–617 (2003)
2. Ma, B., et al.: Gap junction coupling confers isopotentiality on astrocyte syncytium. Glia **64**(2), 214–226 (2016)
3. Nett, W.J., et al.: Hippocampal astrocytes in situ exhibit calcium oscillations that occur independent of neuronal activity. J. Neurophysiol. **87**(1), 528–537 (2002)

---

[1] https://github.com/UNN-VMK-Software/astro-analysis.

4. Sun, M.Y., et al.: Astrocyte calcium microdomains are inhibited by Bafilomycin A1 and cannot be replicated by low-level Schaffer collateral stimulation in situ. Cell Calcium **55**(1), 1–16 (2014)
5. Fiacco, T.A., et al.: Intracellular astrocyte calcium waves in situ increase the frequency of spontaneous AMPA receptor currents in CA1 pyramidal neurons. J. Neurosci. **24**(3), 722–732 (2004)
6. Asada, A., et al.: Subtle modulation of ongoing calcium dynamics in astrocytic microdomains by sensory inputs. Physiol. Rep. **3**(10), e12454 (2015). https://www.ncbi.nlm.nih.gov/pubmed/26438730
7. Nakayama, R., et al.: Subcellular calcium dynamics during juvenile development in mouse hippocampal astrocytes. Eur. J. Neurosci. **43**(7), 923–932 (2016)
8. Shigetomi, E., et al.: Probing the complexities of astrocyte calcium signaling. Trends Cell Biol. **26**(4), 300–312 (2016)
9. Bindocci, E., et al.: Three-dimensional Ca^{2+} imaging advances understanding of astrocyte biology. Am. Assoc. Adv. Sci. **356**(6339), eaai8185 (2017). http://science.sciencemag.org/content/356/6339/eaai8185
10. Krizhevsky, A., et al.: ImageNet classification with deep convolutional neural networks. In: Advances in Neural Information Processing Systems (NIPS 2012), vol. 25, pp. 1097–1105 (2012)
11. Szegedy, C., et al.: Going deeper with convolutions (2014). http://arxiv.org/abs/1409.4842
12. Simonyan, K., et al.: Very deep convolutional networks for large-scale visual recognition (2014). http://www.robots.ox.ac.uk/~vgg/research/very_deep
13. Redmon, J., et al.: You only look once: unified, real-time object detection (2016). https://arxiv.org/abs/1506.02640
14. Redmon, J., Farhadi, A.: YOLO9000: better, faster, stronger (2016). https://arxiv.org/abs/1612.08242
15. Liu, W., et al.: SSD: single shot multibox detector (2016). https://arxiv.org/abs/1512.02325
16. Wu, Y.W., et al.: Spatiotemporal calcium dynamics in single astrocytes and its modulation by neuronal activity. Cell Calcium **55**(2), 119–129 (2014)
17. Dabov, K., et al.: Image denoising by sparse 3D transform domain collaborative filtering. IEEE Trans. Image Process. **16**(8), 2080–2095 (2007)
18. Ji, H., et al.: Robust video denoising using low rank matrix completion. In: IEEE Computer Society Conference on Computer Vision and Pattern Recognition, pp. 1791–1798 (2010)
19. Burges, C.J.C., et al.: Adaptive multi-column deep neural networks with application to robust image denoising. In: Advances in Neural Information Processing Systems (NIPS 2013), vol. 26, pp. 1493–1501 (2013)
20. Lefkimmiatis, S.: Non-local color image denoising with convolutional neural networks. In: IEEE International Conference on Computer Vision and Pattern Recognition (2017)
21. Lefkimmiatis, S.: Universal denoising networks: a novel CNN architecture for image denoising. In: IEEE International Conference on CVPR (2017)
22. Kao, J.P.Y., et al.: Photochemically generated cytosolic calcium pulses and their detection by fluo-3. J. Biol. Chem. **264**(14), 8179–8184 (1989)
23. Ester, M., et al.: A density-based algorithm for discovering clusters in large spatial databases with noise. In: Proceedings of the Second International Conference on Knowledge Discovery and Data Mining (KDD 1996), pp. 226–231 (1996)

Copy-Move Detection Based on Different
Forms of Local Binary Patterns

Andrey Kuznetsov[✉] [ID]

Samara National Research University, Samara, Russia
Kuznetsoff.andrey@gmail.com

Abstract. An obvious way of digital image forgery is a copy-move attack. It is quite simple to carry out to hide important information in an image. Copy-move process contains three main steps: copy the fragment from one place of an image, transform it by some means and paste to another place of the same image. Nowadays researchers develop a lot of copy-move detection solutions though the achieved results are far from perfect. In this paper, it is proposed a comparison of different local binary patterns (LBP) forms in the task of copy-move detection: geometric local binary patterns (GLBP), binary gradient contours (BGC), local derivative patterns (LDP) and simple LBP forms. All these LBP-based solutions are used to create local features that are robust to contrast enhancement, additive Gaussian noise, JPEG compression, affine transform. All these solutions are different in the number of transforms and transform parameters range that can be detected by the algorithm. Another advantage of these features is low computational complexity. Conducted experiments show that GLBP-based features can be used to detect all 4 transforms with a wide range of transforms parameters. The proposed solution showed high precision and recall values during experimental research for wide ranges of transform parameters. Thus, it showed a meaningful improvement in detection accuracy.

Keywords: Forgery · Copy-move · Local binary pattern
Binary gradient contour · Geometric local binary pattern
Local derivative pattern

1 Introduction

Providing information to expand on a topic or to prove some facts is impossible without digital images. For this purpose, people exchange digital photos using social networks (i.e., Facebook) and image services (Instagram, Tumblr, etc.). The number of shared digital images per day for the most popular web applications is incredible: nearly 350 million images on Facebook [1] and 80 million photos in Instagram [2]. These data are not secure at all, nobody checks its originality.

Protection from digital image forgery is a challenging problem nowadays. Everyone can make changes in a digital image using modern software tools on PC or any mobile device in several minutes. But no one can predict the consequences of these changes: whether they will be harmful or not. Conventionally such changes are called attacks.

One of the most frequently used digital image attack is copy-move (Fig. 1). An image with copy-move forgery contains two or more fragments with similar content

© Springer Nature Switzerland AG 2018
W. M. P. van der Aalst et al. (Eds.): AIST 2018, LNCS 11179, pp. 180–190, 2018.
https://doi.org/10.1007/978-3-030-11027-7_18

where one or more distinct fragments are copied and pasted to one or more places in the same image to modify important information. The copied regions are called duplicates. The forgery is usually harder to detect when the copied content is transformed using a pre-processing procedure such as contrast enhancement, noise addition, JPEG compression or affine transform (rotation and scaling) to make it fit the surrounding image area. The purpose of these modifications is to make the tampering not visible.

Fig. 1. Copy-move attack example and its detection result

Two main approaches to copy-move detection are usually used. The first one concerns analysis of all host image pixels (dense-field approach), whereas the second one (keypoint-based approach) is based on keypoints analysis. A small set of pixels is selected from the host image while using the keypoint-based approach [3–5]. This approach is faster in comparison with entire image analysis. A survey provided by Christlein et al. [6] showed performance results for keypoint-based approaches. It was stated that the main problem of this approach is considerably low detection accuracy for duplicates containing smooth regions.

However, the benchmarking paper [6] also showed high performance for dense-field approach. One of the main problems of this approach is high computational complexity due to sliding window analysis scheme. Block-based analysis scheme is usually used to improve analysis speed [6–10]. The second problem, which is a background of many papers, is robustness to transformations. Consequently, the experimental study showed that dense-field approach is more reliable than the keypoint-based approach.

The above comparison compels to focus on development an algorithm corresponding to the dense-field approach. In this paper, we propose a novel copy-move detection algorithm based on different complex forms of Local Binary Pattern (LBP). They are used for texture classification and show robustness to different transformations. For copy-move detection purposes features based on some forms of LBP (GLBP, LDP, BGC) show extremely high detection accuracy for a wide range of transformations parameters. We use LBP-based features in the following three-step copy-move detection scheme: feature extraction, feature matching and post-processing.

In Sect. 2 we briefly describe the three-step copy-move detection scheme. Binary gradient contours (BGC) are presented in Sect. 3. In Sect. 4 Geometric local binary patterns are described. Section 5 provides information about local derivative patterns.

In Sect. 6 we carry out a detailed investigation of the LBP-based algorithm robustness to contrast enhancement, JPEG compression, additive noise and affine transform. Conclusions are drawn in Sect. 7.

2 Copy-Move Detection Scheme

2.1 Feature Extraction

Let f be a host image with size $N_1 \times N_2$, $W_{k,l}$ be a processing block with size $M_1 \times M_2$. Let $t \in \mathbf{N}$ be the block $W_{k,l}$ shift. The number of blocks where features are extracted is calculated as follows:

$$L = \left(\left\lfloor \frac{N_1 - M_1}{s} \right\rfloor + 1 \right) \cdot \left(\left\lfloor \frac{N_2 - M_2}{s} \right\rfloor + 1 \right).$$

One Q-dimensional feature vector $\mathbf{h} \in \mathbf{R}^Q$ corresponds to one block $W_{k,l}$. Features are extracted using one of the algorithms [7–10] and are stored column-wise in the following matrix $\mathbf{FM}^{L \times Q} = [\mathbf{h}_0, \dots, \mathbf{h}_{L-1}]^T$. In this paper we will use different forms of LBP to create feature vectors of local blocks.

2.2 Feature Matching

The second step of the algorithm concerns nearest features matching. The obvious way to find the nearest neighbors is a pairwise comparison of all features, which is computationally inefficient $(O(N^2))$. To improve the speed of nearest features matching we use one of the binary search tree structures – k-d tree.

We use $\mathbf{FM}^{L \times Q}$ to generate the tree with *Matlab* k-d tree implementation – *createns*. There is used Euclidean distance as distance measure. The number of the nearest neighbors K was set to 8. Matched features indexes are stored in the matrix $\mathbf{D}^{L \times K}$, created at the end of tree constructing procedure. Its elements $d_{ij} = \arg(\mathbf{h}_{d_{ij}}), d_{ij} \in \mathbf{N}^+_{[0,L-1]}/\{i\}$ correspond to the indexes of matched features that are the nearest neighbors to $\mathbf{h}_i$.

2.3 Post Processing

The last step of the copy-move detection scheme is made to filter the results of features matching from false positives. During this step, we select such only features pairs $(\mathbf{h}_i, \mathbf{h}_j)$, which meet the following conditions:

1. $\forall i \in [0, L-1], j \in [0, Q-1], \left\| \mathbf{h}_i - \mathbf{h}_{d_{ij}} \right\| \leq 1$, where $\|\bullet\|$ is a Euclid norm;
2. $\forall i \in [0, L-1], j \in [0, Q-1], d\left(\mathbf{h}_i, \mathbf{h}_{d_{ij}}\right) \geq 10$, where $d\left(\mathbf{h}_i, \mathbf{h}_j\right)$ is a distance between $\mathbf{h}_i$ and $\mathbf{h}_j$.

For each pair meeting these constraints we calculate the difference between the corresponding block positions and their frequency. The most frequently occurred features are marked as copy-move in the result image.

3   Binary Gradient Contours

BGC were proposed by Fernández et al. [11]. BGC belong to the class of descriptors based on pairwise comparison of pixel values selected in a clockwise order from a neighborhood using some pre-defined route. It is obvious that the number of different methods of BGC construction depends on the number of different routes for the neighborhood. Three ways of BGC calculation were proposed [11], which will be denoted later as *BGC_1*, *BGC_2* and *BGC_3*.

The first two methods use one route for the whole neighborhood of the central pixel to generate the code. The first method is described by the following sequence $\{0, 1, 2, 3, 4, 5, 6, 7, 0\}$, where '0' denotes to the index of the upper left pixel in the neighborhood. In this case the BGC code is calculated as follows:

$$BGC^1(x_c, y_c) = \sum_{i=0}^{7} I\left(f_i - f_{(i+1)\bmod 8} \geq 0\right) \cdot 2^i. \tag{1}$$

The second method is based on the route of the form $\{0, 3, 6, 1, 4, 7, 2, 5, 0\}$. In this case the BGC code is calculated as follows:

$$BGC^2(x_c, y_c) = \sum_{i=0}^{7} I\left(f_{3i\bmod 8} - f_{3(i+1)\bmod 8} \geq 0\right) \cdot 2^i. \tag{2}$$

Both methods generate $2^8 - 1$ different code values.

The third method of BGC code calculation is different from the previous ones. It is based on two different routes – $\{1, 3, 7, 5, 1\}$ and $\{0, 2, 4, 6, 0\}$. Every route usage leads to $2^4 - 1$ different codes and the whole BGC code is calculated as follows:

$$BGC^3(x_c, y_c) = \left(2^4 - 1\right) \cdot \sum_{i=0}^{3} I\left(f_{2i} - f_{2(i+1)\bmod 8} \geq 0\right) \cdot 2^i$$

$$+ \sum_{i=0}^{3} I\left(f_{2i+1} - f_{(2i+3)\bmod 8} \geq 0\right) \cdot 2^i - 2^4. \tag{3}$$

4   Geometric Local Binary Pattern

The GLBP technique [12] characterizes changes of intensities for a pixel neighborhood by evaluating neighboring points on circles with different radii around the center pixel (Fig. 2). We take 7×7 neighborhood of a pixel and calculate intensities of points lying on three circles with radii: $r_1 = 0.7071$, $r_2 = 1.7725$, and $r_3 = 0.30536$. These points are equally spaced by an angle $2\pi/N$, $N \in \{4, 12, 20\}$. Consequently, we have 36 interpolated points, which values will be used further to compute GLBP codes.

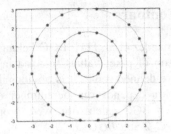

Fig. 2. 36 interpolated points.

The intensity of every interpolated point is computed through bilinear interpolation. If we consider calculating the intensity value of point $P(x,y)$, we use the following equation:

$$f_P = (1-x \quad x) \cdot \begin{pmatrix} f_{i,j} & f_{i,j+1} \\ f_{i+1,j} & f_{i+1,j+1} \end{pmatrix} \cdot \begin{pmatrix} 1-y \\ y \end{pmatrix},$$

where $x, y \in [0,1]$ are Cartesian coordinates within 2×2 neighborhood of $P(x,y)$. The coordinates (x,y) are calculated as follows:

$$x = r_i \cdot \cos((\pi + 2\pi n)/N), y = r_i \cdot \sin((\pi + 2\pi n)/N),$$

where $n \in [0, N-1], i = 1..3$.

When all the locations and intensity values of the interpolated points are calculated, we proceed to GLBP code evaluation. First, the intensity values of the points are compared. The intensity value of every point is compared to the intensity value of the nearest neighbor on an inner circle. The intensity value of a point of the circle with radius r_1 is compared to the intensity value of the center pixel. Some points on the circle with radius r_3 are compared to the average of the intensity values of their two nearest neighbors. Figure 3 shows circle points comparison pattern: the arrows show points to compare.

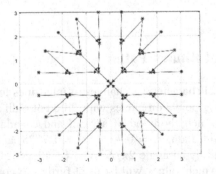

Fig. 3. Intensity values comparison pattern.

Second, we form the matrix $\mathbf{B}^{6\times6}$ that contains comparison results of intensity values. Matrix element is assigned one if the corresponding point intensity value is bigger than the intensity value to which it is compared, otherwise it is assigned zero.

Finally, using these bit values, we calculate 8 6-bit sequences according to the scheme on Fig. 4 and form a 8×6 bit matrix, which is called the GLBP code.

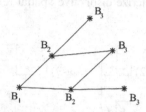

Fig. 4. Bit sequence generating scheme.

5   Local Derivative Pattern

LBP code is a result of applying the first-order derivative of the neighborhood of the center pixel. For a more informative description of the image the operator LDP was proposed in [13]. This operator allows encoding the information obtained as high-order derivatives calculation.

The first-order derivatives along $0°, 45°, 90°$ and $135°$ directions are denoted as $f'_\alpha(x_0, y_0)$, where $\alpha = 0°, 45°, 90°, 135°$. The first-order derivatives at (x_0, y_0) position can be written as follows:

$$f'_0(x_0, y_0) = f_0 - f_4, \quad f'_{45}(x_0, y_0) = f_0 - f_3,$$
$$f'_{90}(x_0, y_0) = f_0 - f_2, \quad f'_{135}(x_0, y_0) = f_0 - f_1. \tag{4}$$

The second-order directional LDP along the α direction at (x_0, y_0) is defined as follows:

$$LDP^2_\alpha(x_0, y_0) = I_2\left(f'_\alpha(x_0, y_0), f'_\alpha(x_1, y_1)\right),$$
$$I_2\left(f'_\alpha(x_0, y_0), f'_\alpha(x_2, y_2)\right),$$
$$\ldots \tag{5}$$
$$I_2\left(f'_\alpha(x_0, y_0), f'_\alpha(x_8, y_8)\right),$$

where I_2 is a predicate defined as follows:

$$I_2(m, n) = \begin{cases} 0, & m \cdot n \geq 0 \\ 1, & m \cdot n < 0 \end{cases}. \tag{6}$$

Finally, the second-order LDP is defined as the concatenation of the four 8-bit directional LDPs:

$$LDP^2(x,y) = \{LDP_\alpha^2(x,y) | \alpha = 0°, 45°, 90°, 135°\}. \tag{7}$$

As it can be seen from the equations above, LDP operator assigns a code to each pixel of the image by comparing two derivatives along the same direction for two adjacent pixels. Then obtained codes are concatenated into a single sequence of 32 bits. Derivative equations along α direction are performed using 16 primitive patterns (Fig. 5). These patterns characterize distinctive spatial relationships in the local area.

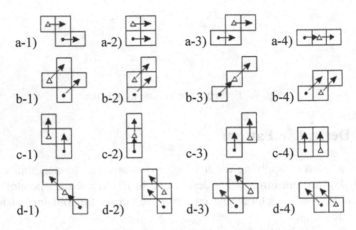

Fig. 5. Primitive patterns used in LDP.

Each of the 16 primitive patterns in Fig. 5 belongs to one of the two groups: a 3-point pattern or a 4-point pattern. A 3-point pattern is assigned to the value '0', if it increases or decreases monotonically (Fig. 5c). In other case, a pattern is assigned to the value '1' (Fig. 5a). Similarly, 4-point pattern is assigned to the value '1' if there is a change of the derivative sign (Fig. 5b), and '0' if it increases or decreases monotonically (Fig. 5d). Thus, the binary code of the image area is formed using the local transitions based on the primitive patterns.

6   Experiments

To carry out research we used a standard PC (Intel Core i5-3470 3.2 GHz, 8 GB RAM). Existing copy-move datasets do not contain all the types of cope-move blocks transformations to do research. This is why we selected 10 grayscale images with size 512×512 and generated forgeries using a self-developed automatic copy-move generation procedure. There were used five types of transforms to create copy-move regions: linear contrast enhancement, JPEG compression, additive noise, rotation and scale.

Four sets of test images were constructed with different copy moved region's size, $48 \times 48, 64 \times 64, 96 \times 96$, and 128×128. In each set, the images were forged by copying a square region at a random location and pasting it to a random non-overlapping position in the same image. There were created 250 forgeries in a set: 25 images for each of the 10 initial images (5 images for each transform type), which were further analyzed using the proposed detection algorithm. Examples of generated images are presented in Fig. 6.

Fig. 6. Copy-move forgeries examples (from left to right, row-wise): contrast enhancement (baboon eyes), additive noise, rotation (small pepper to the left) and scale (waterfall right part) transforms of duplicates

To evaluate the performance of the proposed copy-move forgery detection method *F1_Score (F1)* measure is used. It is calculated based on *Precision (p)* and *Recall (r)* values. They are calculated using true positive (TP), false positive (FP) and false negative (FN) values. In the sense of copy-move detection task TP means the number of correctly detected duplicates, FP – the number of duplicates that have been erroneously detected as copy-move and FN – the falsely missed duplicates. Therefore, p, r and $F1$ values are defined as follows:

$$p = \frac{TP}{TP+FP}, r = \frac{TP}{TP+FN}, F1 = \frac{2 \cdot p \cdot r}{p+r}.$$

Further we compare the proposed solution based on LBP features with the leading dense-field approaches, provided in [6]. It should be mentioned that the proposed LBP-based features use lead to significantly less computational complexity than many other features (PCA, DCT, Zernike, etc.).

Table 1. F1 detection quality measure values for linear contrast enhancement of duplicates.

y = a * x + b	LBP	Uniform LBP	BGC_1	GLBP	LDP
a = 0.9; b = −25	0.84	0.82	0.85	0.91	0.90
a = 0.7; b = −20	0.84	0.75	0.86	0.91	0.89
a = 0.5; b = −15	0.83	0.75	0.85	0.92	0.88
a = 0.3; b = −10	0.51	0.45	0.59	0.90	0.89
a = 0.2; b = −5	0.06	0.04	0.29	0.89	0.88
a = 0.2; b = −5	0.40	0.41	0.46	0.90	0.88
a = 0.3; b = 10	0.74	0.65	0.78	0.91	0.90
a = 0.5; b = 15	0.91	0.87	0.88	0.93	0.91
a = 0.7; b = 20	0.90	0.88	0.89	0.92	0.91
a = 0.9; b = 25	0.92	0.90	0.88	0.93	0.91

Table 2. F1 detection quality measure values for JPEG compression of duplicates.

Q	LBP	Uniform LBP	BGC_1	GLBP	LDP	KPCA [6]	DCT [6]
95	0.85	0.84	0.84	0.93	0.83	0.95	0.88
90	0.78	0.64	0.69	0.90	0.81	0.89	0.84
85	0.56	0.00	0.36	0.91	0.76	0.83	0.78
80	0.00	0.00	0.03	0.88	0.70	0.82	0.80
75	0.00	0.04	0.03	0.85	0.60	0.71	0.73
70	0.00	0.00	0.03	0.85	0.54	0.71	0.73
65	0.00	0.00	0.03	0.82	0.46	0.66	0.66
60	0.00	0.00	0.03	0.82	0.30	0.65	0.66
55	0.00	0.00	0.03	0.79	0.21	0.63	0.65
50	0.00	0.00	0.03	0.82	0.10	0.64	0.63

Table 3. F1 detection quality measure values for duplicates with additive Gaussian noise.

Noise STD	LBP	Uniform LBP	BGC_1	GLBP	LDP	DCT [6]	PCA [6]
0.02	0.28	0.31	0.58	0.52	0.48	0.63	0.62
0.04	0.00	0.05	0.03	0.19	0.20	0.55	0.48
0.06	0.00	0.00	0.06	0.00	0.10	0.50	0.34
0.08	0.00	0.03	0.02	0.00	0.05	0.40	0.26
0.10	0.00	0.04	0.04	0.02	0.00	0.30	0.21

Experimental results presented in Tables 1, 2, 3, 4 and 5 show F1 values for different transforms (distortions) of duplicates embedded in an image. Conducted research shows that LBP-based features (GLBP, LDP and BGC) can be effectively used for copy-move detection on a wide range of distortion parameters. It can be clearly seen that the proposed features show better results in comparison with the best solutions from the survey [6]. However, for extreme values of transformations parameters LBP-based features show low quality measure values and more research needed in this field.

Table 4. F1 detection quality measure values for rotation transform of duplicates.

Angle	LBP	Uniform LBP	BGC_1	GLBP	LDP	Zernike [6]	Bravo [6]
1	0.97	0.90	0.99	0.94	0.48	0.89	0.83
2	0.95	0.87	0.97	0.91	0.20	0.72	0.45
5	0.81	0.65	0.88	0.75		0.72	0.42
7	0.71	0.62	0.74	0.58	0.10	0.68	0.38
10	0.49	0.45	0.54	0.45		0.65	0.36
12	0.43	0.31	0.36	0.42	0.05	0.60	0.36
15	0.32	0.31	0.29	0.42	0.00	0.45	0.35

Table 5. F1 detection quality measure values for scaling transform of duplicates.

Scaling factor	LBP	Uniform LBP	BGC_1	GLBP	LDP	Zernike [6]	KPCA [6]
0.75	0.00	0.00	0.03	0.60	0.54	0.29	0.30
0.8	0.00	0.00	0.20	0.66	0.60	0.42	0.36
0.85	0.29	0.03	0.22	0.63	0.67	0.60	0.58
0.9	0.57	0.07	0.38	0.84	0.82	0.74	0.70
0.95	0.61	0.39	0.51	0.86	0.88	0.77	0.74
1.05	0.52	0.25	0.47	0.67	0.82	0.75	0.77
1.1	0.55	0.27	0.48	0.78	0.79	0.70	0.75
1.15	0.37	0.01	0.28	0.61	0.71	0.55	0.56
1.2	0.23	0.03	0.23	0.45	0.53	0.30	0.35
1.25	0.04	0.02	0.03	0.38	0.46	0.20	0.24

7 Conclusion

In this paper, we proposed a copy-move detection algorithm based on the different forms of LBP. The developed solution showed considerable increase in Precision and Recall values in comparison with state-of-the-art features [6] for wide ranges of transform parameters. We could gain better performance for duplicates transformed with JPEG compression, rotation and scale. Further we will carry out research in developing an algorithm using different LBP-based features fusion techniques to increase performance results, because some experiments showed better performance for BGC features for certain transform parameters ranges.

Acknowledgements. This work was supported by the Federal Agency of Scientific Organization (Agreement 007-3/43363/26) in parts "Copy-move detection scheme" and "Binary Gradient Contours" and by the Russian Foundation for Basic Research (no. 19-07-00138) in parts "Geometric Local Binary Pattern", "Local Derivative Pattern" and "Experiments".

References

1. https://www.brandwatch.com/blog/47-facebook-statistics-2016/
2. https://www.brandwatch.com/blog/37-instagram-stats-2016/
3. Bayram, S., Sencar, H.T., Memon, N.: An efficient and robust method for detecting copy-move forgery. In: Proceedings of the IEEE International Conference on Acoustics, Speech and Signal Processing, pp. 1053–1056 (2009)
4. Bravo-Solorio, S., Nandi, A.K.: Exposing postprocessed copy–paste forgeries through transform-invariant features. IEEE Trans. Inf. Forensics Secur. **7**, 1018–1028 (2012)
5. Cao, Y., Gao, T., Fan, L., Yang, Q.: A robust detection algorithm for copy-move forgery in digital images. Forensic Sci. Int. **214**, 33–43 (2012)
6. Christlein, V., Riess, C., Jordan, J., Angelopoulou, E.: An evaluation of popular copy-move forgery detection approaches. IEEE Trans. Inf. Forensics Secur. **7**, 1841–1854 (2012)
7. Huang, Y., Lu, W., Sun, W., Long, D.: Improved DCT-based detection of copy-move forgery in images. Forensic Sci. Int. **206**, 178–184 (2011)
8. Mahdian, B., Saic, S.: Detection of copy–move forgery using a method based on blur moment invariants. Forensic Sci. Int. **171**, 180–189 (2007)
9. Muhammad, G., Hussain, M., Bebis, G.: Passive copy move image forgery detection using undecimated dyadic wavelet transform. Digit. Invest. **9**, 49–57 (2012)
10. Kang, X., Wei, S.: Identifying tampered regions using singular value decomposition in digital image forensics. In: Proceedings of the 2008 IEEE International Conference on Computer Science and Software Engineering, pp. 926–930 (2008)
11. Fernández, A., Álvarez, M.X., Bianconi, F.: Image classification with binary gradient contours. Opt. Lasers Eng. **49**(9–10), 1177–1184 (2011)
12. Orjuela Vargas, S.A., Yañez Puentes, J.P., Philips, W.: The geometric local textural patterns (GLTP). In: Brahnam, S., Jain, L.C., Nanni, L., Lumini, A. (eds.) Local Binary Patterns: New Variants and Applications. Studies in Computational Intelligence, vol. 506, pp. 85–112. Springer, Heidelberg (2014). https://doi.org/10.1007/978-3-642-39289-4_4
13. Zhang, B., Gao, Y., Zhao, S., Liu, J.: Local derivative pattern versus local binary pattern: face recognition with high-order local pattern descriptor. IEEE Trans. Image Process. **19**(2), 533–544 (2010)

Emotion Recognition of a Group of People in Video Analytics Using Deep Off-the-Shelf Image Embeddings

Alexander V. Tarasov[✉] and Andrey V. Savchenko[iD]

National Research University Higher School of Economics, Nizhny Novgorod, Russia
alxndrtarasov@gmail.com, avsavchenko@hse.ru

Abstract. In this paper we address the group-level emotion classification problem in video analytic systems. We propose to apply the MTCNN face detector to obtain facial regions on each video frame. Next, off-the-shelf image features are extracted from each located face using preliminary trained convolutional neural networks. The features of the whole frame are computed as a mean average of image embeddings of individual faces. The resulted frame features are recognized with an ensemble of state-of-the-art classifiers computed as a weighted sum of their outputs. Experimental results with EmotiW 2017 dataset demonstrate that the proposed approach is 2–20% more accurate when compared to the conventional group-level emotion classifiers.

Keywords: Group emotion recognition · Video-analytic system Convolutional neural network · Face detection

1 Introduction

Emotion recognition can potentially become a foundation for a lot of business applications in such video-analytic fields as smart advertising, education, surveillance, human computer interaction, etc. [1]. As video analytic systems constantly observe multiple persons, group-level analysis [2] raises another issue of decision-making. There exist several known solutions, such as temporal multimodal fusion [3], combining recurrent neural network and 3D convolutional neural networks (CNNs) [4], Bayesian classifiers [5], transfer learning [6,7], were used. The challenges of contemporary video emotion classification methods usually include requirements of real-time processing and low quality of video content provided by typical cameras. Therefore, the goal of our research is to provide an algorithm which is able to process group-level emotion recognition in videos recorded in low resolution, containing several faces. To address the issue of low video quality we locate the facial regions on the frames using Multi-task Cascaded Convolutional Networks (MTCNN) face detector [8]. We examine two simple CNN architectures pre-trained for emotion recognition in low resolution

© Springer Nature Switzerland AG 2018
W. M. P. van der Aalst et al. (Eds.): AIST 2018, LNCS 11179, pp. 191–198, 2018.
https://doi.org/10.1007/978-3-030-11027-7_19

facial photos in order to extract feature vectors of detected faces. The facial feature vectors are aggregated and recognized using a weighted sum of outputs of several classifiers.

The rest of the paper is organized as follows. In Sect. 2 we introduce the proposed algorithm. Experimental results are presented in Sect. 3 group-level emotion recognition sub-challenge of EmotiW 2017 (Emotion Recognition in the Wild) [2]. Concluding comments are given in Sect. 4.

2   Materials and Methods

In this paper we focus on prediction of the average mood of a group of people given an input sequence of video frames $\{X(t)\}, t = 1, 2, ..., T$ [9]. Here t is a number of frame and T is the total number of frames in an observed video. For simplicity, it is assumed the video clip can be characterized with only one emotion. We consider the classification task, in which the video is associated with one of $C > 1$ emotional categories. The classes are defined with the aid of $N \geq C$ reference group-level photos, and the class label $c_n \in \{1, ..., C\}$ of each n-th reference image is known.

Let us assume that there is no video data in a training set. Hence, instead of considering a temporal coherence of sequential frames, we predict emotions of each frame independently. If the size N of the training set is not very large, a deep CNN cannot be trained from scratch. As the most valuable information about emotion is contained in facial expressions of observed persons, we detect all the faces on each frame. The resolution of videos is rather low, and each face in a group photo can be very small. Hence, we implement the CNN-based method, e.g., the MTCNN [8] or the Tiny face detector [10] instead of traditional Viola-Jones cascade classifier of the Histogram of Oriented Gradients (HOG) detector.

Next, the features of each detected face should be computed. Nowadays it is typical to use domain adaptation techniques with a deep CNN, which was pre-trained for a similar task. Many existing emotion recognition methods [6,7] transfer the knowledge from face identification task and use such pre-trained facial descriptors as VGG-Face, VGGFace2, ResFace, CenterFace, etc. However, such descriptors require large resolution of an input facial image (usually, 224×224) and are not so accurate for tiny facial regions. Moreover, the inference time of large CNN is also too large to implement the video frame processing in real-time. Thus, we decided to use an existing small CNN trained to recognize emotion of low resolution facial photos. In particular, we used FER-2013 dataset with 48×48 images. The facial features are computed as the L2-normed outputs of the last but one ("embedding") layer. The whole frame is associated with the mean of individual facial embeddings. The same feature extraction procedure is repeated for all N reference images from the training set. Next, we use the set of the feature vectors to train $K \geq 1$ traditional classifiers, e.g., SVM, Random Forest, Gradient boosting, etc. A final decision for the resulted feature vector of the input frame is made as a weighted sum of the outputs (scores or posterior probabilities) of these classifiers, in which the weights w_k of the k-th classifier are chosen proportionally to its validation accuracy.

The complete proposed procedure is presented in Algorithm 1. It was implemented in a stand-alone Python application using OpenCV, scikit-learn libraries with Keras and TensorFlow frameworks. This application allows dynamical changing the ensemble composition and method of final labeling of single frames. Also, it is quite easy to add new pre-trained classifiers into set by providing the application with serialized objects of classifiers. Our software predicts one of $C = 3$ emotion labels (Positive, Neutral or Negative) learned on the EmotiW 2017 training set [2].

Algorithm 1. Proposed procedure of video-based group emotion recognition

---

Require: N training images of group-level emotions, observed video frames
  $\{X(t)\}, t = 1, 2, ..., T$
Ensure: Emotion label of the input video
1: Train a CNN to classify emotions using an external dataset, i.e., FER-2013
2: **for** each training instance $n \in \{1, ..., N\}$ **do**
3:     Detect facial regions, extract embeddings using pre-trained CNN and obtain an average feature vector $\mathbf{x}_n$
4: **end for**
5: Learn K classifiers using set of feature vectors $\{\mathbf{x}_n\}$, estimate their validation accuracy and normalize all K values in order to obtain weights w_k
6: Initialize the vector of final scores $s_c := 0, c = 1, 2, ..., C$
7: **for** each frame $t \in \{1, ..., T\}$ **do**
8:     Obtain $R \geq 0$ facial regions using, e.g., MTCNN face detector
9:     **for** each class facial region $r \in \{1, ..., R\}$ **do**
10:         Extract embeddings $\mathbf{x}_r(t)$ using pre-trained CNN
11:     **end for**
12:     Compute the frame feature vector $\mathbf{x}(t)$ as an average of embeddings $\mathbf{x}_r(t)$ and normalize it
13:     Initialize the vector of scores $s_c(t) := 0, c = 1, 2, ..., C$
14:     **for** each classifier in an ensemble $k \in \{1, ..., K\}$ **do**
15:         Predict emotion label c^* for an input vector $\mathbf{x}(t)$ using classifier k
16:         Assign $s_{c^*}(t) := s_{c^*}(t) + w_k$
17:     **end for**
18:     **if** $R > 0$ **then**
19:         Assign $s_c := s_c + s_c(t)/R$ for each $c = 1, 2, ..., C$
20:     **end if**
21: **end for**
22: **return** class label with the maximal score s_c

In Fig. 1 we present a brief example of group-level emotion recognition. Here the most accurate (70.9% validation accuracy) Random Forest classifier returns a wrong Neutral emotion for video clip labeled as positive. According to our Algorithm 1, the neutral label gets a score of 0.709. However, when Linear SVM (70.3% validation accuracy) and RBF (Radial Basis Function) SVM (69.3% accuracy) are included into an ensemble, we are able to obtain the correct category, since the positive label gets score 1.396 (0.703 + 0.693) because it is chosen by both SVMs. Hence, our weighted ensemble makes it possible to correct a wrong decision of the most accurate classifiers.

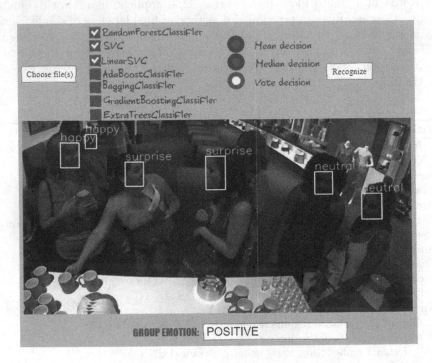

Fig. 1. An example of group-level emotion recognition in the developed application.

3   Experimental Results

As there is no publicly available dataset of videos of a group of people, we provided here an experimental study for a group-level emotion recognition in single images. To validate the accuracy of group emotion recognition during the experimental study we used EmotiW 2017 dataset [2], in which the training/validation sets have been taken from the Group Affect Database 2.0. The training set consists of 3630 photos for $C = 3$ emotion classes and the validation set includes 2065 photos.

We were able to locate 19291 faces on the images from both sets using HOG descriptor from DLIB library. As there were many small unrecognized faces we used MTCNN Face Detector [8]. Though the false positive rate of both detectors was identical, MTCNN Face detector located 2.8-times more (54053) faces. We examined two known CNNs for feature extraction trained to classify 7 emotions (Angry, Disgust, Fear, Happy, Sad, Surprise, Neutral) from the FER-2013 dataset. The first one (hereinafter "CNN-5") includes 5 convolution layers, Global Average Pooling, PReLU activation layer and 3 sequential fully connected layers [11]. The second one is a fully-convolutional CNN architecture (hereinafter "FCN") composed of 10 convolution layers, ReLUs, BatchNorm and Global Average Pooling[1]. The accuracies on the FER-2013 validation subset are equal to 68% and 66% for CNN-5 and FCN, respectively. Selected architectures show comparatively high accuracy among CNN with low computational difficulty [11].

In order to compute final feature vectors, we aggregated either posterior probabilities of 7 basic emotion classes at the softmax layer or the embeddings computed at the last but one layer of both CNNs. We trained a set of conventional classifiers implemented in the scikit-learn library, namely, AdaBoost, RBF SVM, Linear SVM, Bagging with decision trees base classifier, Random Forest, Gradient Boosting and Extra Trees. We used the grid search to tune their parameters and PCA (Principal Component Analysis) to transform group feature vectors into a set of principal components. For example, the usage of 80 out of 112 embeddings from FCN made it possible to increase accuracy by approximately 2%.

Different combinations of classifiers were combined into ensembles. We tested three following ways to aggregate individual classifiers decisions, namely, mean, median and our weighted voting. The weighted voting achieves 1–4% higher accuracy when compared to ensembles based on mean/median decisions. It was experimentally noticed that ensemble of Random Forest, RBF SVM and Linear SVM shows the highest accuracy in comparison with other combinations.

Validation accuracies of individual classifiers and the proposed approach (Algorithm 1) with weighted outputs of Random Forest, RBF SVM and Linear SVM are presented in Table 1. The usage of CNN-5 [11] leads to 10–13% lower accuracy when compared to FCN. The features from the softmax layer also performs 1–2% worse than embedding one. Confusion matrix of our approach with the FCN's embeddings (Fig. 2) shows that our algorithm performs better on Positive images, due to the large number of positive images in a training set.

As we mentioned in introduction, another important issue is the recognition time. The average inference time running on Intel Core i5-5200U CPU, 64-bit machine with NVIDIA GeForce 940M is equal to 32 ms and 25 ms for CNN-5 and FCN, respectively. Our ensemble performs classification in approximately 11 ms, which is 8 ms higher when compared to the fastest individual classifier (Random Forest). The most time-consuming part of our algorithm is face detection. The MTCNN [8] runs at 8–12 FPS on the high-resolution images from the EmotiW dataset, while the Tiny face detector [10] requires several seconds to process one group photo. Also, we do not use any techniques for obtaining global context features [12], since it is time-consuming operation and video-analytic cameras set in one place shoot a video with stable global context almost all the time.

---

[1] https://github.com/LamUong/FacialExpressionRecognition

Table 1. Accuracy (%) on EmotiW 2017 validation set.

Classifiers	CNN-5		FCN	
	Softmax	Embeddings	Softmax	Embeddings
RBF SVM	57.0	60.5	68.5	69.3
Gradient Boosting	61.3	61.9	67.2	70.1
Extra Trees	59.1	61.9	69.6	70.2
Linear SVM	59.6	61.5	69.8	70.3
Bagging	59.0	61.5	68.8	70.7
Random Forest	58.0	59.9	69.0	70.9
AdaBoost	57.4	58.1	69.5	70.9
Proposed approach (all classifiers)	61.5	62.2	70.7	72.3
Proposed approach (Random Forest, RBF SVM and Linear SVM)	61.9	64.2	73.2	75.5

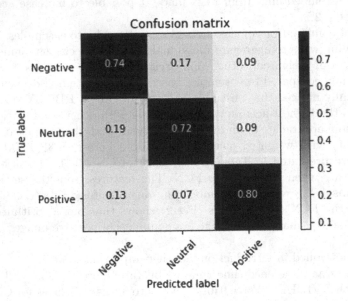

Fig. 2. Confusion matrix of the proposed approach, EmotiW 2017 validation set.

4   Conclusion

In this paper we propose the novel algorithm for the group-level emotion recognition in videos (Algorithm 1). We implemented a special software application that can be useful for emotion recognition in video-analytic and advertising systems. Source code of our solution is made publicly available[2].

---

[2] https://github.com/alxndrtarasov/GrEmRec

Our algorithm outperforms the known methods, namely, the baseline with SVM for CENTRIST descriptors [2], fine-tuning of VGG-16 for the whole photos without face detection and the deep CNNs with Bayesian classifiers [5] by 23%, 11% and 8%, respectively. Our top accuracy (75.5%) is approximately 9% higher when compared to our previous implementation of the algorithm [7] that includes the Tiny Face Detector, but extracts features with the VGGFace CNN. Moreover, we increase the validation accuracy by 0.2% over another known method [7] with a fusion of 4 Random Forests classifiers for various transformations of the VGGFace embeddings, VGG-16 descriptor and specially designed landmark features. What is more important this heavy approach is approximately 20-times slower than the proposed Algorithm 1.

The EmotiW 2017 winner [12] obtained 80.05% validation accuracy, which is 4.5% higher than our best result (Table 1). Their final solution contains an ensemble of 7 models [12] trained on the scene, faces and skeletons separately. Very deep ResNet-152, VGG-FACE and VGG-16 CNNs are used to extract features from skeletons, faces and scene, respectively. As a result, in contrast to the proposed Algorithm 1, such architecture is unsuitable for real-time video analytic systems because of its extremely high run-time complexity.

In future research, we are planning to test our proposed approach for the EmotiW 2018 dataset, which is 3-times larger when compared to the EmotiW 2017 training set. In addition, it would be important to extend the developed algorithm for use in a case where people experience different emotions simultaneously and when emotions change over time. Finally, it is important to analyze the whole image if the faces were not detected, when appropriate global descriptor of the whole scene should be used [7]. In this case it will be important to compress the large CNNs [13] in order to make it suitable for embedded systems.

Acknowledgements. The article was prepared within the framework of the Academic Fund Program at the National Research University Higher School of Economics (HSE) in 2017 (grant №17-05-0007) and by the Russian Academic Excellence Project "5-100".

References

1. Krakovsky, M.: Artificial (emotional) intelligence. Commun. ACM **61**(4), 18–19 (2018)
2. Dhall, A., Goecke, R., Ghosh, S., Joshi, J., Hoey, J., Gedeon, T.: From individual to group-level emotion recognition: EmotiW 5.0. In: 19th ACM International Conference on Multimodal Interaction (ICMI), pp. 524–528. ACM (2017)
3. Vielzeuf, V., Pateux, S., Jurie, F.: Temporal multimodal fusion for video emotion classification in the wild. In: 19th ACM International Conference on Multimodal Interaction (ICMI), pp. 569–576. ACM (2017)
4. Fan, Y., Lu, X., Li, D., Liu Y.: Video-based emotion recognition using CNN-RNN and C3D hybrid networks. In: 18th ACM International Conference on Multimodal Interaction (ICMI), pp. 445–450. ACM (2016)

5. Surace, L., Patacchiola, M., Sönmez, E.B., Spataro, W., Cangelosi, A.: Emotion recognition in the wild using deep neural networks and Bayesian classifiers. In: 19th ACM International Conference on Multimodal Interaction (ICMI), pp. 593–597. ACM (2017)
6. Kaya, H., Gürpınar, F., Salah, A.A.: Video-based emotion recognition in the wild using deep transfer learning and score fusion. Image Vis. Comput. **65**, 66–75 (2017)
7. Rassadin, A., Gruzdev, A., Savchenko, A.: Group-level emotion recognition using transfer learning from face identification. In: 19th ACM International Conference on Multimodal Interaction (ICMI), pp. 544–548. ACM (2017)
8. Zhang, K., Zhang, Z., Li, Z., Qiao, Y.: Joint face detection and alignment using multitask cascaded convolutional networks. IEEE Signal Process. Lett. **23**(10), 1499–1503 (2016)
9. Savchenko, A.V., Belova, N.S., Savchenko, L.V.: Fuzzy analysis and deep convolution neural networks in still-to-video recognition. Opt. Mem. Neural Netw. (Inf. Opt.) **27**(1), 23–31 (2018)
10. Hu, P., Ramanan, D.: Finding tiny faces. In: IEEE Conference on Computer Vision and Pattern Recognition (CVPR), pp. 1522–1530. IEEE (2017)
11. Arriaga, O., Valdenegro-Toro, M., Plöger, P.: Real-time convolutional neural networks for emotion and gender classification. arXiv preprint arXiv:1710.07557 (2017)
12. Guo, X., Polania, L., Barner, K.: Group-level emotion recognition using deep models on image scene, faces, and skeletons. In: 19th ACM International Conference on Multimodal Interaction (ICMI), pp. 603–608. ACM (2017)
13. Rassadin, A.G., Savchenko, A.V.: Compressing deep convolutional neural networks in visual emotion recognition. In: Proceedings of the International Conference on Information Technology and Nanotechnology (ITNT). Session Image Processing, Geoinformation Technology and Information Security Image Processing (IPGTIS), CEUR-WS, vol. 1901, pp. 207–213 (2017)

General Topics of Data Analysis

Extraction of Visual Features
for Recommendation of Products
via Deep Learning

Elena Andreeva[1,3](✉), Dmitry I. Ignatov[1], Artem Grachev[1],
and Andrey V. Savchenko[2]

[1] National Research University Higher School of Economics, Moscow, Russia
maerville@yandex.ru
[2] National Research University Higher School of Economics,
Nizhniy Novgorod, Russia
[3] ex Mail.Ru Group, Moscow, Russia
http://www.hse.ru

Abstract. In this paper (The first author is the 1st place winner of
the Open HSE Student Research Paper Competition (NIRS) in 2017,
Computer Science nomination, with the topic "Extraction of Visual Features for Recommendation of Products", as alumni of 2017 "Data Science" master program at Computer Science Faculty, HSE, Moscow), we
describe a special recommender approach based on features extracted
from the clothes' images. The method of feature extraction relies on pre-
trained deep neural network that follows transfer learning on the dataset.
Recommendations are generated by the neural network as well. All the
experiments are based on the items of category Clothing, Shoes and Jewelry from Amazon product dataset. It is demonstrated that the proposed
approach outperforms the baseline collaborative filtering method.

Keywords: Visual feature · Recommender system · Fine tuning
Collaborative filtering

1 Introduction

Conventional content-based recommender systems use only a textual information
about the product, while in terms of clothes it is not enough. Text categories and
annotations cannot fully describe the appearance of the clothes, and these categories and annotations are difficult to get as well. That is why visual features can
be helpful. Usually, a user browses through several visually similar items before
choosing the one to buy. By visually similar we mean clothes with same colors or
forms or pattern for example. As we said before, text-based systems cannot fully
estimate those parameters (how many different patterns can one manually define
in a store?), while collaborative filtering methods does not pay any attention to
product's appearance at all, as they rely only on click or view activity.

The trend of using neural networks as part of the recommender system is
gaining in popularity [2,5,6].

© Springer Nature Switzerland AG 2018
W. M. P. van der Aalst et al. (Eds.): AIST 2018, LNCS 11179, pp. 201–210, 2018.
https://doi.org/10.1007/978-3-030-11027-7_20

In this paper, we address two problems. The first one deals with extraction of visual features to find clothes that are visually similar for a product (non-personalized recommender). The second problem is to make a personalized recommender system based on visual features as well.

The paper is organized as follows. Section 2 briefly overviews several related papers. Section 3 describes the introduced solutions based on pre-trained neural networks and fine-tuning schemes for product recommendation taking into account the dataset peculiarities. Section 4 contains short dataset description and the obtained results for seven compared models. Section 5 specially pays attention to visual similarities and the perceived quality of the models' outputs. Section 6 concludes the paper.

2   Related Work

A considerable number of papers about extracting and usage of visual features appeared in last few years [7,11]. Not all of these papers refer to a clothes images and products, but they propose similar methods based on a transfer learning. The main differences are in the architecture of a pre-trained net, learning method, and training data. In Pinterest [12], different pre-trained architectures were used, such as VGG16 [10], ResNet101, ResNet152 [4], etc. As a fine-tuning method classification-based scheme was chosen; text search queries that match images' annotations were used as targets to classify images. In [9], the architecture named VisNet was proposed, which consists of parallel Shallow Nets and deep VGG16 net. The net was fine-tuned like a siamese net, taking triplets $\langle q, p, n \rangle$ as an input, where q is a query image, p is a similar image and n is a negative example.

In [1], the dataset contained users' marks of images similarity (links from scene photo to products' stand-alone images). Thus, collecting products images from scenes required crowdsourcing.

3   Methods

3.1   Extraction of Visual Features

In our work, we use an architecture inspired by VisNet. In fact, we take a pre-trained ResNet101 and add a parallel shallow net, which consists of only two layers: convolution and maxpooling, as shown in Fig. 1. As a method for fine-tuning a classification-based scheme is used. The Amazon product data is chosen as a dataset for experiments[1]. One of the main reasons for choosing this dataset is a presence of users' behavior data (information about users purchases). In the Amazon product dataset [8], there are no any "similar" marks as in [1], or user annotations and text queries as in [12]. But still we do have some metadata on products, such as title, brand, and categories. We have decided to use the words as targets, supposing that visually similar items can be described by similar (or the same) words. So, we have collected the words from categories, brands,

---

[1] http://jmcauley.ucsd.edu/data/amazon/.

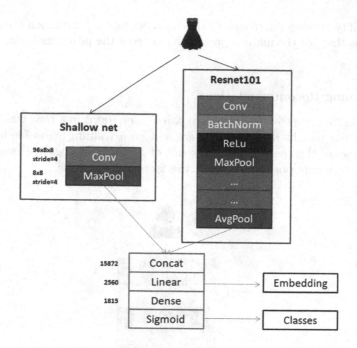

Fig. 1. The architecture of the deep convolutional net

and titles of products and counted them. All the words are normalized, stop-words are removed. Some words are considered as synonyms, so their counts are summed. We keep top-1815 words as targets for classification (it is guaranteed, that every word is related to at least 100 products in the dataset). Thus, the task becomes a multi-label classification problem, but slightly modified. In order to make the Net to pay attention to all the existing classes, despite their frequency, normally the re-balancing is used. However, it is not easy in case of multi-label classification. So the trick with re-weighting of the impact of each class in the loss function is used. The final loss between the target, t_i, and output, o_i is defined as

$$loss(o, t) = -\frac{1}{n} \sum_i w_i(t_i \cdot log(o_i) + (1 - t_i) \cdot log(1 - o_i)), \text{ where}$$

w_i is the weight of i-th class (word) and defined as

$$w_i = (\max_{j \in C}(f_j) + 0.01) - f_i, \text{ where}$$

f_i is a frequency of a i-th word and C is the set of all classes.

A simple dense neural net (see Fig. 2) is used to generate recommendations. As an input to the net we use a so-called "users embedding", which is simply the mean vector of all extracted visual features from the products that a user purchased. The last activation is *softmax*, so as output we get for each product its probability of being purchased by this user. So, we can recommend to a person those items that got the highest probability (except the products he has already bought).

3.2 Forming Recommendations

We train this net by providing purchase pairs $\langle user, item \rangle$ in a row, which means if the user has bought n products, there will be n training pairs for him. One of advantages of this method is an absence of necessity of negative samples, for training we use only positive ones, i.c. the facts of a purchase.

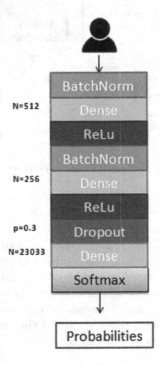

Fig. 2. The architecture of the recommender net (the layers sizes and drop-out rate are given along the left side of the network diagram)

4   Experiments

4.1   Data

All the experiments are carried out on Amazon product data since it is open and large dataset with the information about user's behaviour suitable for making personalized recommendations [8].

The dataset contains reviews on bought products from 24 non-intersected categories, meta-information about products along with their features extracted by VGG16. In addition to the full set of reviews for each category the authors provide their subsets, the so-called "5-cores": it is guaranteed that in such subset each user left no less than five reviews and each product has no less than five reviews as well.

In this paper, we use mainly "5-core" of the category "Clothing, Shoes and Jewelry". For certain experiments extra products have been added from the same category independently of the number of respective reviews.

After removal of duplicate reviews, the final dataset contains 278,677 reviews, 39,387 users, and 23033 products.

Each review is interpreted as a bought item independently of the obtained rating since each user-item interaction is an important signal: the user was interested in the item, saw and bought it.

4.2   Evaluation

In this section, we describe the evaluation of the proposed models against several baselines. The fine-tuned convolutional net indeed extracts good visual features: as shown in Fig. 3, the nearest neighbours (in terms of $L1$ distance) for products are really visually similar. But the main task is generation of personalized recommendations. We have compared the methods in terms of recall@k proposed in [3]. The first model (baseline one) is collaborative filtering, trained via Alternating Least Squares ($rank = 200$) (CF in Fig. 4). The second and the third models are our simple dense neural nets, where features extracted by not fine-tuned ResNet101 from purchased products are used (Net (ResNet) and Net (VGG) respectively in Fig. 4). The fourth and the fifth ones are again simple dense nets, but their user embeddings are based on vectors produced by fine-tuned ResNet (ResNet+FT in Fig. 4); the latter of them is paired with a shallow net in parallel (ResNet+FT+conv in Fig. 4). For benchmarking purposes we have included

Fig. 3. Examples of visually similar products

two models based on logistic regression over visual features of products that a target user bought (positive examples) or ignored (negative examples): models LogRegr (VGG) and LogRegr (ResNet) in Fig. 4.

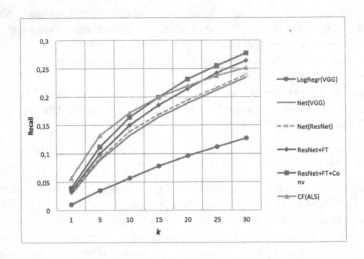

Fig. 4. Recall@k for seven models

As wee see in Fig. 4, fine-tuning improves the result of ResNet. Our final model outperforms a baseline model starting from $k = 15$.

We use 5-fold cross-validation during the experiments taking the 5-core part of the dataset.

Among the two models that use logistic regression, ResNet-based model has a slightly better performance. However, this experiment rather shows that logistic does not perform well in case of absence explicit negative examples (two times worse than collaborative filtering via ALS).

As follows from the results in Table 1, neural network as a classifier is reasonably good to capture the "peculiarities" of each user and the distributions

Table 1. The results of comparison in terms of Recall@30 and AUC

Model	Recall@30	AUC
CF (ALS)	0.251 ± 0.005	5.470 ± 0.042
LogRegr (VGG)	0.1272 ± 0.001	2.210 ± 0.010
LogRegr (ResNet)	0.1273 ± 0.0005	2.344 ± 0.006
Net (VGG)	0.234 ± 0.002	4.524 ± 0.016
Net (ResNet)	0.239 ± 0.001	4.667 ± 0.007
ResNet+FT	0.2642 ± 0.005	5.097 ± 0.034
ResNet+FT+Conv	0.277 ± 0.007	5.508 ± 0.033

of purchase probability: the figures are close to collaborative filtering and two times higher than for logistic regression as a top-layer model (see Net (VGG) and Net (ResNet) in the table). Since ResNet demonstrates better results, it has been selected as a single model for further fine tuning. As one can see, final improvement in recall and AUC is reached by fine-tuned ResNet model with convolutional shallow net in parallel.

5 Results Analysis by Visual Inspection

In this section we delve into short visual inspection of particular recommendations provided by four studied recommenders: the collaborative filtering model, trained via Alternating Least Squares ($rank = 200$); the second and the third ones are based on features extracted with VGG16 and ResNet101 dense neural nets, with a final recommender scheme relying on logistic regression over purchased products in terms of those features (LogRegr (VGG) and LogRegr (ResNet) in Fig. 4); the fourth one is a dense net over user embeddings based on vectors produced by fine-tuned ResNet (FT ResNet+conv in Fig. 4).

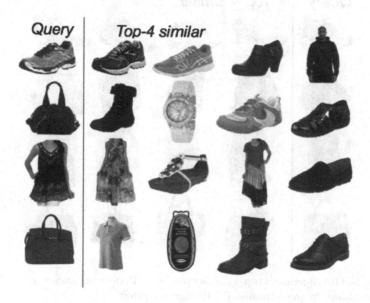

Fig. 5. Collaborative filtering results: queries vs top-4 similar products

As expected, collaborative filtering could only capture visually similar by chance since it is heavily based on users' shopping behaviour (Fig. 5).

In Figs. 6 and 7 one can see examples of products that are similar to each other in the space of visual features learned by each model respectively. It is clear that VGG16 takes into account color and shape while ResNet101 concentrates

Fig. 6. VGG16: queries vs top-4 similar products. (Color figure online)

Fig. 7. ResNet101: queries vs top-4 similar products. Perfect matches w.r.t. to labelling scheme are shown in green frames. (Color figure online)

on more abstract features (similarity by colors have been overlooked due to the excessive net's depth).

Subjectively, ResNet performed better since it does not demonstrates rather rude mismatches like closeness of sport bags and ladies' bags or high similarity of a one color t-shirt to a tunic with a pattern.

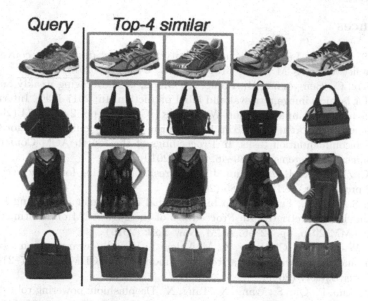

Fig. 8. Trained Resnet101 with a shallow convolutional net in parallel: queries vs top-4 similar products (Color figure online)

Let us have a look at the results of the fourth model in Fig. 8. It is clear that the net shows the best results in terms of visual similarity. All the running shoes have the same pattern though of different color. In the second row, the closest bag is almost a copy of the original bag w.r.t. color, shape and pockets. There are no black bags in the top-4 of the most similar bags to the original red one.

6 Conclusion and Future Work

In this paper, we have proposed a method for extracting visual features of products and an algorithm for making recommendations based on these extracted features. The proposed algorithm outperforms collaborative filtering by recall@k metric with $k \geq 15$. Extracted visual features may also be used for generating a non-personalized recommendations of similar items. In future we plan to collect alternative datasets possibly with assessors' marks and experiment with other methods of fine-tuning such as siamese nets.

Acknowledgements. This study is supported by Russian Federation President grant MD-306.2017.9. Andrey V. Savchenko is supported by the Laboratory of Algorithms and Technologies for Network Analysis, National Research University Higher School of Economics. The work of Dmitry I. Ignatov (contributed to Sects. 1, 2, 3 and 6) was supported by the Russian Science Foundation under grant 17-11-01294 and performed at National Research University Higher School of Economics, Russia. The authors would like to thank Dmitry Soloviev from Mail.ru and Gleb Gusev from Yandex team for their piece of advice and relevant literature on the problem.

References

1. Bell, S., Bala, K.: Learning visual similarity for product design with convolutional neural networks. ACM Trans. Graph. (TOG) **34**(4), 98 (2015)
2. Corbière, C., Ben-younes, H., Ramé, A., Ollion, C.: Leveraging weakly annotated data for fashion image retrieval and label prediction. In: 2017 IEEE International Conference on Computer Vision Workshops (ICCVW), pp. 2268–2274 (2017)
3. Cremonesi, P., Koren, Y., Turrin, R.: Performance of recommender algorithms on top-N recommendation tasks. In Proceedings of the fourth ACM Conference on Recommender Systems, pp. 39–46. ACM (2010)
4. He, K., Zhang, X., Ren, S., Sun, J.: Deep residual learning for image recognition. arXiv preprint arXiv:1512.03385 (2015)
5. Jiang, S., Wu, Y., Fu, Y.: Deep bi-directional cross-triplet embedding for cross-domain clothing retrieval. In: Proceedings of the 2016 ACM on Multimedia Conference, MM 2016, pp. 52–56. ACM, New York (2016)
6. Kang, W.-C., Fang, C., Wang, Z., McAuley, J.: Visually-aware fashion recommendation and design with generative image models. In: ICDM, pp. 207–216. IEEE Computer Society (2017)
7. Liu, Z., Luo, P., Qiu, S., Wang, X., Tang, X.: Deepfashion: powering robust clothes recognition and retrieval with rich annotations. In: CVPR, pp. 1096–1104. IEEE Computer Society (2016)
8. McAuley, J., Targett, C., Shi, Q., Van Den Hengel, A.: Image-based recommendations on styles and substitutes. In: Proceedings of the 38th International ACM SIGIR Conference on Research and Development in Information Retrieval, pp. 43–52. ACM (2015)
9. Shankar, D., Narumanchi, S., Ananya, H.A., Kompalli, P., Chaudhury, K.: Deep learning based large scale visual recommendation and search for e-commerce. arXiv preprint arXiv:1703.02344 (2017)
10. Simonyan, K., Zisserman, A.: Very deep convolutional networks for large-scale image recognition. CoRR abs/1409.1556 (2014)
11. Veit, A., Kovacs, B., Bell, S., McAuley, J., Bala, K., Belongie, S.: Learning visual clothing style with heterogeneous dyadic co-occurrences. In: International Conference on Computer Vision (ICCV), Santiago, Chile (2015)
12. Zhai, A., et al.: Visual discovery at pinterest. In: Proceedings of the 26th International Conference on World Wide Web Companion, pp. 515–524. International World Wide Web Conferences Steering Committee (2017)

Regression Analysis with Cluster Ensemble and Kernel Function

Vladimir Berikov[1,2] and Taisiya Vinogradova[1(✉)]

[1] Novosibirsk State University, Novosibirsk 630090, Russia
t.vinogradova@g.nsu.ru
[2] Sobolev Institute of Mathematics SBRAS, Novosibirsk 630090, Russia
berikov@math.nsc.ru

Abstract. In this paper, we consider semi-supervised regression problem. The proposed method can be divided into two steps. In the first step, a number of variants of clustering partition are obtained with some clustering algorithm working on both labeled and unlabeled data. Weighted co-association matrix is calculated using the results of partitioning. It is known that this matrix satisfies Mercer's condition, so it can be used as a kernel for a kernel-based regression algorithm. In the second step, we use the obtained matrix as kernel to construct the decision function based on labelled data. With the use of probabilistic model, we prove that the probability that the error is significant converges to its minimum possible value as the number of elements in the cluster ensemble tends to infinity. Output of the method applied to a real set of data is compared with the results of popular regression methods that use a standard kernel and have all the data labelled. In noisy conditions the proposed method showed higher quality, compared with support vector regression algorithm with standard kernel.

Keywords: Regression analysis · Cluster analysis
Ensemble clustering · Kernel methods

1 Introduction

Regression task consists in detecting a hidden regularity in a set of data objects expressed in some value which is commonly referred to as a *response variable*.

Consider a general population X of objects where each object has a number of features and a set of responses of objects Y. There is unknown target function $f^* : X \to Y$ that needs to be approximated. Let all objects be described by real-valued features and responses also be real-valued.

In the problem of semi-supervised regression [4], we have an input data set $X_N = \{x_1, \dots, x_N\}$ of objects—a random sample from X. Objects from X_N can be divided into two parts:

- $X_c = \{x_1, \dots, x_k\}$ are objects with known values of Y in the corresponding set $Y_c = \{y_1, \dots, y_k\}$, $y_i = y(x_i)$ is a response variable;

© Springer Nature Switzerland AG 2018
W. M. P. van der Aalst et al. (Eds.): AIST 2018, LNCS 11179, pp. 211–220, 2018.
https://doi.org/10.1007/978-3-030-11027-7_21

– $X_u = \{x_{k+1}, \ldots, x_N\}$ are objects with unknown values of Y.

In this paper we consider that it is required to perform so-called *transductive learning*, i.e., build a decision function $f : X_u \to Y$ optimizing a certain quality criterion.

There are many approaches to solving regression problems. Both parametric and nonparametric approach can be used. Parametric methods such as Generalized Least Squares are based on the assumption that the function has a known form with unknown parameters. Parameters need to be chosen to minimize the standard deviation that is an example of quality criterion [1]. Nonparametric approaches do not impose such restrictions on the target function [2]. Traditional parametric and nonparametric methods usually work well if their underlying assumptions are true, but can give misleading results otherwise. Robust regression methods are designed to be used in case of violations of assumptions by the underlying data-generating process [3].

However, the development of methods that give effective and robust solutions in cases characterized by complex data structures, nonlinearity of dependencies and noisiness, remains an urgent task.

Since we identify an object with a point of a finite-dimensional feature space, some characteristics of real structure of the object set can become rather difficult to distinguish in such a projection of potentially infinite-dimensional space of data properties onto the feature space. Thus the linkage between an object and the corresponding value of a response variable can also be blurred.

Sometimes in that case it is useful to consider the objects from a different point of view and to characterise the object set with another set of features and points in a new, possibly very high dimensional, feature space. Kernel-based methods are able to limit the consideration of this new feature space to computing the dot product of feature vectors in it without even knowing what this space is. By identifying the appropriate kernel, we can restore the regularities of data structure.

Previous kernels are usually able to detect certain data structures and one needs to guess what kernel should be used in a particular case. The more dimensions a feature space has, the harder it is to assess its structure. Algorithms of cluster analysis joint into ensemble in this paper are expected to form the universal proximity metrics that helps one to choose the kernel that corresponds to the structure of the set of data. Moreover, standard regression analysis methods are usually very unstable to noise. Our task is to propose a method that overcomes mentioned difficulties.

2   Methodology

Suppose that the set of objects identified with X is divided into several classes $C_1^X, \ldots, C_p^X$, directly unobservable, and belonging to a certain class C_i^X reflects the value of Y. There can exist latent variables $h_1, \ldots, h_p$ such that if $x_i \in C_k^X$ then $y_i = y(x_i) = h_k + \epsilon$, where ϵ is a random value with mean $E[\epsilon] = 0$

and variance $var[\epsilon] = const$. Forms of the dependence of y_i on x_i can also differ depending on the class of x_i. For example, the condition of the patient's cardiovascular system (sick or healthy) manifests itself through such observable characteristics as upper and lower blood pressure, heart rate. If this assumption is fulfilled, the belonging of objects to one class reflects the proximity of their responses.

Semi-supervised regression using ensemble clustering can be divided into two basic parts. This method first generates different partitions on all data objects using clustering algorithms, and then calculates the averaged co-association matrix using different partitions. In the second step, it uses the averaged co-association matrix so obtained previously as the kernel for the existing kernel based methods such as Support Vector Regression (SVR) and Nearest Neighbor Regression (NNR) for training these algorithms on X_c. Finally, we use this trained machine learning model for regression task on the unmarked data X_u.

2.1 Cluster Ensembles and Co-association Matrices

Ensemble clustering is one of the implementations of the collective methodology in machine learning. There are many ways of combining algorithms into an ensemble [5,6]. One of the approaches is *evidence accumulation* [7]. Following it, the decision is found in three steps. First of all, a number of clustering variants are obtained. For each partition, the co-association boolean $N \times N$ matrix is calculated, where N is number of objects in X. The matrix elements indicate if the pair belongs to the same cluster or not. In the second step, the weighted co-association matrix is found taking into account all the obtained variants of partition; the weights can be equal or reflect some quality and diversity estimates of the partitions. In the final step, the matrix elements are used as new similarity measures between data points to construct the final partition (if clustering task is considered) or to build the regression function (in regression analysis case).

Using a cluster ensemble usually helps to construct robust and effective solution, especially in case of noisy data or uncertainty in data model. Generally, the metrics given by a co-association matrix reflects the data structure better than the metrices used in each individual algorithm.

Consider a set of various clustering algorithms $\mu_1, \ldots, \mu_m, \ldots, \mu_M$ (such as k-means, spectral clustering, hierarchical partitioning, etc.). Let algorithm μ_m work several times with different parameter settings. In each lth run, the algorithm generates a partition composed of $K_{l,m}$ groups, $l = 1, \ldots, L_m$, where L_m is total number of runs for μ_m. To calculate the averaged co-association matrix H, we consider clustering results for all available objects $X_N = \{x_1, \ldots, x_N\}$. Elements of the matrix are:

$$h(i,j) = \sum_{m=1}^{M} \alpha_m \frac{1}{L_m} \sum_{l=1}^{L_m} h_{l,m}(i,j), \tag{1}$$

where $i, j \in \{1, \ldots, N\}$ are the object numbers $(i \neq j)$, $\alpha_m \geq 0$ are given weights, $\sum_{m=1}^{M} \alpha_m = 1$; $h_{l,m}(i,j) = 0$ if the pair (i,j) belongs to different clusters in l-th variant of partition obtained by algorithm μ_m, otherwise it equals one. The weights α_m can be the same or, for example, reflect the correspondence between the partition of objects formed by μ_m and real groups of their responses.

2.2 Combination of Cluster Ensemble and Regression Algorithms

In the field of kernel-based pattern analysis the following theorem is in the basis:

Theorem 1 *(Mercer). A function $K(x, x')$ is a kernel function (i.e., it defines inner product in a certain feature space), iff for any set of points $\{x_i\}_{i=1}^N$ in R^d and real numbers $\{c_i\}_{i=1}^N$ the matrix $\mathbf{K} = (K(i, j)) = (K(x_i, x_j))_{i,j=1}^N$ is nonnegativity definite:* $\sum_{i,j=1}^N c_i c_j K(i, j) \geq 0$.

This property guaranties the validity of a "kernel trick" [8]. It is also known that function H defined in (1) satisfies Mercer's conditions and thus can be used as a kernel [9].

3   Co-associated Algorithms

In this paper, we propose two novel semi-supervised regression algorithms using averaged co-association matrix as the kernel matrix for machine learning algorithms, named CASVR and CANNR. These methods are both using the idea of kernel mentioned above but differ in the regression algorithms used in the second step. In CASVR we use Support Vector Regression algorithm, and in CANNR Nearest Neighbor Regression method is used. Notice, that other kernel methods of regression analysis can also be applied.

The outline of CANNR and CASVR algorithms is summarized as follows.

Algorithm *CANNR* (*Co-associative NNR*):
Input: objects X_c with known responses Y_c and objects X_u; number of clustering algorithms M, number of partition variants L_m for each algorithm $\mu_m, m = 1..., M$
Output: values of responses for objects from X_u.
1. Perform clustering of objects $X_c \cup X_u$ with algorithms $\mu_1, \ldots, \mu_M$, having received L_m variants of partition from each algorithm.
2. Calculate matrix H on $X_c \cup X_u$ under the formula (1).
3. Use *NNR* method: for each unmarked object $x \in X_u$ find a response of the most similar (with respect to H) marked object $x' \in X_c$.

Algorithm *CASVR* (*Co-associative SVR*):
Input: objects X_c with known responses Y_c and objects X_u; number of clustering algorithms M, number of partition variants L_m for each algorithm $\mu_m, m = 1..., M$
Output: responses of objects from X_u.
1. Perform clustering of objects $X_c \cup X_u$ with algorithms $\mu_1, \ldots, \mu_M$, having

received L_m variants of partitioning for each algorithm.
2. Calculate matrix H on $X_c \cup X_u$ under the formula (1).
3. Train SVR on the marked data X_c using matrix H as kernel.
4. With the help of SVR predict responses for objects in X_u.

4   CANNR Convergence

Let us consider the convergence properties of CANNR algorithm within the following probabilistic model:

– For simplicity, let us suppose that $M = 1$, i.e. the cluster ensemble consists of the variants of partitions obtained by single algorithm μ with random parameters $\Omega_1, \ldots, \Omega_L$ considered as i.i.d. values taken from some allowable set of algorithm's parameters.

– For fixed $x_i, x_j \in X_u$ we denote $L_1(i,j) = \sum_{l=1}^{L} h_l(x_i, x_j)$, and $L_0(i,j) = L - L_1(i,j)$. Quantity L_1 equals to the number of variants of clustering in which the algorithm voted to merge the pair x_i, x_j, and L_0 is the number of variants voted against the union. Let $Y(x)$ be directly unobserved value of the response for object $x \in X_u$.

– Following nearest neighbor regression rule, for all $x_i \in X_u$ we assign the decision y_i, where y_i is the response of the nearest to x_i object $x'_j, j \in X_c$ (in the sense of the measure of similarity defined by matrix H): $x'_j = arg \max_{x_{j''} \in X_c} H(x_i, x_{j''})$.

– Suppose that X is divided into several classes $C_1^X, \ldots, C_p^X$, each class C_i^X corresponding to a certain class C_i^Y of responses from Y due to the properties of data distribution.

– Consider a random value

$$z(Y(x_i), Y(x_j)) = \begin{cases} 1, & \text{if } Y(x_i) \text{ and } Y(x_j) \text{ are from the same class,} \\ 0, & \text{otherwise} \end{cases}$$

which corresponds to the true status of the responses of the pair (x_i, x_j).

Under these conditions, the following property is valid:

Theorem 2. *Let $\forall l \in \{1, \ldots, L\}$ and $\forall x_i, x_j \in X_u$ the inequalities be hold:*

$$P[h_l(x_i, x_j) = 1 | z(Y(x_i), Y(x_j)) = 1] > \frac{1}{2},$$

$$P[h_l(x_i, x_j) = 0 | z(Y(x_i), Y(x_j)) = 0] > \frac{1}{2}$$

and $L_0(i,j) = const$. Then for any object $x_i \in X_u$, the probability that CANNR algorithm assigns to x_i the response which does not belong to the class that contains true value of $Y(x_i)$, $P_{er}(x_i) = P[Y(x_i) \in C_j^Y, y' \notin C_j^Y]$, converges to zero as $L \to \infty$.

Proof. Let us fix an object $x \in X_u$. Further, let x' be the nearest neighbor of x and $y' = Y(x')$ be its response. Denote

$$q_1 = P[h_l(x, x') = 1 | z(Y(x), Y(x')) = 1] > \frac{1}{2}$$

$$q_0 = P[h_l(x, x') = 0 | z(Y(x), Y(x')) = 0] > \frac{1}{2}.$$

Let us calculate the probability that our algorithm assigns a response from a wrong class to x:

$$P_{er}(x) = P[Y(x) \in C_j^Y, y' \notin C_j^Y]$$

$$= P[Y(x) \in C_j^Y, y' \notin C_j^Y \mid x, x', h_1(x, x') = h_1^0, .., h_L(x, x') = h_L^0]$$

$$= P[z(Y(x), Y(x')) = 0 \mid h_1(x, x') = h_1^0, .., h_L(x, x') = h_L^0]$$

$$= \frac{P[z(Y(x), Y(x')) = 0, h_1(x, x') = h_1^0, .., h_L(x, x') = h_L^0]}{P[h_1(x, x') = h_1^0, .., h_L(x, x') = h_L^0]}$$

$$= \frac{P[h_1(x, x') = h_1^0, .., h_L(x, x') = h_L^0 \mid z(Y(x), Y(x')) = 0] P[z(Y(x), Y(x')) = 0]}{\prod_l P[h_l(x, x') = h_l^0]}$$

$$= \frac{\prod_{l=1}^{L_0} P[h_l(x, x') = 0 \mid z(Y(x), Y(x')) = 0] \prod_{l=1}^{L_1} P[h_l(x, x') = 1 \mid z(Y(x), Y(x')) = 0] P[z(Y(x), Y(x')) = 0]}{\prod_{l=1}^{L_0} P[h_l(x, x') = 0] \prod_{l=1}^{L_1} P[h_l(x, x') = 1]}$$

$$= \frac{P[z(Y(x), Y(x')) = 0] q_0^{L_0} q_1^{L_1}}{(P[h(x, x') = 0])^{L_0} (P[h(x, x') = 1])^{L_1}}.$$

Denote

$$p_{00} = P[z(Y(x), Y(x')) = 0, h(x, x') = 0],$$
$$p_{01} = P[z(Y(x), Y(x')) = 0, h(x, x') = 1],$$
$$p_{10} = P[z(Y(x), Y(x')) = 1, h(x, x') = 0],$$
$$p_{11} = P[z(Y(x), Y(x')) = 1, h(x, x') = 1].$$

Then the random vector (h, z) follows two-dimensional Bernoulli distribution $Ber(p_{00}, p_{01}, p_{10})$ and

$$q_0 = \frac{p_{00}}{p_{00} + p_{01}}, \quad q_1 = \frac{p_{11}}{p_{11} + p_{10}},$$

$$P[h(x, x') = 0] = p_{00} + p_{10},$$

$$P[z(Y(x), Y(x')) = 0] = p_{00} + p_{01}.$$

In this case,

$$P_{er}(x) = (p_{00} + p_{01}) \times$$

$$\frac{p_{00}^{L_0} p_{01}^{L_1}}{[(p_{00} + p_{01})(p_{00} + p_{10})]^{L_0}[(p_{00} + p_{01})(1 - p_{00} - p_{10})]^{L_1}}$$

$$= \frac{(p_{00} + p_{01})^{1-L_1} p_{00}^{L_0} p_{01}^{L_1}}{(p_{00} + p_{10})^{L_0}(1 - p_{00} - p_{10})^{L_1}}$$

$$= \frac{(p_{00} + p_{01})^{1-L_0} p_{00}^{L_0}}{(p_{00} + p_{10})^{L_0}} \frac{p_{01}^{L_1}}{(p_{00} + p_{01})^{L_1}(1 - p_{00} - p_{10})^{L_1}}$$

$$= A(L_0) \frac{p_{01}^{L_1}}{(p_{00} + p_{01})^{L_1}(1 - p_{00} - p_{10})^{L_1}}$$

$$= A(L_0)[\frac{p_{01}}{(p_{00} + p_{01})(1 - p_{00} - p_{10})}]^{L_1}$$

$$= A(L_0)[\frac{p_{01}}{(p_{00} + p_{01})(p_{01} + p_{11})}]^{L_1}$$

$$= A(L_0)[\frac{P[z(Y(x), Y(x')) = 0, h(x, x') = 1]}{P[z(Y(x), Y(x')) = 0]P[h(x, x') = 1]}]^{L_1} \xrightarrow[L \to \infty]{} 0,$$

where $A(L_0) = \frac{(p_{00}+p_{01})^{1-L_0} p_{00}^{L_0}}{(p_{00}+p_{10})^{L_0}} = const.$

The Theorem is proved.

5   Real Data Processing

Consider the application of a semi-supervised regression method with co-associative kernel to a real dataset. We represent the conducted experiments using the following case.

Concrete is the most important material in civil engineering. The concrete compressive strength is a highly nonlinear function of age and ingredients. These ingredients include cement, blast furnace slag, fly ash, water, superplasticizer, coarse aggregate, and fine aggregate. Let us examine the dataset [10] having 9 real-valued features (8 dimensions in feature space and one response variable we need to predict—compressive strength) and 1030 instances—concrete constructions. The actual concrete compressive strength (MPa) for a given mixture under a specific age (days) was determined from laboratory [11]. Basic properties of the data are given in Table 1.

The set of objects was divided into two equal parts: 515 objects in training sample and 515 objects in the test one. CASVR and CANNR algorithms were applied as follows.

To construct the co-association matrix, k-means algorithm was used. K-means was chosen because its complexity is inear with respect to the number of objects and it works fast. Different numbers of clusters were examined, and it turned out that the most acceptable is to run the algorithm about a hundred

Table 1. Dataset parameters

Name	Data type	Measurement	Description
Cement (component 1)	Quantitative	kg in a m3 mixture	Input variable
Blast furnace slag (component 2)	Quantitative	kg in a m3 mixture	Input variable
Fly ash (component 3)	Quantitative	kg in a m3 mixture	Input variable
Water (component 4)	Quantitative	kg in a m3 mixture	Input variable
Superplasticizer (component 5)	Quantitative	kg in a m3 mixture	Input variable
Coarse aggregate (component 6)	Quantitative	kg in a m3 mixture	Input variable
Fine aggregate (component 7)	Quantitative	kg in a m3 mixture	Input variable
Age	Quantitative	Day (1–365)	Input variable
Concrete compressive strength	Quantitative	MPa	Output variable

times and separate the set into $30 + q$ clusters on every q-th iteration. It was mentioned that the algorithm shows good results when we consider the cluster partition uses not the entire space of feature variables, but its two-dimensional random subspace (this remark was especially important during the work with datasets of higher dimension). In addition, the initial seeds for clusters were chosen at random for every run. Thus, co-association matrices for every iteration were built and then the final matrix was calculated as weighted sum of the matrices for distinct algorithms with equal weights (averaged co-association matrix).

In the second step, SVR and NNR methods were trained using the first part of the object set and then applied to the second one as to test sample.

Further, some noise was added to the data. A part of the general set was chosen at random and the values of its features were "spoiled". To spoil the i-th object, for every value $a_{i,j}$ of its j-th feature $a_{i,j}$ was replaced with $a_{i,j} + \epsilon$ where ϵ is a value of the standard normally distributed quantity.

Finally, the result of the calculation was compared with the output of the Support Vector Regression and Nearest Neighbor Regression algorithms that use standard kernels and have the whole set of the responses as the input.

Determination coefficient is the fraction of the variance of the dependent variable explained by the considered regression model, that is, explanatory variables. The results of the comparison of the algorithms expressed in the determination coefficients are shown in Tables 2 and 3.

Table 2. Determination coefficients (SVR)

Kernel type	Without noise	10% of noise	30% of noise	50% of noise
Linear	0.750	0.518	0.420	0.310
RBF	0.882	0.880	0.864	0.849
Co-associative	0.882	0.882	0.880	0.869

Table 3. Determination coefficients (NNR)

Kernel type	Without noise	10% of noise	30% of noise	50% of noise
Linear	0.670	0.450	0.300	0.216
RBF	0.784	0.685	0.573	0.476
Co-associative	0.780	0.700	0.674	0.620

RBF-kernel is the kernel formed with the radial basis function. For fixed objects x and x' it is defined as

$$K(x, x') = exp(-\gamma |||(x - x')|||^2),$$

where $\gamma > 0$ is a parameter that must be chosen. In this experiment we put $\gamma = 1$.

P percent of noise means that $\lfloor \frac{515 \cdot P}{100} \rfloor$ of objects were chosen at random and their features were "spoiled" so as mentioned above.

Finally, absolute values of the errors were investigated for the considered kernels and compared with each other. Statistical tests of the significance of the differences were conducted. Both parametric (pairwise two-tailed t-test together with Shapiro-Wilk normality test) and nonparametric (sign test) criteria were applied and showed that the hypothesis which assumes no significant difference should be rejected with significance level of 0.06.

6   Conclusion and Future Work

In this paper, the new method of solving semi-supervised regression task is presented. The convergence property is proved for one of its variants within the certain probabilistic model. The theorem proposed proves that if the algorithms of cluster analysis applied to construct co-association matrix partition the set of objects according to some directly unobservable class splitting, then increasing the number of elements in the ensemble leads to a decrease of the error probability. If the condition of the interconnection between the classes of objects and corresponding classes of their responses is satisfied, this means that the responses obtained by the algorithm are close to the true values.

In addition, the experiments on real data are described. The proposed method tends to show better results in comparison with the algorithms that use standard kernels. The difference in quality becomes even more noticeable in noise conditions.

In the future, we plan to investigate the issues connected with the weights of the summands in the co-associative matrix and to find out the weights that are optimal in the sense of their accuracy estimates. Convergence properties of algorithms of this type should also be studied in more detail in order to generalize the mentioned results. In particular, it is planned to consider other kernel methods instead of NNR and SVR and to adapt the proposed method for processing data of any type.

Acknowledgment. The article was prepared according to the scientific research program "Mathematical methods of pattern recognition and prediction" in the Sobolev Institute of mathematics SB RAS. The research was partly supported by RFBR grant 18-07-00600 and partly by the Russian Ministry of Science and Higher Education under the 5-100 Excellence Programme. We also want to express our gratitude to the reviewers, as their comments helped us to fill the missing and outlined areas for further research.

References

1. Amemiya, T.: Generalized least squares theory. In: Advanced Econometrics. Harvard University Press (1985)
2. Henderson, D.J., Parmeter, C.F.: Applied Nonparametric Econometrics. Cambridge University Press, New York (2015)
3. Maronna, R., Martin, D., Yohai, V.: Robust Statistics: Theory and Methods. Wiley, Hoboken (2006)
4. Zhu, X.: Semi-supervised learning literature survey. Technical report, Department of Computer Science, University of Wisconsin, Madison, no. 1530, pp. 35–36 (2008)
5. Ghosh, J., Acharya, A.: Cluster ensembles. Wiley Interdiscip. Rev.: Data Min. Knowl. Discov. **1**(5), 305–315 (2011)
6. Vega-Pons, S., Ruiz-Shulclope, J.: A survey of clustering ensemble algorithms. IJPRAI **25**(3), 337–372 (2011)
7. Fred, A., Jain, A.: Combining multiple clusterings using evidence accumulation. IEEE Trans. Pattern Anal. Mach. Intell. **27**, 835–850 (2005)
8. Shawe-Taylor, J., Cristianini, N.: Kernel Methods for Pattern Analysis. Cambridge University Press, Cambridge (2004)
9. Berikov, V., Karaev, N., Tewari, A.: Semi-supervised classification with cluster ensemble. In: Proceedings of 2017 International Multi-conference on Engineering, Computer and Information Sciences (SIBIRCON), Novosibirsk, Russia, 18–22 September 2017, pp. 245–250. IEEE (2017)
10. Dua, D., Karra Taniskidou, E.: UCI machine learning repository. School of Information and Computer Science, University of California, Irvine, CA (2017). http://archive.ics.uci.edu/ml
11. Yeh, I.-C.: Modeling of strength of high performance concrete using artificial neural networks. Cem. Concr. Res. **28**(12), 1797–1808 (1998)

Tree-Based Ensembles for Predicting the Bottomhole Pressure of Oil and Gas Well Flows

Dmitry I. Ignatov[1]([✉]) [iD], Konstantin Sinkov[2], Pavel Spesivtsev[2], Ivan Vrabie[1], and Vladimir Zyuzin[2,3]

[1] National Research University Higher School of Economics, Moscow, Russia
dignatov@hse.ru, vrabie93@mail.ru
[2] Schlumberger Moscow Research, Moscow, Russia
{KSinkov,PSpesivtsev,VZyuzin}@slb.com
[3] Moscow Institute of Physics and Technology, Moscow, Russia

Abstract. In this paper, we develop a predictive model for the multiphase wellbore flows based on ensembles of decision trees like Random Forest or XGBoost. The tree-based ensembles are trained on the time series of different physical parameters generated using the numerical simulator of the full-scale transient wellbore flows. Once the training is completed, the ensemble is used to predict one of the key parameters of the wellbore flow, namely, the bottomhole pressure. According to our recent experiments with complex wellbore configurations and flows, the normalized root mean squared error (NRMSE) of prediction below 5% can be achieved and beaten by ensembles of decision trees in comparison to artificial neural networks. Moreover, the obtained solution is more scalable and demonstrate good noise-tolerance properties. The error analysis shows that the prediction becomes particularly challenging in the case of highly transient slug flows. Some hints for overcoming these challenges and research prospects are provided.

Keywords: Multi-phase flow · Tree-based regression · Time series Bottomhole pressure

1 Introduction

The problem of multiphase flows is encountered in many industrial applications. In nuclear energetics, the gas-liquid flows with phase transitions are formed in the process of reactor cooling. In the oil and gas industry, the three-phase flows are formed in the wells and pipelines. Such flows are often transient with all parameters varying in time and in space. For example, in the oil and gas wells the highly transient severe or terrain-induced slugging can be formed in the system even at the steady-state boundary conditions [1]. Typically, in the analysis of such problems one is interested in evaluation of parameters at certain locations (e.g., at the well bottom). This is challenging because all the parameters evolving

© Springer Nature Switzerland AG 2018
W. M. P. van der Aalst et al. (Eds.): AIST 2018, LNCS 11179, pp. 221–233, 2018.
https://doi.org/10.1007/978-3-030-11027-7_22

in time according to the complex physical phenomena. In the classical approach, the transient multiphase flows are studied using mathematical modeling and numerical simulations [2].

In this work, the alternative approach based on machine learning is explored. Use of machine learning approaches to study the processes taking place in wellbores is a relatively new area. In [3–6] Artificial Neural Networks were used to predict steady-state bottomhole pressure (BHP) in wells operated under conditions of multiphase flow. The authors assumed that the BHP is a single-valued characteristic of a particular well that depends on wellhead pressure (WHP) and surface flow rates that are, in turn, also single-valued characteristics of this particular well. However, during the essentially transient process of well startup, which we are focused on, the BHP is affected not only by the current values of rates and the WHP, but by the states of wellbore at the previous moments of time. In other words, the instant value of BHP is history dependent. In contrast to previous studies, we address specifically the transient problem of predicting the time-dependent BHP. To predict this characteristic, we must use the history of WHP and surface rates, which are also represented as the time series. An interested reader may also refer to similar case-study with Neural Networks as a machine learning tool [7] and oil recovery prediction by tree-based ensembles [8].

The paper is organized as follows. In Sect. 2, we briefly describe the problem statement. Section 3 explains the data generation process and provides the summary of prepared datasets. In Sect. 4, we mention the tree-based models used in our experiments along with short summary of their best performance in the conducted experiments; more details on the experiments and the obtained results are given in Sect. 5. Section 6 concludes the paper.

2   Problem Statement

The proposed approach is developed for a particular problem relevant to the oil and gas industry. To reach for the hydrocarbon bearing underground formations the deep wells are drilled. The depth of wells can reach several thousands of meters. Based on economical considerations, some wells are equipped with the measurement devices installed at the well bottom (e.g. pressure gauges) or even provided with the distributed measurement systems. However, there is a significant number of "cheaper" wells having only surface gauges at best. Nevertheless, to better understand and control the behavior of these wells, especially if they demonstrate tendencies for transient behavior, the evaluation of parameters at the well bottom is still required. Hence, the idea of the proposed approach is to train machine learning algorithms on the set of wells containing data and then provide predictions for the wells where no gauges were installed. To evaluate the feasibility of the approach, the transient numerical multiphase wellbore flow simulator is used [1]. The parameter of interest is the bottomhole pressure. This parameter is transient and affected significantly by the following physical phenomena captured by the simulator:

– Dynamics of volume fraction distributions of phase in the wellbore,

- Velocity difference between phases (also referred to as slip),
- Compressibility of phases,
- Release of gas in the wellbore as pressure drops below the bubblepoint value,
- Friction.

The simulator is operated such that the produced results are representative of the field data that can be recorded by the gauges.

3   Data Generation and Datasets

As it was stated above, to generate the dataset that could be used for training and for validation, a transient multiphase wellbore flow simulator is used [1].
  The behavior of a well is defined by the following set of principal parameters:

- **Well geometry.** The well is represented by a set of segments. Each segment i is described by three principal numbers: measured depth $MD(i)$, true vertical depth $TVD(i)$, and diameter $d(i)$.
- **Inflow.** Several volumetric source terms (also referred to as sources) are placed along the well at different points. The source term is a simplified representation of a perforation where the fluids from the underground formation enter the wellbore. Each source i is described by its position expressed in terms of measured depth $MD_s(i)$ and oil flow rate time series $q_o(t; i)$, water flow rate time series $q_w(t; i)$, and gas flow rate time series $q_g(t; i)$. The sources are sorted by their depth, so the source with $i = 1$ is the closest to the surface.
- **Wellbore control.** The wellbore is controlled with using the wellhead pressure time series $WHP(t)$.

Other parameters are not listed here because they either are kept constant during all the experiments or are not relevant to the considered problem formulation (e.g., roughness of pipe walls). Using this list of parameters, the simulator models the transient multiphase wellbore flow over a given duration and provides the results in terms of the following time series:

- $BHP(t)$—BHP (measured at the position of the source closest to the surface),
- $Q_o(t)$—surface oil volumetric flow rate at standard conditions,
- $Q_w(t)$—surface water volumetric flow rate at standard conditions,
- $Q_g(t)$—surface gas volumetric flow rate at standard conditions.

Often, in a real field situation, the available data are the parameters known before the well is started to flow and the time series that can be measured at the surface during the flow. Therefore, we take the following list of input parameters for our machine learning model: the $WHP(t)$ and rates $(Q_o(t), Q_w(t), Q_g(t))$, the well geometry $(MD(i), TVD(i), d(i))$, and the measured depth of the source closest to the surface that usually coincides with the position of the topmost perforation $(MD_s(1))$. The output of the model is the $BHP(t)$, important and often unknown wellbore flow characteristic.

Massive simulations with different samples are carried out to form data suitable for training machine learning algorithms. The input wellbore flow parameters are varied in a unique (random) way for each sample. The variation is carried out as follows:

– Ten different geometry configurations are used. For each sample, the geometry is picked randomly. There are five types of well trajectories (Fig. 1) with different diameter parameters (see [7] for details).
– There are six sources with positions generated randomly for each sample. To avoid positioning of sources too close to the surface, the sources are allowed to be located starting from the second wellbore segment. There is no limitation imposed on spacing between sources.
– For each source, the $q_w(t)$ and $q_o(t)$ are given as

$$q_w(t) = C_w e^{-\frac{(t-A)^2}{B}},$$

$$q_o(t) = \begin{cases} 0, & \text{when } t < A; \\ C_o \left(1 - e^{-\frac{(t-A)^2}{B}}\right), & \text{when } t \geqslant A. \end{cases}$$

The parameters A and B of these functions are shared between water and oil sources; the amplitudes of water and oil sources C_w and C_o are chosen independently. All parameters are generated randomly affecting amplitude, time of maximum water flow rate, and time of oil rate buildup (Fig. 2).
– The $q_g(t)$ is fixed to zero in the current study. However, oil contains dissolved gas so that the resulting gas rate at the outlet $Q_g(t)$ is non-zero.
– The WHP(t) is generated as a random piecewise constant/linear/quadratic functions with parameters t_i, A_i, B_i, and C_i, $i = \overline{0, k}$ defined as

$$\text{WHP}(t) = \begin{cases} A_0 t^2 + B_0 t + C_0, & \text{when } t < t_1; \\ A_1 t^2 + B_1 t + C_1, & \text{when } t_1 \leqslant t < t_2; \\ \dots, & \\ A_k t^2 + B_k t + C_k, & \text{when } t_k \leqslant t. \end{cases}$$

In the majority of cases, the parameters were chosen so that piecewise constant and linear functions are frequently generated. Also, no specific conditions are imposed on the continuity and monotonicity of these functions.

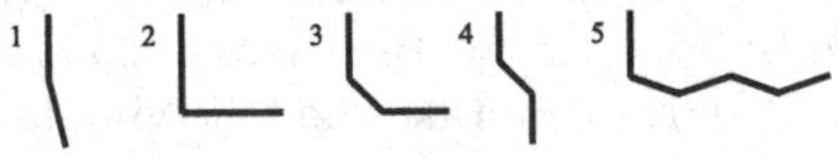

Fig. 1. Schematic representation of the five types of geometry.

An example of one sample with graphs of WHP(t) and BHP(t) functions and surface flow rates $Q_o(t)$, $Q_w(t)$, and $Q_g(t)$ are shown in Fig. 3. Note that in this

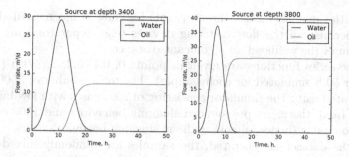

Fig. 2. Examples of the source functions for a selected scenario. Only two out of six sources are shown for brevity.

figure negative values of surface flow rates can be observed. This is due to the backflow across the outlet boundary of the computation domain of the numerical simulation that can be caused by rapid changes of WHP or may be due to the formation of a slug flow regime in the wellbore. Usually, the periods of backflow at surface are small compared to the overall simulation time and have no significant influence on the overall system behavior. In reality, backflow across the oil and

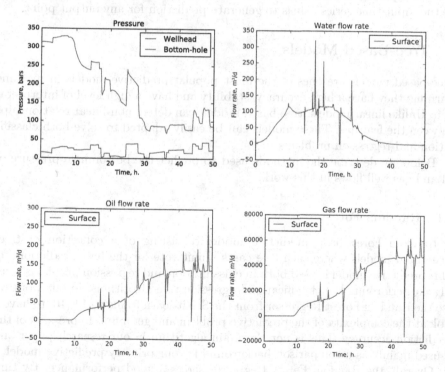

Fig. 3. Example of a sample with graphs of WHP and BHP functions and the surface flow rates $Q_o(t)$, $Q_w(t)$, and $Q_g(t)$.

gas well outlet may also exist. However, this backflow is not measured and not recorded because all the flow metering devices (e.g., separators) are installed downstream of the wellhead, at significant distance.

All time-series functions are given in points 0, 0.1, 0.2, ..., 50 h (total 501 points over 50 h simulated for each sample). The total number of 3500 samples were generated using the simulator consisting of 2000 cases with low inflow rates of sources (and, therefore, pronounced slugging behavior) and 1500 cases with higher inflow rates (hence, almost without slugging behavior).

After the dataset is generated, the samples are randomly mixed to avoid formation of biased patterns. After this, the dataset is split into a set of objects and the correct answers on them. This subset is often referred to as the training set. In our analysis, the training set comprised 80% of samples. The remaining part of samples constitutes the test dataset. Similarly to the training set, the testing dataset consists of the set of objects and the correct answers on them; however, they are not given to our learning algorithms to train on. This set is used to evaluate the quality of a machine learning algorithm. The machine learning problem on the generated dataset is formulated as follows. Given the time series data at the surface, geometry parameters, and measured depth of the source closest to the surface, predict the $BHP(t)$ time series for a new case that was not a part of the training dataset. The model is assumed to have access to all the input time series values to generate prediction for any output point.

4   Tree-Based Models

Tree-based models are ones of the most popular predictive models in machine learning: they have a high accuracy, stability and have a high level of interpretation. Unlike linear models, tree-based models can detect non-linear relationships between the features. These models can be easily adapted to solve both classification and regression problems.

Below we describe the methods used to predict the bottomhole pressure of oil and gas well flows in this work.

4.1   Random Forest

A Random Forest is a predictive model consisting of a collection of tree-structured models where each tree casts a unit vote for the best prediction [9]. This predictive model is used both in classification and regression problems, usually outperforming the classification/regression tree algorithms. In our work we use the Random Forest Regressor from the Scikit-learn framework[1] [10] to have a look at the complexity of the predictive problem and get a first impression of the prediction accuracy. This is not the main algorithm in our research paper and is used mainly as a comparison background for our primary predictive model.

Overall, the Random Forest Regressor shows a good performance. By tuning some of the most important parameters of the model, we have achieved

---

[1] http://scikit-learn.org.

the following results: mean normalized root mean squared error (NRMSE) is equal to 5.44% and worst NRMSE = 17.33% (Fig. 4). Training on a set of 1000 multidimensional time series (along with their derived features, see Sect. 5) and predicting 500 separate series, this method performs slightly worse than the Neural Networks (mean NRMSE = 4.0% and worst NRMSE = 17%, cf. [7]).

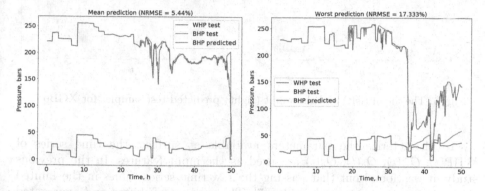

Fig. 4. The mean (left) and the worst (right) predicted test samples for Random Forest

4.2 Boosting Trees

Boosting refers to a class of learning algorithms that fit models by combining many simpler models [11]. These simpler models are typically referred to as base models and are learnt using a base learner or weak learner. These simpler models tend to perform slightly better than a random guess, but when selected carefully and aggregated using a boosting algorithm, they form a relatively more accurate model. This is sometimes referred to as an ensemble model as it can be viewed as an ensemble of base models.

In our work we use the extension of tree boosting algorithms called XGBoost[2]. To be more precise, we use the XGBRegressor algorithm that solves the regression problem using the "boosting trees" approach. This tree boosting system gives state-of-the-art results on a wide range of problems in the classification, regression, and ranking tasks [12].

By tuning the parameters of the model we achieved the following results: mean NRMSE = 3.83% and worst NRMSE = 15.24% (Fig. 5). As we can notice, the boosting algorithm outperforms Neural Networks [7] and Random Forest.

5 Experiments and Error Analysis

5.1 Impact of Feature Generation

In machine learning, feature generation is a technique that can potentially improve model performance and detect unseen relations between features.

---

[2] http://xgboost.readthedocs.io.

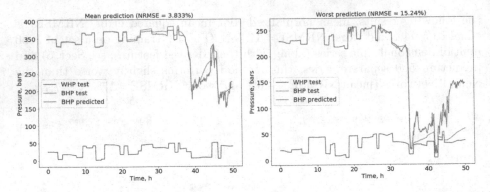

Fig. 5. The mean (left) and the worst (right) predicted test samples for XGBoost

To characterize the parameters available at surface, the time series of WHP(t), $Q_w(t)$, $Q_o(t)$, $Q_g(t)$ are used as the input features. In the previous study, it was found out that passing the raw time series values in the limited time window produces good results [7]. The used set of shifted and aggregated values consists of the following features:

- WHP$(t+i)$, $i = \overline{-2, 2}$, i.e. values of WHP at time point t, when BHP is predicted, two preceding and two subsequent to t moments of WHP recording
- $Q_w(t+i)$, $i = \overline{-5, 5}$, similar for oil and gas
- $(Q_w(t+i) + Q_w(t+1+i))/2$, $i = \overline{6, 12}$, similar for oil and gas
- $(Q_w(t+i) + Q_w(t-1+i))/2$, $i = \overline{-12, -6}$, similar for oil and gas.

All the geometry and flow rates features give, in total, a vector of 74 values. In our work we used the PolynomialFeatures function from the Scikit-learn library [10] to artificially generate new features as polynomials of the features we already have. Unfortunately, this method makes the model and its interpretation more complex, as from 74 features we obtain more than 2000 features. For our problem, this way of generating new features have not improved the accuracy of the predictions. We have obtained the following performance of the XGBoost model with the extended feature space: mean NRMSE $= 4.07\%$ and worst NRMSE $= 16.89\%$.

Another way to generate new features is through the help of domain experts. Discussing with the experts the lack of features that might explain better some of the oil and gas flow properties, we ended up generating features as a ratio of the gas flowrate to the water flowrate. The model with new features has been slightly improved in the worst case scenario: mean NRMSE $= 3.75\%$ and worst NRMSE $= 14.98\%$ (Fig. 6).

5.2   Grid Search by Cross-Validation

In order to get the best possible hyperparameters of the XGBRegressor method, a cross-validation was performed on a separate dataset with 300×500

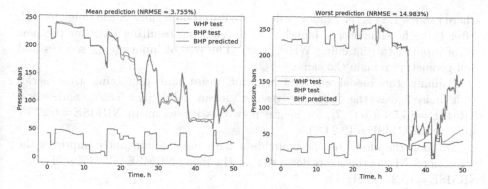

Fig. 6. The mean (left) and the worst (right) predicted test samples for XGBoost over the enhanced dataset with the domain-driven features

(multidimensional series × measurements) using the following set of tuning parameters: *max_depth, learning rate, n_estimators, min_child_weight, subsample* and *colsample_bytree*. The number of grid search experiments was 2400 for different combinations of the provided values of tuned parameters.

Surprisingly, the set of parameters that the cross-validation test suggested was not better than our best manual parameter settings. Thus the result of the cross-validation test gave the following values of the XGBRegressor parameters: *max_depth = 10, learning rate = 0.07, n_estimators = 950, min_child_weight = 23, subsample = 0.8,* and *colsample_bytree = 0.75*. The model performance using the above set of values is the following: mean NRMSE = 3.81% and worst NRMSE = 16.41%. By manually adjusting the parameters of the model we get the next values: *max_depth = 13, learning rate = 0.04, n_estimators = 800, min_child_weight = 15, subsample = 0.9,* and *colsample_bytree = 0.8*. The XGBRegressor method with the above settings performs as follows: mean NRMSE = 3.81% and worst NRMSE = 15.24%.

5.3 Noisy Data

The performance of the model was also tested on noisy version of the input data with the same parameters as in [7].

Typically, the wellhead has a hydraulic connection with the separator installed downstream. In the separator, the phases recovered from the wellbore are split and sent further (e.g., to production facilities). The pressure in the separator might oscillate, and the oscillations might propagate backwards and, in turn, affect the WHP. Hence, if the separator pressure is affected by noise, then the WHP may become noisy, too. Assuming this, we regenerated the datasets by running the simulator on the samples with the WHP affected by noise. The original WHP generated as described above was modified by adding a random noise:

$$\text{WHP}_{\text{noise}}(t) = \text{WHP}(t)[1 + 0.25 \cdot \text{rand}(-1, 1)],$$

where $\mathrm{rand}(-1,1)$ is the random value distributed uniformly in the range $[-1,1]$. After this, the simulator is launched to provide the resulting BHP for the new set of input data containing $\mathrm{WHP}_{\mathrm{noise}}(t)$. The rest of inputs (e.g., sources and well geometry) remain the same.

Training the model on data without noise and predicting time series with noise gives the following results: mean $\mathrm{NRMSE} = 5.89\%$ and worst $\mathrm{NRMSE} = 19.59\%$ (cf. [7], for neural networks it was mean $\mathrm{NRMSE} = 6.226\%$ and worst $\mathrm{NRMSE} = 19.24\%$).

Training the model on the input data with noise significantly improves the predictions driving the errors down so that the mean $\mathrm{NRMSE} - 3.71\%$ and worst $\mathrm{NRMSE} = 15.21\%$ (Fig. 7).

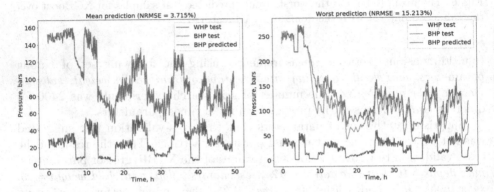

Fig. 7. The mean (left) and the worst (right) predicted test samples for "noisy" training and "noisy" test data

6 Conclusions

6.1 Current Results

The proposed tree-based solutions for multiphase wellbore flows predicts the BHP using the problem-specific input parameters: wellhead pressure, surface flow rates, and wellbore geometry. As the transient aspects of behavior are well captured and allow us to make predictions of the change in variables over time one can conclude that the model does identify physical correlations intrinsic to the considered phenomena.

The presented approach uses the multiphase wellbore flow data generated by physics-based simulator of [1] as the training dataset. Hence, the developed machine learning approach has a potential to process the multiphase wellbore flow data recorded in field operations directly to yield the BHP predictions. The presented approach can be used as an alternative to traditional solutions of the inverse problem of determining the BHP from surface measurements for the transient multiphase wellbore cleanup problem.

In addition to the previous study [7], we observe the following:

- The model can benefit from expert-driven features like the ratio of gas to water flowrate.
- It was earlier demonstrated that the model is capable of predicting the BHP even in the slug flow regime. Some of the samples in this regime can result in significant prediction errors independently of the regression model; it seems that tree-based models, as well as neural network based ones, need more such rare examples.
- In most considered cases, the mean NRMSE value can be further reduced less than 4%.
- The model trained on original data without noise was applied to predict the BHP for the test dataset containing noise. Although the average error increased, the model was capable of capturing the basic features of the noisy transient BHP behavior.
- On an average modern computer (CPU Intel Core i7, 4 cores 2.9 GHz, RAM size, GPU: NVidia 5200M, and HDD disk type), even without GPU usage, tree-based models are capable to provide comparable and slightly better results trained over the same sets within 15–40 mins rather than within 15–25 h in comparison with neural networks (two hidden layers with 50 and 100 neurons).

6.2 Error Analysis

In order to better understand the nature of observed errors, we performed an analysis of the test set to determine the physical characteristics of the cases that were the most challenging for prediction. To do this, we plotted the geometry type of every experiment in the test set and the overall flow rate of every observation vs NRMSE (Fig. 8).

As one may observe, the cases with the geometry type of "2a" or "2b" in combination with a low total flowrate result in the highest prediction error. We believe that there are few cases with this combination of parameters in the training set, thus, leading to the poor predictions on the test set.

6.3 Future Work

The slug flows are a challenging problem for both data-driven predictions and classical physics-based modeling and simulation. The slug flows are characterized by periods of time when there is no liquid production rate observed at the surface at all. In this study, we have sorted out the cases where the model performed well and the challenging samples as follows. All the samples with NRMSE higher than 9.8% have been identified and manually inspected (10 samples out of 500 samples used as the test set). Several peculiarities of those time series have been identified. According to the proposed working hypothesis, the rare combination of certain events resulted in low presence of such cases in the generated data. So, the straightforward approach of further improvements is to generate enough number

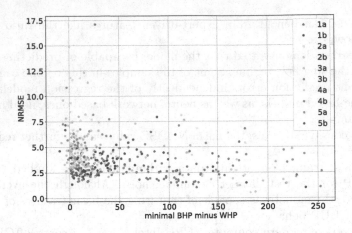

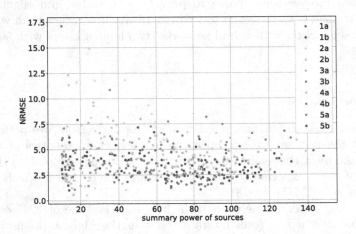

Fig. 8. Geometry type (top) and total flowrate (bottom) of the observations

(up to several hundreds) of samples similar to those rare cases for training and test the newly trained models on the original test set.

The detailed study of the models' noise tolerance and work with real data is left for future experiments. Another interesting venue for further studies is to use global dependencies found as latent factors for the whole times series for one input sample as its digital signature in addition to local features computed for a single time step measurement and related shifts. This can be done via regularized matrix factorization for time series [13] even in case of missing values in the input samples.

Acknowledgments. The paper was partially prepared within the framework of the Basic Research Program at the National Research University Higher School of Economics (HSE) and supported within the framework of a subsidy by the Russian Academic Excellence Project '5-100'. The authors are grateful to Schlumberger and to Marina Bulova, Dean Willberg, Bertrand Theuveny in particular for supporting this work. Dmitry Ignatov was supported by the Russian Foundation for Basic Research, grants no. 16-01-00583 and 16-29-12982.

References

1. Spesivtsev, P., Kharlashkin, A., Sinkov, K.: Study of the transient terrain-induced and severe slugging problems by use of the drift-flux model. SPE J. **22**(5) (2017). https://doi.org/10.2118/186105-PA
2. Theuveny, B., et al.: Integrated approach to simulation of near-wellbore and wellbore cleanup. In: SPE Annual Technical Conference and Exhibition, Society of Petroleum Engineers (2013)
3. Osman, E.A., Ayoub, M.A., Aggour, M.A.: An artificial neural network model for predicting bottomhole flowing pressure in vertical multiphase flow. In: SPE Middle East Oil and Gas Show and Conference. Society of Petroleum Engineers (2005)
4. Jahanandish, I., Salimifard, B., Jalalifar, H.: Predicting bottomhole pressure in vertical multiphase flowing wells using artificial neural networks. J. Pet. Sci. Eng. **75**(3), 336–342 (2011)
5. Ashena, R., Moghadasi, J.: Bottom hole pressure estimation using evolved neural networks by real coded ant colony optimization and genetic algorithm. J. Pet. Sci. Eng. **77**(3), 375–385 (2011)
6. Ebrahimi, A., Khamehchi, E.: A robust model for computing pressure drop in vertical multiphase flow. J. Nat. Gas Sci. Eng. **26**, 1306–1316 (2015)
7. Spesivtsev, P., et al.: Predictive model for bottomhole pressure based on machine learning. J. Pet. Sci. Eng. **166**, 825–841 (2018)
8. Krasnov, F., Glavnov, N., Sitnikov, A.: A machine learning approach to enhanced oil recovery prediction. In: van der Aalst, W.M.P., et al. (eds.) AIST 2017. LNCS, vol. 10716, pp. 164–171. Springer, Cham (2018). https://doi.org/10.1007/978-3-319-73013-4_15
9. Breiman, L.: Random forests. Mach. Learn. **45**(1), 5–32 (2001)
10. Pedregosa, F., et al.: Scikit-learn: machine learning in Python. J. Mach. Learn. Res. **12**, 2825–2830 (2011)
11. Friedman, J.H.: Greedy function approximation: a gradient boosting machine. Ann. Stat. **29**(5), 1189–1232 (2001)
12. Chen, T., Guestrin, C.: XGBoost: a scalable tree boosting system. In: Proceedings of the 22nd ACM SIGKDD International Conference on Knowledge Discovery and Data Mining, KDD 2016, pp. 785–794. ACM, New York (2016)
13. Yu, H., Rao, N., Dhillon, I.S.: Temporal regularized matrix factorization for high-dimensional time series prediction. In: Advances in Neural Information Processing Systems: Annual Conference on Neural Information Processing Systems 2016, Barcelona, Spain, 5–10 December 2016, vol. 29, pp. 847–855 (2016)

Lookup Lateration: Mapping of Received Signal Strength to Position for Geo-Localization in Outdoor Urban Areas

Andrey Shestakov[1], Danila Doroshin[2], Dmitri Shmelkin[2],
and Attila Kertész-Farkas[1(✉)]

[1] School of Data Analysis and Artificial Intelligence, Faculty of Computer Science,
National Research University Higher School of Economics (HSE), Moscow, Russia
akerteszfarkas@hse.ru
[2] Huawei Technologies Co. Ltd., Moscow, Russia

Abstract. The accurate geo-localization of mobile devices based upon received signal strength (RSS) in an urban area is hindered by obstacles in the signal propagation path. Current localization methods have their own advantages and drawbacks. Triangular lateration (TL) is fast and scalable but employs a monotone RSS-to-distance transformation that unfortunately assumes mobile devices are on the line of sight. Radio frequency fingerprinting (RFP) methods employ a reference database, which ensures accurate localization but unfortunately hinders scalability.

Here, we propose a new, simple, and robust method called lookup lateration (LL), which incorporates the advantages of TL and RFP without their drawbacks. Like RFP, LL employs a dataset of reference locations but stores them in separate lookup tables with respect to RSS and antenna towers. A query observation is localized by identifying common locations in only associating lookup tables. Due to this decentralization, LL is two orders of magnitude faster than RFP, making it particularly scalable for large cities. Moreover, we show that analytically and experimentally, LL achieves higher localization accuracy than RFP as well. For instance, using grid size 20 m, LL achieves 9.11 m and 55.66 m, while RFP achieves 72.50 m and 242.19 m localization errors at 67% and 95%, respectively, on the Urban Hannover Scenario dataset.

Keywords: Non-line-of-sight
Outdoor mobile device geo-localization · Fingerprinting

1 Introduction and Literature Review

The localization of wireless mobile devices [8,20] has become a key issue in emergency cases [1,6], surveillance, security, family tracking, radio network planning and optimization, et cetera. GPS-based localization, while very accurate, is not

W. M. P. van der Aalst et al. (Eds.): AIST 2018, LNCS 11179, pp. 234–246, 2018.
https://doi.org/10.1007/978-3-030-11027-7_23

preferred because of its high energy consumption; thus, a wide range of technologies has emerged for localization based on received signal strength (RSS), time of arrival (ToA), and angle of arrival (AoA). The two latter methods, ToA and AoA, require additional equipments, such as an antenna array to time synchronization between the transmitter and the receiver, while AoA requires some array to identify the signal's angle. However, RSS measurements are ubiquitous and readily available in almost all wireless communication systems; hence, it seems plausible to gain information about mobile devices' position.

The main problem is characterized as follows: We assume that we are given a set of antennas $A = \{a_i = (a_{i_x}, a_{i_y}) \in R^2\}$ and several RSS measurements for a single mobile phone: $M_{RSS} = [r_{a_1}, r_{a_2}, ..., r_{a_{|A|}}]$, where r_{a_i} denotes RSS from antenna a_i, measured in dBm if it is observed; NaN otherwise. Antenna a_i and its location will be used interchangeably in this article. The main task is to determine the mobile phone's location, where its RSS measurements were observed. A straightforward method to carry out a simple localization is the triangular lateration (TL) that calculates the mobile phone's location by solving the following optimization problem:

$$\min_{(x,y)} \sum_{i:r_{a_i} \in M_{RSS}} \left((a_{i_x} - x)^2 + (a_{i_y} - y)^2 - (d(r_{a_i}))^2\right)^2, \tag{1}$$

where $d(.)$ denotes a distance function. For monotone functions $d(.)$ and for at least three measurements, the optimization problem becomes convex and avoids the local minima phenomenon. However, there are two main issues with approach. First, it requires a well-calibrated signal-to-distance conversion function $d(.)$. For urban areas, the COST-HATA models were developed by the COST European Union Forum based on various field experiments and research [13]. Second, it assumes mobile devices on line of sight (LoS), which is hardly met in practice in urban areas because signal propagation is hindered by obstacles, concrete constructions, churches, tunnels, underpasses, etc. In this case, the observed distance can be decomposed as $d(r_{a_i}) = L_i + e_i + NLOS_i$, where L_i is the true distance and e_i is the receiver noise assumed to be a zero mean Gaussian random variable. Note that, $NLOS_i \gg e_i$ according to field test results.

The quantity $NLOS_i$ denotes the error caused by obstacles on the signal propagation path. These obstacles interfere with the signal by either weakening it or changing its path a.k.a multipath propagation. In either case, the antenna receives weaker signals indicating mobile phones farther away than they are as illustrated in Fig. 1A. The NLOS error can be modeled by a probabilistic distributions [5] or can be considered a deterministic error [10,12]. These methods show promising results when few obstacles can be found on the landscape, though how these methods will work in a dense urban area remains to be seen.

To characterize the NLoS error, we analyzed the Urban Hannover Scenario dataset [15], which contains approximately 22 Gb of localized RSS measurements from Hannover. For further details, see results section or [15]. In our analysis we first selected tower ID = 183 randomly and identified mobile phones located at the fourth degree from the tower's azimuth. The data are shown in Fig. 1C, where

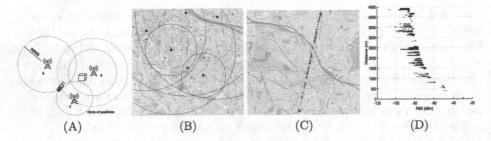

(A)   (B)   (C)   (D)

Fig. 1. (A) Triangular lateration hindered by obstacles. D(RSS) denotes the distance determined by RSS. The dashed circle denotes the true distance of a mobile phone from tower B, which appears farther because of the obstacle (represented by a box) between them. (B) Lateration in Hannover dataset using the six strongest RSSs. The red dot denotes the mobile device's true location, the black dots denote antennas, and the circles denote the circle of positions. Data (User ID = 10048 and Time = 0.1) is from [15]. (C) Location of mobiles phones (marked blue dots) from antenna ID = 183 (marked blue triangle at the bottom) at 4th degree to its azimuth (clockwise). (D) Distance vs. RSS for phones from panel (C). (Color figure online)

the blue triangle denotes the tower's location and where the blue dots indicate the mobile phones' locations. The scatter plot of the measured RSSs versus the true distance is shown in Fig. 1D. The relationship indicates an exponential-like correlation. However, the weaker the signal is, the noisier the correlation becomes. As signal strength decreases, the conversion becomes more ambiguous. For instance, an observed −90 dBm signal could indicate a mobile phone located somewhere between 2.4 km and 4.5 km. It is also worth noticing that the phones are not distributed evenly in this range; rather, they are concentrated at certain distances. For instance, no phone is located between 2.5 km and 3.0 km in this direction. This gap can be seen at the railways shown in Fig. 1C.

The radio frequency fingerprinting (RFP) [2,7] approach is designed to overcome the NLoS problem. This is a two-phase algorithm. The first step, called offline training phase or surveying, is to construct a reference dataset that consists of a collection of localized RSS measurement vectors $M_{XY} = [r_{a_1}, r_{a_2}, \ldots, r_{a_{|A|}}]$ all over the area, where x, y denotes its true position. In the second phase, localization phase, the query device's position producing M_{RSS} is estimated based on the nearest dataset member via a k-Nearest Neighbor (kNN) approach. In other words, the location x, y is determined by solving the following search problem:

$$\tilde{x}, \tilde{y} = \operatorname*{argmin}_{x,y} d(M_{RSS}, M_{XY}), \qquad (2)$$

where the distance function $d(.,.)$ can be, say, an Euclidean distance or cosine distance [9], etc., which simply omit NaN values. One drawbacks of RFP is the lack of scalability, which means that it can be considerably slow over large areas.

Grid systems are often utilized to mitigate data redundancy, making localization faster with less memory consumption in which measurements replaced by their grid centers. Therefore, RFP can be formulated as follows: localize query

M_{RSS} in the center of grid $\tilde{x}y$ calculated by $\tilde{x}\tilde{y} = \text{argmin}_{xy \in \mathcal{G}}\, d(M_{RSS}, \mu_{xy})$, where μ_{xy} is the sample mean observed in grid xy, $\mathcal{G}$ denotes the set of grids and $d(.,.)$ is a suitable distance function. This reduces search time and the size of the reference datasets by a couple of magnitudes at the expense of localization accuracy, and the trade-off can be controlled with the grid's size. To make localization even faster, Campos et al. [3] have developed filtering procedures to reduce the search space.

Besides RFP, which is based on a k-NN, several classic machine learning methods have been evaluated and tested on localization problems. For instance, Wu et al. [18] have tried support vector machines, Nuño-Barrau et al. [14] have used linear discriminant analysis, and Campus et al. and Magro et al. [4,11] have used genetic algorithms for positioning measurements. Artificial neural networks are also popular; they can learn a regression function to map a M_{RSS} to its corresponding location $\tilde{x}, \tilde{y}$ [16,17,19]. No matter which methods are used, they all need to learn highly nonlinear mapping like in Fig. 1D to provide a bona fide approximation. In our opinion, this seems quite challenging in practice.

In this paper, we present a new method called lookup lateration (LL) for the rapid and accurate geo-localization of mobile phones based on RSS. The underlying idea is based on the decentralization of the reference dataset. For every antenna tower $a_i \in A$, LL builds a lookup table to store associating reference locations, that is, $\tau_{a_i}(r_{a_i}) = \{(XY) \mid r_{a_i} \in M_{XY}\}$. Therefore, the localization of a given query M_{RSS} can be carried out by simply determining the common locations in the associating lookup tables as follows: $\tilde{x}, \tilde{y} = \bigcap_{r_{a_i} \in M_{RSS}} \tau_{a_i}(r_{a_i})$. The main advantages of LL compared to RFP is that LL does not need to search the whole reference dataset, only two-six tables associated to query measurements. This results in a great acceleration; LL is around 100 times faster than RFP on the Urban Hannover Scenario dataset, and we believe LL can be even faster in very big cities.

LL resembles TL to some extent. Every lookup table can be considered an RSS-to-distance nonlinear mapping $d(.,.)$ (more precisely, a relation) w.r.t. a given antenna tower. However, instead of solving a nonconvex optimization problem, LL carries out the localization by determining common elements in the lookup tables. Therefore, LL does not involve local minima problems, but it may result in multiple locations, which need to be addressed.

2   Lateration Using RSS-to-Location Lookup Tables

In the previous section, we concluded that the relationship between received RSS and true distance is nonmonotone in dense urban areas. Using any nonmonotone, continuous function $d(.)$ in Eq. 1 would result in a nonconvex optimization problem. Here, we introduce a new procedure that we termed lookup lateration (LL).

The procedure is provided a collection of M_{XY} measurements annotated with their true locations as training data. Then lookup tables $\tau_a(r)$ contain mobile locations with respect to RSS r measured by antenna a. This procedure is shown by Algorithm 1. To avoid redundant locations in lookup tables and reduce their size, nearby locations can be grouped using clustering algorithms, and the center of clusters can be stored in lookup tables. Because RSS measurements r are real valued and hindered by some measurement noise, we simply applied binning techniques to group measurements together. The corresponding bin b of a measurement r using bin size s is calculated $b = \lfloor r/s \rfloor$. In most of our experiments, we simply used bin $s = 1$; thus, all r were rounded down to the closest integer. For instance, -64.7 dBm is rounded down to -65. In the rest of the paper, $s = 1$, unless it is specified otherwise.

Figure 2A shows an example, where data were taken from the Urban Hannover Scenario and green dots mark the locations stored in $\tau_{88}(-50)$, while blue triangle (bottom) indicates the tower location (ID $= 88$).

Algorithm 1. **Construction of lookup tables.** The input is a list of localized RSS measurement $M = \{(x_i, y_i, r_{a_i})\}_{i=1}^{N}$. Triplet (x_i, y_i, r_{a_i}) denotes location x_i, y_i, where RSS r_{a_i} is measured from tower a_i. D is the maximal allowed diameter of a cluster.

```
1: procedure CONSTRUCTLOOKUPTABLES(M,D)
2:     for all antenna tower a do
3:         for all RSS r do
4:             S ← {(x_i, y_i) | r_{a_i} = r, a_i = a}
5:             τ_a(r) ← Clustering(S, D)    ▷
           Group mobiles nearby.
6:         end for
7:     end for
8:     return τ
9: end procedure
```

Algorithm 2. **Lookup lateration.** The input $M_{RSS} = \{r_{a_1}, r_{a_2}, ..., r_{a_n}\}$ is a list of RSS measurements in decreasing order. Returns a location estimation for M_{RSS}.

```
1: procedure LOOKUPLATERATION(M_RSS)
2:     Remove r_{a_i} ∈ M_RSS from M_RSS if
     τ_{a_i}(r_{a_i}) is empty
3:     return some default location if M = ∅
4:     k ← 1
5:     C^k ← τ_{a_k}(r_{a_k})
6:     for k = 1 ... |M_RSS| do
7:         break if C^k is ambiguous
8:         k ← k + 1
9:         C^k ← {c_i|c_i ∈ C^{k-1}, c_j ∈
         τ_{a_k}(r_{a_k}), d(c_i, c_j) ≤ T}
10:    end for
11:    C^k ← C^{k-1} if C^k is empty
12:    return mean of locations in C^k for query
     M_RSS
13: end procedure
```

Now the next step is to determine the location of a given measurement $M_{RSS} = [r_{a_1}, r_{a_2}, ..., r_{a_{|A|}}]$. Here, the principle is that it can be carried out by determining the common locations in the corresponding tables $\tau_{a_i}(r_{a_i})$, that is, M_{RSS} is annotated by the location $\bigcap_i \tau_{a_i}(r_{a_i})$ for $r_{a_i} \neq NaN$ and $\tau_{a_i}(r_{a_i}) \neq \emptyset$. In practice, this could lead easily to an empty set or unambiguous locations. Our method, shown in Algorithm 2, is slightly different as, in a greedy manner, it takes into account that stronger signals provide more reliable information. Let $M = \{r_{a_{i_1}}, r_{a_{i_2}}, ..., r_{a_{i_n}}\}$ be a list of the observed RSS from M_{RSS} in decreasing order. Our algorithm starts with the set of candidate locations $C^1 = \tau_{a_1}(r_{a_1})$ provided by the strongest signal. In subsequent iterations, in the while loop at line 6, candidate location $c_i \in C^1$ is eliminated from C^1 if it does not appear as a candidate location from another antenna a_k $(k = i_2, ..., i_n)$ within a tolerance T. This iteration terminates if either all measurements are processed or the candidate locations in C^k have a smaller variance than a predefined threshold. The iteration also terminates when C^k is emptied. Figure 2 shows an example of how this algorithm works.

$$(A) \qquad\qquad\qquad (B) \qquad\qquad\qquad (C)$$

Fig. 2. Illustration of lookup lateration for query $M_{RSS} = \{88{:}{-}50;\ 27{:}{-}55;\ 135{:}{-}55\}$. Candidate locations are marked by green dots, and three antennas (ID $= 88$ [bottom], ID $= 27$ [top], ID $= 135$ [right hand side]) are marked by blue triangles. (A) Initial locations in $C^1 = \tau_{88}(-50)$ at the beginning of Algorithm 2. (B) Next, locations not present in $\tau_{27}(-55)$ are removed, and the resulting locations in C^2 are shown. (C) Points not present in $\tau_{135}(-55)$ are also removed, yielding an ambiguous location estimation in C^3 for M_{RSS}. In this example, the lookup lateration stops after three iterations. (Color figure online)

3   Error Estimation in Grid Systems

The LL method can also be used with grid systems, which can also lead to further simplifications of the algorithm. In our experiments, Algorithm 1 simply stored unique grid references in $\tau_a(r)$. Thus, clustering algorithms were not needed. Moreover, filtering condition $(d(c_i, c_j) < T)$ in line 9 of Algorithm 2 was replaced with a Kronecker delta $(\delta_{c_i,c_j} \overset{?}{=} 1)^1$, which is more quickly evaluated.

Now we can compare the error obtained by fingerprinting and lookup lateration methods under two assumptions: (1) completeness and (2) unambiguousness. By completeness, we assume that there are enough data observed in each grid. This ensures that error is not induced by data sparsity, poor design, or the localization is not carried out in, say, Paris if site survey was taken in New York, etc. By unambiguousness we assume that identical observations cannot be found in different grids. This assumption ensures that any observation can be determined unambiguously. Now we can claim the following:

Theorem 1. *Let* $\mathbb{E}[\mathcal{E}_{LL}]$ *and* $\mathbb{E}[\mathcal{E}_{RFP}]$ *be the expected error obtained with LL and RF, respectively. Under the conditions mentioned above,* $\mathbb{E}[\mathcal{E}_{LL}] \leq \mathbb{E}[\mathcal{E}_{RFP}]$.

Proof. First, let us consider the fingerprinting method. Let $\mathcal{B} \subset R^{|A|}$ be the measurement space, $\mathcal{G}$ be the grid system, P_{xy} be the density distribution of measurements belonging to grid $xy \in \mathcal{G}$, and μ_{xy} be the mean vector of P_{xy}. The distance function d in RFP localization implicitly specifies a Voronoi partition of $\mathcal{B}$: $\{\mathcal{R}_{xy}\}_{xy \in \mathcal{G}}$, where

$$\mathcal{R}_{xy} = \{r \in \mathcal{B} \mid d(r, \mu_{xy}) \leq d(r, \mu_{uv})\ \forall uv \neq xy \in \mathcal{G}\}. \tag{3}$$

---

[1] Defined by Kronecker delta $(\delta_{a,b} = 1 \iff a = b,$ otherwise 0).

Second, let $M_{RSS} \in \mathcal{B}$ denote a query measurement observed at position M_{xy}, which is in grid $xy \in \mathcal{G}$.

If $M_{RSS} \in \mathcal{R}_{xy}$ and $M_{RSS} \notin \mathcal{R}_{uv}$ for any other $uv \neq xy \in \mathcal{G}$, then RFP localizes M_{RSS} in the center of grid xy. The localization error is $\epsilon_1 = d(M_{xy}, c_{xy})$, that is, the distance between M_{xy} and the center of the grid c_{xy}. Let $\mathcal{E}_1$ be this type of error and let $\mathbb{E}_{xy}[\epsilon_1]$ be the expected value of this error type in grid xy.

If $M_{RSS} \notin \mathcal{R}_{xy}$, then $\exists uv \in \mathcal{G}$ in which RFP localizes M_{RSS}. The error is $\epsilon_2 = d(M_{xy}, c_{uv})$. Let $\mathbb{E}_{xy}[\epsilon_2]$ denote the expected error of this type with respect to grid xy. Let $\mathcal{E}_2$ denote this type of error. Note that error ϵ_2 is always greater than error ϵ_1, which implies that

$$\mathbb{E}_{xy}[\epsilon_1] < \mathbb{E}_{xy}[\epsilon_2]. \tag{4}$$

If $M_{RSS} \in \mathcal{R}_{xy}$ and $M_{RSS} \in \mathcal{R}_{uv}$ for $uv \neq xy \in \mathcal{G}$, then that means either M_{RSS} is on the border of two regions $\mathcal{R}_{xy}$ and $\mathcal{R}_{uv}$ or $\mu_{xy} = \mu_{uv}$. In the first case, RFP has to choose between the two grids. Let us give some advantage to RFP and assume for the sake of simplicity, it always chooses the correct grid somehow. In the second case, we have $\mathcal{R}_{xy} = \mathcal{R}_{uv}$. In our opinion this case happens rarely in practice, so we omit this type of error, and we assume that $\mu_{ab} \neq \mu_{cd}$ for any $ab, cd \in \mathcal{G}$ in the rest of this proof.

The probability that RFP localizes M_{RSS} in its correct grid is

$$p_{xy}(\mathcal{E}_1) = \int_{\mathcal{R}_{xy}} P_{xy}(r_{a_1}, r_{a_2}, \ldots, r_{a_{|A|}}) \, dr_{a_1} \ldots dr_{a_{|A|}}, \tag{5}$$

and it is indicated by the white area under P_{xy} in Fig. 3. The probability that RFP localizes M_{RSS} in a grid incorrectly is

$$p_{xy}(\mathcal{E}_2) = \int_{\mathcal{B} \setminus \mathcal{R}_{xy}} P_{xy}(r_{a_1}, \ldots, r_{a_{|A|}}) \, dr_{a_1} \ldots dr_{a_{|A|}} = 1 - p_{xy}(\mathcal{E}_1). \tag{6}$$

and it is indicated by the gray area under P_{xy} in Fig. 3.

Last, the total expected error for the fingerprinting method can be summarized as follows:

$$\mathbb{E}[\mathcal{E}_{RFP}] = \sum_{xy \in \mathcal{G}} p(xy) \left[p_{xy}(\mathcal{E}_1) \mathbb{E}_{xy}[\epsilon_1] + p_{xy}(\mathcal{E}_2) \mathbb{E}_{xy}[\epsilon_2] \right], \tag{7}$$

where $p(xy)$ denotes a priori probability of receiving an RSS from grid xy.

Now we examine LL, and let us consider $M_{RSS} = [r_{a_1}, r_{a_2}, \ldots, r_{a_{|A|}}]$ measured in grid xy. Because of the first and second assumptions, we have $xy \in \tau_{a_i}(r_{a_i})$ and $\bigcap_i \tau_{a_i}(r_{a_i}) = \{xy\}$, respectively. This means when Algorithm 2 terminates, the set C^k unambiguously contains only one candidate grid, where the measurement was observed. Hence, the total expected error for lookup lateration can be summarized as follows:

$$\mathbb{E}[\mathcal{E}_{LL}] = \sum_{xy \in \mathcal{G}} p(xy) \mathbb{E}[\mathcal{E}_1], \tag{8}$$

Following from Eq. 4 we obtain $\mathbb{E}[\mathcal{E}_{LL}] \leq \mathbb{E}[\mathcal{E}_{RFP}]$, which proves our claim. ∎

To illustrate the proof of Theorem 1, let us consider a scenario in which there is one antenna tower A and two nearby grids G_{xy}, G_{uv}. Furthermore, let us assume that A receives $-53, -55, -57, -59$, and -61 dBm from G_{xy} and $-58, -60, -62, -64$, and -66 dBm from G_{uv}. The mean values are $\mu_{xy} = -57$ and $\mu_{uv} = -62$ and $\mathcal{R}_{xy} = \{x \mid x \geq -59.5\}$ and $\mathcal{R}_{uv} = \{x \mid x \leq -59.5\}$. RFP locates measurement $M = -61$ dBm in grid G_{uv} incorrectly because $M \in \mathcal{R}_{uv}$, that is, M closer to μ_{uv} than to μ_{xy}. However, LL constructs a table τ_A in which $\tau_A(-66) = \tau_A(-64) = \tau_A(-62) = \tau_A(-60) = \tau_A(-58) = \{G_{uv}\}$ and $\tau_A(-61) = \tau_A(-59) = \tau_A(-57) = \tau_A(-55) = \tau_A(-53) = \{G_{xy}\}$. Therefore, LL will find $G_{xy} \in \tau_A(-61)$ and localize M in it correctly.

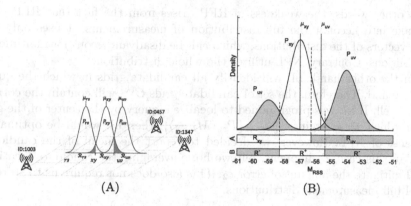

(A)   (B)

Fig. 3. (A) Distribution of measurements for three grids rs, xy, and uv in fingerprinting scenario. Decision boundaries between $\mathcal{R}_{rs}, \mathcal{R}_{xy}$, and $\mathcal{R}_{uv}$ are halfway between distribution means. (B) Decision boundaries in case of multi-modal density distributions. The decision boundary defined with RFP is located between μ_{xy}, μ_{uv} and the corresponding regions $\mathcal{R}_{xy}, \mathcal{R}_{xy}$ denoted on horizontal bar A. Decision boundaries defined by the NPL non-continuous and are shown on bar B.

One may argue that Theorem 1 is based on two conditions that are hard to ensure in practice. We may agree, but in our opinion, Theorem 1 shows the conceptual limitation of the fingerprinting method. However, if we easy up on the conditions and if the unambiguousness is not required, then we claim

Theorem 2. *Fingerprinting method is suboptimal in the sense of the Neyman-Pearson criterion.*

Proof. Localization using grid systems can be considered a classification problem in which grids are considered classes, and a localization method has to decide whether to classify a query M_{RSS} in a given class xy. Let $P^+(M_{RSS}) = P_{xy}(M_{RSS})$ and $P^-(M_{RSS}) = \frac{1}{Z}\sum_{uv \neq xy} P_{uv}(M_{RSS})$ be likelihood functions, where Z is an appropriate normalization factor. According to the Neyman-Pearson Lemma (NPL), the highest sensitivity can be achieved with the following likelihood-ratio rule:

$$\text{Localize } M_{RSS} \text{ in grid } xy \text{ if } \quad \frac{P^+(M_{RSS})}{P^-(M_{RSS})} \geq \eta, \qquad (9)$$

where η is a trade-off parameter among false positive, false negative error, and statistical power. Let $\eta = 1$ for sake of simplicity. The region $\mathcal{R}^+ = \{M_{RSS} \mid P^+(M_{RSS}) \geq P^-(M_{RSS})\}$ in which a query $M_{RSS} \in \mathcal{R}^+$ is localized in grid xy and defined by the likelihood ratio can be noncontinuous region. However, Fig. 3B illustrates that the region $\mathcal{R}_{xy}$ defined by RFP in Eq. 3 is always continuous and different from $\mathcal{R}^+$. Therefore, RFP yields less or equal sensitivity than it could be achieved with NPL, where equality holds iff all distribution belonging to grids are symmetric and unimodal. ∎

In other words, the weakness of RFP arises from the fact that RFP does not take into account the full distribution of measurements, it uses only the mean vectors of the distributions, which can be disadvantageous for multimodal distributions. Contrary, NPL utilizes the whole distribution.

On the other hand, LL will identify all candidate grids in which the query can be found. Therefore, the set of candidate grids $C^{(k)}$ will contain the correct grid as well. If LL was programmed to localize a query by the center of the grid $\tilde{x}\tilde{y}$ for which $\tilde{x}\tilde{y} = \text{argmax}_{xy \in C^{(k)}} \{P_{xy}(M_{RSS})\}$, then LL would be optimal in the sense of NPL. However, we decided to report the mean of the candidate grids. In this case, $\mathcal{E}_2$ happens, but we hope averaging the candidate locations would mitigate the amount of error ϵ_2. This also does not require us to store or model full measurement distributions.

4   Results and Discussion

We have carried out our experiments on the Urban Hannover Scenario dataset [15]. This dataset contains approximately 22 Gb of RSS measurements simulated in Downtown Hannover, along with a reference x-y location. The reference point (0-0) for the coordinate system is the lower left corner of the scenario. The data is the result of a prediction with a calibrated ray tracer using 2.5D building information. For each mobile phone, the 20 strongest RSS measurements are provided. For further details, we refer to [15]. Data was split into training and test sets randomly, and experiments were repeated ten times. Results were averaged. The variance in results was very small because of the dataset's huge size; therefore, standard deviation is not shown for the sake of simplicity. Algorithms were implemented in Python programming language and executed on a PC equipped with a 3.4 GHz CPU and a 16 GB RAM.

First, we investigated the localization accuracy of RFP and LL methods using grid sizes 5, 10, 20, and 50 m. The full dataset was split to 10% training and 90% test data. The results shown in Fig. 4 clearly tell us LL and RFP perform nearly the same when the grid size is small (Figs. 4A–B). The performance of LL remains roughly the same as the size of the grid grows (Figs. 4C–E), while the performance of RFP decreases quickly. In our opinion the dramatic drop in the performance of RFP is in accordance with Theorems 1 and 2. Large grids cover

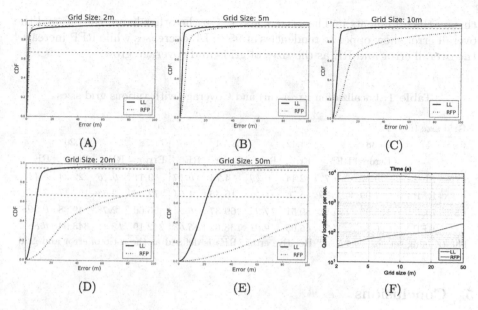

Fig. 4. Performance of RFP and LL using various grid sizes. Dashed lines corresponds to errors at 67% and 95%.

a large area containing various obstacles on the ray propagation path, which result in wide, multimodal measurement distributions w.r.t. grids. This is not taken into account by RFP. Next, we analyzed the speed of these methods, and the results are summarized in Fig. 4F. LL seems to be extremely fast, around two magnitudes faster compared to the RFP method. This improvement stems from the fact that LL processes only a few lookup tables related to a query. On the other hand, RFP iteratively processes all grids while seeking the most similar RSS vector pattern. It is also surprising, the execution time of LL does not depend on the grid size, while RFP quadratically becomes slower as the grid size decreases.

Next, we investigated how RFP and LL methods perform with improper site surveying. We calculated a site coverage defined as the ratio of grids that contain training data. For instance, 35% of coverage means that the training data belong to 35% of the total grids, and all queries would be localized in one of these grids. The coverage is driven by two factors: (i) grid size and (ii) the size of the training data. First, we took 5% of the training data and calculated the coverage and the localization error at 67% and 95% using various grid sizes. The results shown in Table 1 tell us a larger grid size results in larger coverage, and both methods yield a larger localization error at 67%. However, if we take a closer look and calculate a relative error (RE) as the ratio of the error and the grid size, we can observe opposite tendencies. The RE obtained with LL decreases as the grid size grows, and we think this is the result of increased coverage. However, the RE obtained with RFP increases in spite of increased coverage, and we explain this by the arguments in Theorems 1 and 2. On the other hand, the RE and the overall error show opposite tendencies at 95%. LL decreases while RFP increases the overall error and RE as the sizes of grids and coverage grow.

Table 1. Localization Error (m) and Coverage with various grid sizes

Grid size (m)		1		5		10		20		50	
Cov. (%)		26.38		54.42		66.74		71.78		87.20	
		Error	RE[a]	Error	RE	Error	RE	Error	RE	Error	RE
67%	LL	0.77	0.77	2.44	0.49	4.74	0.47	9.11	0.46	22.24	0.44
	RFP	1.09	1.09	5.76	1.15	16.36	1.64	72.50	3.63	157.17	3.14
95%	LL	125.98	125.98	89.51	17.9	69.37	6.94	55.66	2.78	49.88	0.998
	RFP	6.32	6.32	45.05	9.01	138.62	13.68	242.19	12.11	343.94	6.88

The training set size was 5%. [a] Relative error (RE) is defined as the ratio of error and grid size.

5   Conclusions

In this article, we have presented a new method called lookup lateration for localization problems in densely populated urban areas using received signal strength (RSS) data. Our method combines the advantages from both triangular lateration and fingerprinting.

Lookup lateration outperforms triangular lateration and fingerprinting-based methods in localization accuracy as well. Triangular lateration does not take into account NLOS objects; hence, its performance in urban areas is always limited. We also showed conceptual limitations of the fingerprinting method in Theorems 1 and 2. Briefly, the problem is that RSS observations are aggregated in the same grid position, and their variances are not taken into account. As a consequence, fingerprinting is prone to annotate observations with incorrect grid positions. This situation cannot happen with lookup lateration because it stores all grid positions for any observed RSS values separately. This fact makes LL very robust to larger grid sizes.

Finally, it is very easy to migrate from RFP to LL. Since LL does not need any additional data compared to RFP, lookup tables can be constructed from the site surveying data of RFP. Lookup tables can also be an intelligent reorganization of fingerprinting data, which preserves measurement variance implicitly. Our method is very simple, easy to implement in distributed systems, and inexpensive to maintain, which, we hope, can make it appealing for large-scale industrial applications.

References

1. Directive 2002/58/EC on privacy and electronic communications
2. Bahl, P., Padmanabhan, V.N.: Radar: an in-building RF-based user location and tracking system. In: Proceedings of Nineteenth Annual Joint Conference of the IEEE Computer and Communications Societies, INFOCOM 2000, vol. 2, pp. 775–784. IEEE (2000)
3. Campos, R.S., Lovisolo, L.: A fast database correlation algorithm for localization of wireless network mobile nodes using coverage prediction and round trip delay. In: VTC Spring. IEEE (2009)
4. Campos, R.S., Lovisolo, L.: Mobile station location using genetic algorithm optimized radio frequency fingerprinting. In: Proceedings of ITS 2010 International Telecommunications Symposium (2010)
5. Cong, L., Zhuang, W.: Nonline-of-sight error mitigation in mobile location. IEEE Trans. Wirel. Commun. **4**(2), 560–573 (2005)
6. FCC: Revision of the commission rule to ensure compatibility with enhanced 911 emergency calling system. Technical report RM-8143, Federal Communications Commission (FCC), Washington, DC (2015)
7. Gentile, C., Alsindi, N., Raulefs, R., Teolis, C.: Geolocation Techniques: Principles and Applications. Springer, Heidelberg (2012)
8. Gezici, S.: A survey on wireless position estimation. Wirel. Pers. Commun. **44**(3), 263–282 (2008)
9. He, S., Chan, S.H.G.: Tilejunction: mitigating signal noise for fingerprint-based indoor localization. IEEE Trans. Mobile Comput. **15**(6), 1554–1568 (2016)
10. Ma, C., Klukas, R., Lachapelle, G.: A nonline-of-sight error-mitigation method for TOA measurements. IEEE Trans. Veh. Technol. **56**(2), 641–651 (2007)
11. Magro, M.J., Debono, C.J.: A genetic algorithm approach to user location estimation in UMTS networks. In: The International Conference on "Computer as a Tool", EUROCON 2007, pp. 1136–1139. IEEE (2007)
12. Marco, A., Casas, R., Asensio, A., Coarasa, V., Blasco, R., Ibarz, A.: Least median of squares for non-line-of-sight error mitigation in GSM localization. In: 2008 IEEE 19th International Symposium on Personal, Indoor and Mobile Radio Communications, pp. 1–5, September 2008
13. Neskovic, A., Neskovic, N., Paunovic, G.: Modern approaches in modeling of mobile radio systems propagation environment. IEEE Commun. Surv. Tutorials **3**(3), 2–12 (2000)
14. Nuño-Barrau, G., Páz-Borrallo, J.M.: A new location estimation system for wireless networks based on linear discriminant functions and hidden Markov models. EURASIP J. Appl. Signal Process. **2006**, 159–159 (2006)
15. Rose, D.M., Jansen, T., Werthmann, T., Türke, U., Kürner, T.: The IC 1004 urban hannover scenario - 3D pathloss predictions and realistic traffic and mobility patterns (2013)
16. Takenga, C., Xi, C., Kyamakya, K.: A hybrid neural network-data base correlation positioning in GSM network. In: 2006 10th IEEE Singapore International Conference on Communication Systems, pp. 1–5, October 2006
17. Takenga, C.M., Kyamakya, K.: Location fingerprinting in GSM network and impact of data pre-processing (2006)
18. Wu, C.L., Fu, L.C., Lian, F.L.: WLAN location determination in e-home via support vector classification. In: 2004 IEEE International Conference on Networking, Sensing and Control, vol. 2, pp. 1026–1031. IEEE (2004)

19. Yu, L., Laaraiedh, M., Avrillon, S., Uguen, B.: Fingerprinting localization based on neural networks and ultra-wideband signals. In: 2011 IEEE International Symposium on Signal Processing and Information Technology (ISSPIT), pp. 184–189, December 2011
20. Zekavat, R., Buehrer, R.M.: Handbook of Position Location: Theory, Practice and Advances, 1st edn. Wiley/IEEE Press, Hoboken/Piscataway (2011)

Dropout-Based Active Learning
for Regression

Evgenii Tsymbalov[✉], Maxim Panov, and Alexander Shapeev

Skolkovo Institute of Science and Technology (Skoltech),
Nobel Street 3, Moscow 121205, Russia
{e.tsymbalov,m.panov,a.shapeev}@skoltech.ru

Abstract. Active learning is relevant and challenging for high-dimensional regression models when the annotation of the samples is expensive. Yet most of the existing sampling methods cannot be applied to large-scale problems, consuming too much time for data processing. In this paper, we propose a fast active learning algorithm for regression, tailored for neural network models. It is based on uncertainty estimation from stochastic dropout output of the network. Experiments on both synthetic and real-world datasets show comparable or better performance (depending on the accuracy metric) as compared to the baselines. This approach can be generalized to other deep learning architectures. It can be used to systematically improve a machine-learning model as it offers a computationally efficient way of sampling additional data.

Keywords: Regression · Active learning · Uncertainty quantification
Neural networks · Dropout

1 Introduction

Active learning is crucial for the applications in which annotation of new data is expensive. Selection of samples is usually based on uncertainty estimation, by adding the samples on which the prediction is most uncertain to the training set. The standard techniques include the models that estimate their uncertainty directly, such as Bayesian methods [1], or algorithms that estimate uncertainty indirectly, through the deviation of the outputs of an ensemble of models. Models of the first type become computationally expensive when it comes to a large number of samples ($>10^4$) and input dimensions (>10). Ensemble-like models require full or partial independent training of several models, which is also time-consuming, even though it can be done in parallel.

At the same time, neural networks can easily handle large amounts of data and thus are widely used in different areas of applied machine learning such as computer vision [2], speech recognition [3], as well as physics [4], manufacturing [5], or chemistry [6]. The popularity of deep learning has increased after the regularization techniques like dropout [7] and optimization procedures [8,9] have

© Springer Nature Switzerland AG 2018
W. M. P. van der Aalst et al. (Eds.): AIST 2018, LNCS 11179, pp. 247–258, 2018.
https://doi.org/10.1007/978-3-030-11027-7_24

been developed. However, there is still a lack of theoretical understanding about capabilities of neural networks, in particular, estimating model uncertainty.

In our work, we propose a fast active learning algorithm for neural networks for regression problems. Our approach is based on a stochastic output of the neural network model, which is performed using different dropout masks at the prediction stage and used to rank unlabeled samples and to choose among them the ones with the highest uncertainty. We demonstrate that our approach has a comparable or better performance with respect to the baselines in a series of numerical experiments.

The rest of the paper is organized as follows. We discuss the related work in Sect. 2. In Sect. 3 our MCDUE (Monte-Carlo Dropout Uncertainty Estimation)-based active learning method is described in detail. Section 4 contains numerical experiments on both real-world and synthetic data. Section 5 draws the conclusions to the paper.

2   Related Work

2.1   Active Learning

Active learning [10] is a framework in which a machine-learning algorithm may choose the most informative unlabeled data samples, and ask an external oracle to annotate them. In statistics and engineering applications, such a setup is usually referred to as an *adaptive design of experiments* [11,12]. In this setup, one has a set (finite or infinite) unlabeled examples, the so-called pool. The function ranking the data points is referred to as an *acquisition function* or a *querying function*. Ideally, it should be designed in such a way that adding a relatively small number of data samples helps to improve the performance of the model.

There are numerous approaches to constructing an acquisition function. In the case of Bayesian models, such as the Gaussian Processes [1], model uncertainty may be defined explicitly: various estimates can be adopted to rank data points or even to choose them optimally over the defined region [13,14]. However, the major drawback of such models is that calculations are usually intractable for a large number of input dimensions. To cope with this problem various techniques and analogues were proposed, such as Bayesian Neural Networks [15], but all of the methods still remain expensive to train when it comes to a considerably large train set or a complex model.

Another approach is a committee-based active learning [16], also known as *query-by-committee*, where an ensemble of models is trained on the same or different parts of the data. In this approach the measured inconsistency of the predictions obtained from different models for a given data sample may be considered as an overall ensemble uncertainty. Various investigations in this field include diversification of models [17] and boosting/bagging exploitation [18]. Unfortunately, in the case of neural network ensembles (see [19] for a detailed review) this approach often leads to an independent training of several models, which may be computationally expensive for the large-scale applications.

2.2 Dropout

Dropout [7,20] is one of the most popular techniques used for a neural network regularization. To put it in plain words, it randomly mutes some of the neurons in hidden layers during the training stage, forcing them to output zero regardless of an input. This feature of the technique is taken into account during the backpropagation stage of training. Stochastic by nature and simple in implementation, it allows to efficiently reduce overfitting and thus has paved the way for new state-of-the-art results in almost every deep learning application. First proposed as an engineering, empirical approach to reducing the correlation between weights, later it has obtained its theoretical interpretation as an averaged ensembling technique [20], a Bernoulli realization of the corresponding Bayesian neural network [21] and a latent variable model [22]. It was shown in [23] that using dropout at the prediction stage (i.e., stochastic forward passes of the test samples through the network, also referred to as *MC dropout*) leads to unbiased Monte-Carlo estimates of the mean and the variance for the corresponding Bayesian neural network trained using variational inference. In [23], Gal also proposes and analyses direct estimates for a model uncertainty. We use a lightweight version of this approach, estimating the model uncertainty based on a sample standard deviation (which is, up to a factor, an unbiased model variance estimate) of the stochastic output of the network.

Although there are some applications of the abovementioned approach to active learning setting for classification problems [24,25], to the best of our knowledge, there has been no study of applicability of this approach to the regression task. In this work, we evaluate our MC dropout-based approach and compare it with high-throughput baselines. Our experiments have shown comparable or better accuracy and supremacy in a speed-accuracy trade-off of the proposed approach.

3 Methodology

3.1 Problem Statement

Let
$$y = f(x), \ x \in \mathcal{X} \subset \mathbb{R}^n, \ y \in \mathbb{R}$$
be some unknown function which is to be approximated using the values from the training set

$$D_{\text{train}} = \{x_j^{\text{train}}, f(x_j^{\text{train}}), \ j = 1, \ldots, N_{\text{train}}\}.$$

Suppose we have a model (more specifically, a neural network) $\hat{f} \colon \mathcal{X} \to \mathbb{R}$ trained on D_{train} with a mean squared error

$$L(\hat{f}, D_{\text{train}}) = \sum_{j=1}^{N_{\text{train}}} \left(f(x_j^{\text{train}}) - \hat{f}(x_j^{\text{train}}) \right)^2$$

as a fitting criterion.

Let us now focus on the setting in which we want to decrease the value of loss function on some set of test points D_{test} by extending the training set D_{train} with some set of additional samples and performing an additional training of the model $\hat{f}$. More precisely, we are given another set of points called the "pool"

$$\mathcal{P} = \{x_j, \ j = 1, \ldots, N_{\text{pool}}, \ \mathcal{P} \subset \mathcal{X}\},$$

which represents unlabeled data. Each point $x^* \in \mathcal{P}$ may be annotated by computing $f(x^*)$ so that the pair $\{x^*, f(x^*)\}$ is added to the training set. We suppose that in a practical application the process of annotation is expensive (i.e., it requires additional resources such as computational time or money), hence we need to choose as few additional points from the set $\mathcal{P}$ as possible to achieve the desired quality of the model. To achieve this goal, we use an acquisition function

$$A(\hat{f}, \mathcal{P}, D_{\text{train}}): \mathcal{P} \to \mathbb{R}_+,$$

which ranks the points from $\mathcal{P}$ in such a way that the points with the larger values of A become more appealing for the model to learn on. In the experimental setting of this work, we use a fixed number m of the points to add at each stage of the active learning process.

In most practical applications an acquisition function is related to a model uncertainty, which may be defined in various ways depending on the model and the field. There are also approaches, such as random sampling, that do not use the information from the model $\hat{f}$ (see Sect. 3.3 for further details). As for computationally heavy models with hundreds of thousands of parameters, such as neural networks, these approaches may be considered as more preferable. In the next section, we introduce a Monte-Carlo Dropout Uncertainty Estimation (MCDUE) approach, which enables us to collect uncertainty information from the neural network.

3.2   Monte-Carlo Dropout Uncertainty Estimation

Using dropout at the prediction stage allows us to generate stochastic predictions and, consequently, to estimate the variance of these predictions. Our approach rests on the hypothesis that data samples with higher standard deviations have larger errors of true function predictions. Although this is not always the case (see Fig. 1), concerning a neural network of a reasonable size trained on a reasonable number of samples we have observed a clear correlation between dropout-based variance estimates and prediction errors (see Fig. 2). It should be noted that the result does vary (like any other result of neural network training) depending on several factors: architecture and size of the neural network, samples used for initial training and training hyperparameters, such as regularization, learning rate, and dropout probability. The MCDUE-based active learning algorithm we propose is summarized below.

1. **Initialization.** Choose a trained neural network $\hat{f}(x) = \hat{f}(x, \omega)$, where ω is a vector of weights. Set the dropout probability π. Set the number of stochastic runs T.

2. **Variance estimation.** For each sample x_j from the pool $\mathcal{P}$:

   (a) Make T stochastic runs using dropout of the model $\hat{f}$ and collect outputs $y_k = \hat{f}_k(x_j) = \hat{f}(x_j, \omega_k), k = 1, \ldots, T$, where ω_k are sampled from Bernoulli distribution with parameter π.

   (b) Calculate the standard deviation (as an acquisition function):

$$s_j = A^{\mathrm{MCDUE}}(x_j) = \sqrt{\frac{1}{T-1}\sum_{k=1}^{T}(y_k - \bar{y})^2}, \ \bar{y} = \frac{1}{T}\sum_{k=1}^{T}y_k.$$

3. **Sampling.** Pick m samples with the largest standard deviations s_j.

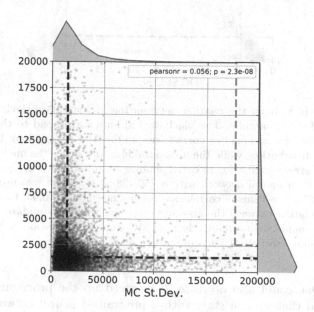

Fig. 1. Scatter plot shows the relation between the MC standard deviation and the absolute error for test samples. The black dashed lines correspond to the medians of distributions, the vertical blue line corresponds to the 0.99 percentile of the MC standard deviation distribution, while the horizontal blue line shows the median percentile of the absolute error distribution of corresponding samples, which is equal to 0.783 in this case. Five-layer neural network with a 256-128-64 structure was used on the Online News Popularity dataset [26]. The Pearson correlation coefficient equals to 0.056, thus showing no linear relation between the absolute error and the MC standard deviation. (Color figure online)

In our experiments we used the dropout probability $\pi = 0.5$, and number of stochastic runs for each sample $T = 25$. We found out that, generally, the decrease in π results in an approximately linear scaling of standard deviations s_j for not too small π. The computational cost per one sample is $O(TN_{\mathrm{pool}})$. However, on modern GPU-based implementations the sampling can be done

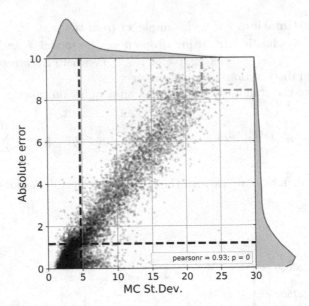

Fig. 2. Scatter plot shows the relation between the MC standard deviation and the absolute error for test samples. The black dashed lines correspond to the medians of distributions, the vertical blue line corresponds to the 0.99 percentile of the MC standard deviation distribution, while the horizontal blue line shows the median percentile of the absolute error distribution of corresponding samples, which is equal to 0.975 in this case. Five-layer neural network with a 256-128-64 structure was used on the CT slices dataset [27]. The Pearson correlation coefficient equals to 0.93, thus showing an almost linear relation between the absolute error and the MC standard deviation, so if we choose a sample with a relatively high MC standard deviation, it will probably have large absolute error. (Color figure online)

in parallel. One could also decrease T to speed up the procedure. It should be emphasized that we can start with a pre-trained neural network from the previous iteration if we train the model on the extended training set; such a method may significantly speed-up retraining.

3.3   Baselines

We use the following baselines to compare the performance.

Random Sampling. This algorithm samples random points x from the pool. Computational cost is $O(1)$ in this case, which makes this algorithm the fastest compared to all the others.

Greedy Max-Min Sampling. To sample a point, this algorithm takes the point from the pool most distant from the training set (in the l_2 sense) and adds it to the training set. This process continues until the required number of points is added to the training set. The acquisition function for max-min sampling is

$$A_{\mathrm{MM}}(x) = \min_{k} \; \|x - x_k^{\mathrm{train}}\|^2.$$

In this case, computational cost can be estimated as $O(N_{train}N_{pool})$.

Batch Max-Min Sampling. Although straightforward and intuitive, the max-min sampling is also computationally expensive in the case of a large number of dimensions and pool/training set size, since it requires that a full distance matrix is calculated on every stage of active learning. We propose the batch version that has the same acquisition function A_{MM} but samples K points which are the most distant from the training set. Although it does seem less optimal, this solution speeds up sampling up to K times assuming K is the number of samples to be sampled on each iteration. In our experiments, we set K equal to 4.

4   Results

4.1   Experimental Setup

We focus on the non-Bayesian methods as comparable in terms of the computational time for large datasets and models and compare the active learning algorithms in the following experimental setup:

1. Initialization of the initial dataset I, training pool $\mathcal{P}$, number of samples added on each step m, the final size of the dataset f, network architecture, and learning parameters.
2. The network is trained on the initial dataset I. Its weights are copied to the networks corresponding to each active learning algorithm.
3. For each active learning algorithm:
   (a) While $|I| < f$:
      i. Obtain the rank r_j for every $x_j \in \mathcal{P}$ using an acquisition function A.
      ii. Sample point set $S \subset \mathcal{P}$, $|S| = m$ with maximal ranks r_j.
      iii. Add S to I: $I := I \bigcup S$.
      iv. Exclude S from the corresponding $\mathcal{P}$: $\mathcal{P} := \mathcal{P}/S$.
      v. Train the neural network on I.
   (b) Calculate the metrics.

   For each experiment, the number of training epochs was set to 10000. We used the l_2-regularization of the weights with the regularization parameter $\alpha = 10^{-5}$, a five-layer fully-connected network with the $256-128-64$ architecture and leaky linear rectifier [28] with leakiness $\beta = 0.01$ as an activation function. We used the Theano library [29] and Lasagne framework [30]. Data points were shuffled and split in the following ratio: 20% on a training set, 60% on a pool, 20% on the test set.

4.2   Metrics

During the neural network training, we optimize the mean squared error (MSE, l_2) metric. However, we also report the mean absolute error (MAE, l_1) and maximum absolute error (MaxAE, l_∞).

We believe that the actual task of regression is more general than optimization of one given metric (e.g., MSE); thus, the choice of a particular metric is merely an operationalization of the real problem behind the regression task. Moreover, several applications exist, in which the maximal error is a much more appropriate accuracy metric (like chemistry or physical simulations) than the mean error. Unfortunately, it is hard to use the l_∞ loss function for training neural network since it is non-differentiable. In case one deals with two algorithms that have a similar MSE and significantly different MaxAE, the algorithm with a smaller maximal error should be preferred.

4.3   Datasets

We took the data from UCI ML repository [31], see the Table 1 for more details. All the datasets represent real-world problems with 15+ dimensions and 30000+ samples. The exception is the synthetic Rosenbrock 2000D dataset, which has 10000 samples and 2000 dimensions.

Table 1. Summary of the datasets used in our experiments.

Dataset name	# of samples	# of attributes	Feature to predict
BlogFeedback [32]	60021	281	Number of comments
SGEMM GPU [33]	241600	18	Median calculation time
YearPredictionMSD [34]	515345	90	Year
Relative location of CT slices [27]	53500	386	Relative location
Online News Popularity [26]	39797	61	Number of shares
KEGG Network [35]	53414	24	Clustering coefficient
Rosenbrock 2000D [36]	10000	2000	Function value

Since the neural network training procedure is stochastic by its nature, we conducted 20 experiments shuffling the dataset and re-initializing the weights each time.

4.4   Ratio Plots

First, we compare our MCDUE-based approach with the baseline approaches using the ratio of errors in various metrics. Figures 3 and 4 show that our active learning approach has a better performance than random sampling in RMSE and MaxAE metrics, and a small accuracy increase as compared to a max-min algorithm. It should be noted that as the number of active learning iterations (new data gathering and learning on the top of it) increases (thus leaving the data pool empty), ratio turns to 1.

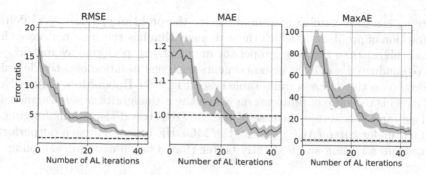

Fig. 3. Comparison of MCDUE-based algorithm and random sampling algorithm: the ratio of errors (training curves) for various metrics on the KEGG Network dataset [35]. Ratio bigger than 1 (dashed line) shows the superiority of the MCDUE-based algorithm. The blue line shows the mean over 25 experiments, the standard deviation is also shown. One can see that the proposed algorithm outperforms the random sampling significantly on RMSE and MaxAE metrics. (Color figure online)

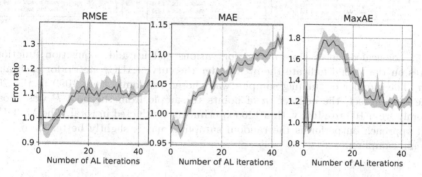

Fig. 4. Comparison of MCDUE-based algorithm and max-min sampling algorithm: the ratio of errors (training curves) for various metrics on KEGG Network dataset [35]. Ratio bigger than 1 (dashed line) shows the superiority of the MCDUE-based algorithm. The blue line shows the mean over 25 experiments, the standard deviation is also shown. One can see that the proposed algorithm slightly outperforms the max-min sampling across all the metrics by up to 20%. (Color figure online)

4.5   Dolan-More Plots

To compare the performance of the algorithms across the different datasets (see Table 1), the initial number of training samples and training samples themselves, we will use Dolan-More curves, which, following [37], may be defined as follows. Let q_a^p be an error measure of the a-th algorithm on the $\mathcal{P}$-th problem. Then, defining the performance ratio $r_a^p = \frac{q_a^p}{\min_x(q_x^p)}$, we can define the Dolan-More curve as a function of the performance ratio factor τ:

$$\rho_a(\tau) = \frac{\#(p : r_a^p \leq \tau)}{n_p}, \tag{1}$$

where n_p is a total number of evaluations for the problem p. Thus, $\rho_a(\tau)$ defines the fraction of problems in which the a-th algorithm has the error not more than τ times bigger than the best competitor in the chosen performance metric.

We conducted a number of experiments with one iteration of active learning performed on the datasets from Table 1, see Fig. 5 for Dolan-More curves. Note that $\rho_a(1)$ is the ratio of problems on which the a-th algorithm performance was the best, and it is always the case of the MCDUE-based algorithm. Judging by the area under curve (AUC) metric, the MCDUE-based approach outperforms the random sampling and is slightly better than a batch max-min sampling.

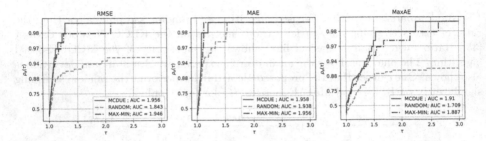

Fig. 5. Log-scaled Dolan-More curves for various metrics and acquisition functions. Figures on the legend indicate the area under the curve (AUC) metric for each algorithm. The number of training samples was chosen randomly from 1000 to the 20% of the training set, the number m of points to sample from the pool $\mathcal{P}$ was chosen randomly from 100 to 1100, with a 140 experiments conducted in total. The MCDUE-based approach outperforms the random sampling and is slightly better than batch max-min sampling.

5 Summary and Discussion

We have proposed an MCDUE-based (Monte-Carlo Dropout Uncertainty Estimation-based) approach to active learning for the regression problems. This approach allows neural network models to estimate self-uncertainty of unlabeled data samples. Numerical experiments on real-life datasets have shown that our algorithm outperforms both random sampling and the max-min baseline. Compared to the latter, our algorithm is faster and relies on the information provided by the model.

Theoretical connections with Bayesian neural networks require further investigations. In this empirical study we propose an *ad hoc*, lightweight approach that may be adopted to strengthen the existing models. In future work, several factors that may affect the results, among which are network architecture, regularization, and dropout probability, require further in-depth study. Another significant improvement would be in a model could estimate the number of samples to acquire on each step.

Acknowledgements. The work was supported by the Skoltech NGP Program No. 2016-7/NGP (a Skoltech-MIT joint project).

References

1. Rasmussen, C.E.: Gaussian processes in machine learning. In: Bousquet, O., von Luxburg, U., Rätsch, G. (eds.) ML 2003. LNCS (LNAI), vol. 3176, pp. 63–71. Springer, Heidelberg (2004). https://doi.org/10.1007/978-3-540-28650-9_4
2. Szegedy, C., et al.: Inception-v4, inception-resnet and the impact of residual connections on learning. In: AAAI, vol. 4. (2017)
3. Sainath, T.N., et al.: Deep convolutional neural networks for LVCSR. In: 2013 IEEE International Conference on Acoustics, Speech and Signal Processing (ICASSP). IEEE (2013)
4. Baldi, P., Sadowski, P., Whiteson, D.: Searching for exotic particles in high-energy physics with deep learning. Nat. Commun. **5**, 4308 (2014)
5. Anjos, O., et al.: Neural networks applied to discriminate botanical origin of honeys. Food Chem. **175**, 128–136 (2015)
6. Schütt, K.T., et al.: Quantum-chemical insights from deep tensor neural networks. Nat. Commun. **8**, 13890 (2017)
7. Hinton, G.E., et al.: Improving neural networks by preventing co-adaptation of feature detectors. arXiv preprint arXiv:1207.0580 (2012)
8. Tieleman, T., Hinton, G.: Lecture 65.-rmsprop: divide the gradient by a running average of its recent magnitude. COURSERA: Neural Netw. Mach. Learn. **4**(2), 26–31 (2012)
9. Kingma, D.P., Ba, J.: Adam: a method for stochastic optimization. arXiv preprint arXiv:1412.6980 (2014)
10. Settles, B.: Active learning. Synth. Lect. Artif. Intell. Mach. Learn. **6**(1), 1–114 (2012)
11. Fedorov, V.: Theory of Optimal Experiments. Elsevier, Amsterdam (1972)
12. Forrester, A., Keane, A.: Engineering Design via Surrogate Modelling: A Practical Guide. Wiley, Hoboken (2008)
13. Sacks, J., et al.: Design and analysis of computer experiments. Stat. Sci. **4**, 409–423 (1989)
14. Burnaev, E., Panov, M.: Adaptive design of experiments based on gaussian processes. In: Gammerman, A., Vovk, V., Papadopoulos, H. (eds.) SLDS 2015. LNCS (LNAI), vol. 9047, pp. 116–125. Springer, Cham (2015). https://doi.org/10.1007/978-3-319-17091-6_7
15. Neal, R.M.: Bayesian Learning for Neural Networks, vol. 118. Springer, New York (2012)
16. Seung, H.S., Opper, M., Sompolinsky, H.: Query by committee. In: Proceedings of the Fifth Annual Workshop on Computational Learning Theory. ACM (1992)
17. Melville, P., Mooney, R.J.: Diverse ensembles for active learning. In: Proceedings of the Twenty-First International Conference on Machine Learning. ACM (2004)
18. Mamitsuka, N.A.H.: Query learning strategies using boosting and bagging. In: Machine Learning: Proceedings of the Fifteenth International Conference (ICML 1998), vol. 1. Morgan Kaufmann Publishers Inc. (1998)
19. Li, H., Wang, X., Ding, S.: Research and development of neural network ensembles: a survey. Artif. Intell. Rev. **49**(4), 455–479 (2018)
20. Srivastava, N., et al.: Dropout: a simple way to prevent neural networks from overfitting. J. Mach. Learn. Res. **15**(1), 1929–1958 (2014)
21. Gal, Y., Ghahramani, Z.: Dropout as a Bayesian approximation: representing model uncertainty in deep learning. In: International Conference on Machine Learning (2016)

22. Maeda, S.: A Bayesian encourages dropout. arXiv preprint arXiv:1412.7003 (2014)
23. Gal, Y.: Uncertainty in Deep Learning. University of Cambridge, Cambridge (2016)
24. Kampffmeyer, M., Salberg, A.-B., Jenssen, R.: Semantic segmentation of small objects and modeling of uncertainty in urban remote sensing images using deep convolutional neural networks. In: CVPRW IEEE Conference (2016)
25. Gal, Y., Islam, R., Ghahramani, Z.: Deep Bayesian active learning with image data. arXiv preprint arXiv:1703.02910 (2017)
26. Fernandes, K., Vinagre, P., Cortez, P.: A proactive intelligent decision support system for predicting the popularity of online news. In: Pereira, F., Machado, P., Costa, E., Cardoso, A. (eds.) EPIA 2015. LNCS (LNAI), vol. 9273, pp. 535–546. Springer, Cham (2015). https://doi.org/10.1007/978-3-319-23485-4_53
27. Graf, F., Kriegel, H.-P., Schubert, M., Pölsterl, S., Cavallaro, A.: 2D image registration in CT images using radial image descriptors. In: Fichtinger, G., Martel, A., Peters, T. (eds.) MICCAI 2011. LNCS, vol. 6892, pp. 607–614. Springer, Heidelberg (2011). https://doi.org/10.1007/978-3-642-23629-7_74
28. Maas, A.L., Hannun, A.Y., Ng, A.Y.: Rectifier nonlinearities improve neural network acoustic models. In: Proceedings of ICML, vol. 30, no. 1 (2013)
29. Al-Rfou, R., et al.: Theano: a Python framework for fast computation of mathematical expressions. arXiv preprint arXiv:1605.02688, vol. 472, p. 473 (2016)
30. Dieleman, S., et al.: Lasagne: first release, August 2015 (2016). https://doi.org/10.5281/zenodo.27878
31. Dua, D., Karra Taniskidou, E.: UCI Machine Learning Repository. University of California, School of Information and Computer Science, Irvine, CA (2017). http://archive.ics.uci.edu/ml
32. Buza, K.: Feedback prediction for blogs. In: Spiliopoulou, M., Schmidt-Thieme, L., Janning, R. (eds.) Data Analysis, Machine Learning and Knowledge Discovery. SCDAKO, pp. 145–152. Springer, Cham (2014). https://doi.org/10.1007/978-3-319-01595-8_16
33. Nugteren, C., Codreanu, V.: CLTune: a generic auto-tuner for OpenCL kernels. In: 2015 IEEE 9th International Symposium on Embedded Multicore/Many-Core Systems-on-Chip (MCSoC). IEEE (2015)
34. Bertin-Mahieux, T., et al.: The million song dataset. In: ISMIR, vol. 2, no. 9 (2011)
35. Shannon, P., et al.: Cytoscape: a software environment for integrated models of biomolecular interaction networks. Genome Res. 13(11), 2498–2504 (2003)
36. Rosenbrock, H.H.: An automatic method for finding the greatest or least value of a function. Comput. J. 3(3), 175–184 (1960)
37. Dolan, E.D., Moré, J.J.: Benchmarking optimization software with performance profiles. Math. Program. 91(2), 201–213 (2002)

Analysis of Dynamic Behavior Through Event Data

Neural Approach to the Discovery Problem in Process Mining

Timofey Shunin[(✉)], Natalia Zubkova, and Sergey Shershakov

Laboratory of Process-Aware Information Systems (PAIS Lab),
Faculty of Computer Science,
National Research University Higher School of Economics, Moscow 101000, Russia
{tmshunin,nszubkova}@edu.hse.ru, sshershakov@hse.ru

Abstract. Process mining deals with various types of formal models. Some of them are used at intermediate stages of synthesis and analysis, whereas others are the desired goals themselves. Transition systems (TS) are widely used in both scenarios. Process discovery, which is a special case of the synthesis problem, tries to find patterns in event logs. In this paper, we propose a new approach to the discovery problem based on recurrent neural networks (RNN). Here, an event log serves as a training sample for a neural network; the algorithm extracts RNN's internal state as the desired TS that describes the behavior present in the log. Models derived by the approach contain all behaviors from the event log (i.e. are perfectly fit) and vary in simplicity and precision, the key model quality metrics. One of the main advantages of the neural method is the natural ability to detect and merge common behavioral parts that are scattered across the log. The paper studies the proposed method, its properties and possible cases where the application of this approach is sensible as compared to other methods of TS synthesis.

Keywords: Process mining · Transition systems · Quality metrics
Recurrent neural networks · Process models synthesis · FSA/FSM

1 Introduction

Neural networks are statistical computing models that are applied today to many practical tasks, e.g. image processing, machine translation and pattern recognition. In *supervised learning*, a neural network trains on a sample of already known objects, i.e. for each input we have a predefined correct answer. The main idea behind training a neural network is to tune such a configuration, so that the model's answers are as close to the correct ones as possible. As for *recurrent neural networks* (RNN), they not only learn on input objects, but also provide the *context* for each following prediction. This helps a neural network preserve

This work is supported by the Basic Research Program of the National Research University Higher School of Economics and funded by RFBR according to the Research project No. 18-37-00438 "mol_a".

the state in which the decision was made. In this paper, we discuss applying RNN methods to process mining problem of *process discovery*.

The task of process discovery consists in deriving a model so that it captures the behavior present in the input data. As it is similar to the task of *pattern recognition* [1], in this paper we focus on solving the process discovery task using a recurrent neural network approach. Regarding an event log as an input training sample, for each event in the log we can train our neural network to predict the next event. Moreover, each state of the neural network can be interpreted as a state in a *finite state machine* (FSM), or, more generally, a *transition system* (TS), that is implicitly present behind the configuration of the neural network. Our ultimate goal is to extract this transition system that represents a model of a process present in the input log. We assume that one of the advantages of the neural method is its ability to detect common behavioral parts that are scattered across the log.

While transition systems themselves are appropriate process models, such models are commonly known as low-level. Thus, a derived TS that includes all behaviors present in the log, e.g. a prefix tree, is comparable in size to the log itself, which could lead to a model that is too complex to analyze. An *adequate* model should provide a certain level of abstraction allowing to study both simple and complicated processes. Another issue is that the nature of these models does not allow them to explicitly represent concurrency. This problem can be eliminated by using appropriate models allowing concurrency such as *BPMN* or *UML activity diagrams*, which are generally modeled based on Petri nets. There exists an approach to synthesize a Petri net from a TS, based on the theory of regions [2–4]. However, direct application of the region-based algorithm to intermediate models is exponentially dependent on the size of a TS, and, correspondingly, on the size of the input log [5]. In this paper we focus on synthesizing transition systems themselves and do not regard the following step of deriving models with concurrency.

Previously, a few works that cover methods of inferring transition systems using neural networks were published [6,7]. At that time, technologies based on neural networks were just starting to appear. Moreover, the computational powers of computers were significantly lower compared to the power of modern ones. Since the field of practical machine learning has experienced a rise in the recent decade, which resulted in rationality and great ease of implementing complex algorithms in practice, we believe that it is now that neural networks could solve practical tasks of process mining and be used to their full potential.

One of the main goals of our study is to analyze the efficiency of the proposed approach, i.e. see how different quality metrics depend on input data and user-specified parameters. There exists several approaches to evaluate the quality of models inferred by different process discovery techniques. In process mining, a number of *quality metrics* [8–11] are used to quantify an extent to which a model describes an input log. We follow the same way, prove that our approach produces *perfectly-fit* models, and compute metrics for the inferred models to assess their quality.

The main contributions are as follows: (1) describing and implementing a method of building transition systems using recurrent neural networks; (2) comparing the quality metrics of models constructed by the described method to quality metrics of models inferred using the *frequency-based reduction algorithm* (FQR) [10] and *prefix trees* [3]; (3) concluding properties of event logs where RNN method shows better results as compared to the FQR method.

The rest of the paper is organized as follows. Section 2 gives a brief overview of related works in the context of process mining and neural networks. In Sect. 3 we describe some relevant concepts needed for the explanation of the RNN approach which is itself described in Sect. 4. Experiments, metrics calculations and results are discussed in Sect. 5. Finally, Sect. 6 concludes the paper and gives some directions for future work.

2   Related Work

The idea of using neural networks to infer FSMs is not completely new. Das and Mozer in [6] described a clustering architecture for finite state machine induction. Their idea was to teach neural network on both positive and negative examples of strings, composed from a two-symbol alphabet, so the network could predict if the input string is a valid example or not. They clustered the inner states of a neural network to extract states for an FSM. This approach was evolved by Cook and Wolf in 1998 [7] in the context of process discovery. They were aiming at inferring grammars for the samples of valid strings. Their extension allowed more complex languages which were not limited by two symbols.

Buijs et al. in [8] formulate key quality metrics for assessing Petri net models in process mining, namely *replay fitness, simplicity, precision* and *generalization*. While their variant of calculating metrics is widely used by the community, it is not the only possible one. Other calculations of simplicity are rather similar to the one proposed in [8], whereas for precision there is no common baseline. In an attempt to bring such a baseline, recently Tax et al. [11] have proposed five axioms a good precision measure should satisfy in order to describe any log or any model consistently.

In [10], authors propose a new approach that allows to infer perfectly fit TSs whose simplicity and precision could be adjusted by configuring the algorithm parameters. The paper describes a method for inferring TS based on the frequency-based reduction algorithm. A new approach for measuring precision of perfectly fit TSs, based on models simulation, is also present there. In this paper, we use that approach to calculate simplicity and precision for comparing models.

3   Preliminaries

3.1   Process Mining Related Concepts

Let A be a set of activities generated by a process. An *event e* is an occurence of an activity in a log. A *token* is an element of a log that is passed to a neural network. We assume an event and a token to be synonyms which are used in different contexts. A set Λ of all unique events of the log is called an input alphabet. By $|\Lambda|$ we denote the cardinality of the input alphabet.

Definition 1 (Trace, Event log). *A trace σ is a finite sequence of events $\sigma = \langle e_1, ..., e_n \rangle \in A^+$, where A^+ is the set of all non-empty finite sequences over A. An event log L is a multiset of traces.*

$L = [\langle a, b, c \rangle, \langle a, b, d \rangle]$ is an example of an event log of two traces.

Definition 2 (Transition System). *A transition system TS is a tuple $TS = (S, E, T, s_0, AS)$, where S is finite space of states, E is a finite set of labels, $T \subseteq S \times E \times S$ is a set of transitions, $s_0 \in S$ is an initial state, and $AS \subseteq S$ is a set of final (accepting) states. By $s_i \in S$ we denote a ith state in a TS.*

3.2   Neural Networks Related Concepts

A *neural network* [12] is a computing model that usually consists of three main parts: an *input layer*, a *hidden layer* and an *output layer*. Each layer is a set of neurons, and each neuron is a computational unit that applies certain *activation functions* to data flowing through it. The layers are linked by weighed connections that transmit signals from a neuron in one layer to one or more neurons in the next layer. The resulting transformation of data passed to the input layer should meet the conditions implied by a problem solved by the network. Such conditions in a neural network are represented by a *loss function*. In order to get this transformation, the weights of the tying connections are configured. The process of configuring weights is called a *learning, or training process*. It consists in optimizing the loss function, or error, using *back propagation* [12]. On each training iteration, the current error is propagated to the previous layer, where it is used to modify the weights in such a way that the error is minimized. In [1], Bishop provides a more complete insight on the meaning and purpose of computations present in a neural network.

The computations could also be influenced by some parameters of a neural network that could not be optimized during the learning process. Such parameters are called *hyperparameters* and they should be defined before the learning process begins. The number of neurons in the hidden layer is an example of such a parameter.

The approach described in this paper is based on *recurrent neural networks* (RNN). Such networks have one or more recurrent connections linking together output and input layers, which allows a network to use its previous computations for the following predictions. The training of an RNN is usually similar to the training of an ordinary neural network and often involves *stochastic gradient descent (SGD)* [12].

4   Description of the Neural Approach

For clarification of the approach we consider the event log $L = [\langle a, b, c, d, e \rangle, \langle a, b, d \rangle]$ as a running example.

The inner computations in a neural network require preprocessing our traces as follows:

1. Adding two new reserved tokens "$" and "#" to each trace indicating the beginning and the end of the trace respectively.
2. Padding all traces with "#" symbol so all traces have the same length.

This way, L transforms into $\tilde{L} = [\langle \$, a, b, c, d, e, \# \rangle, \langle \$, a, b, d, \#, \#, \# \rangle]$, where all traces have the same length of 7. We also encode present tokens with integers from 0 to $|A| + 1$.

In order to build the desired TS we create a neural network that is capable of generating traces from the input log. On each step the network predicts the probabilities of next possible tokens and chooses the next token from the most probable ones. Interpreting the problem of predicting tokens as a task of *classification* allows us to use specific methods and architectures [13] that have proven their efficiency [14]. For each *token* (event) α_i we want to determine a *class of tokens* of the following token α_{i+1}, i.e. an activity represented by it. Moreover, we should take into consideration the history of the trace; otherwise the prediction would only depend on the previous token, which would lead to learning a graph of relations between individual events. That is why a recurrent connection is present, composing a more complex input vector. If the network is capable of generating plausible traces, then the inner state, from which the prediction is made, is credible and represents both the history of the trace and the current token, thus could be used as a state in the TS. To solve this task, we developed a neural network architecture presented in Fig. 1.

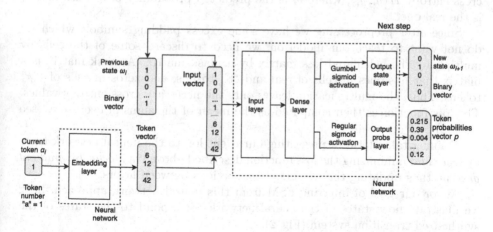

Fig. 1. Neural network architecture with example values

In the *embedding layer*, an integer, that encodes the next token from the trace is transformed into a *token vector*. The *input vector* for the neural network is a concatenation of two vectors: a *binary vector*, which represents a *previous state*

Table 1. Token probabilities for the first trace of $\tilde{L}$

$	0.215
a	0.39
...	...
#	0.12

Table 2. Cross entropy matrix for $\tilde{L}$

$	a	b	c	d	e	#	
2.57	1.45	2.65	1.77	3.72	4.62	0	16.78
$	a	b	d	#	#	#	
2.57	1.45	2.65	2.51	0	0	0	9.18
Average							12.98

of the neural network, and a *token vector*. Then the *input vector* is sent to the *dense (hidden) layer*, which applies a linear transformation to the vector. After that, different activation functions are applied to the computed vector in order to extract two different entities from the output layer. The first one is a new *discrete binary vector*, interpreted as the *new state of the RNN*, and returned as the new parameter for the next iteration. To compute it we use the *Gumbel Sigmoid* as it has proven its efficiency in approximating discrete values [15]. The other is a vector of token probabilities p (see Table 1). Here, we apply regular sigmoid function as its output appertains to the segment of $[0; 1]$. We use the vector of token probabilities to calculate our loss function as follows.

We build a $m \times r$ loss (cross entropy) matrix, where m is the number of traces in the input log and r is the maximum length of a preprocessed trace from the input log. Each row of the matrix relates to a trace in the log, each element relates to a token (event) in a trace. On every step, an element is calculated as cross entropy $H(\hat{y}_i, y_i)$, where $\hat{y}_i$ is the predicted probability of a token and y_i is the real one.

Since after preprocessing we have some excess padding symbols which we do not want to teach our model on, we need to discard some of the cells by multiplying the calculated loss matrix by a mask matrix. A mask matrix is a matrix that has ones in cells of real and "$" tokens, and zeros in cells of "#" tokens. Thus, the multiplication leaves only the necessary cross entropy values. The loss function is then computed as the mean of the sums of every row (see Table 2).

Training takes form of presenting our event log to the neural network trace by trace and minimizing the loss function described above via *stochastic gradient descent*, thus updating the weights of connections between layers.

As for our goal of inferring FSM from this neural network approach, first, we illustrate how states of the neural network correspond to the states of the synthesized transition system (Fig. 2).

Let Ω be a set of states of a neural network represented by the states of neurons in the output layers. By $\omega_i \in \Omega$ we denote a state of a recurrent neural network on step i of our algorithm. Then, we define three bijective functions $\xi : \Omega \to \mathbb{N}$, $\psi : \mathbb{N} \to S$ and $\phi : \Omega \to S$ as follows. As ω_i is a binary vector, ξ interprets it as a binary number and converts it to the equivalent decimal

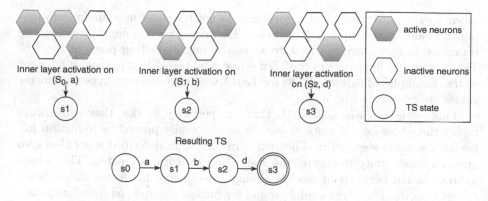

Fig. 2. Inner layer activation patterns on different inputs

number; ψ converts a number j from $\mathbb{N}$ into a state s_j; ϕ is a composition of ψ and ξ; Ω is isomorphic to S through $\mathbb{N}$.

The construction of the $TS = (S, E, T, s_0, AS)$ proceeds according to the following steps. We start by adding state s_0 to the originally empty TS. The initial state ω_0 of the neural network is a vector of zeros. On i-th step the current state of the RNN is ω_i and a new input token is α_i obtained from the *original* input trace. Both compose the input vector of the RNN, and output vectors are ω_{i+1} and p_{i+1}. The current state of TS is $s_i = \phi(\omega_i)$; the following state $s_{i+1} = \phi(\omega_{i+1})$ is added in the TS if it is not present or reused otherwise. These states are connected by the transition $t_i = (s_i, \alpha_i, s_{i+1})$. The described step is repeated for all non-"#" tokens of each trace.

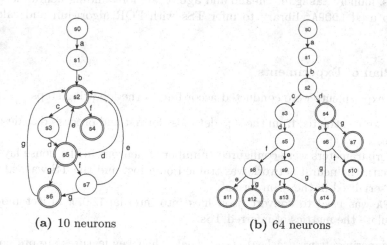

(a) 10 neurons   (b) 64 neurons

Fig. 3. TS inferred for *Log1*

An extracted TS has a unique transition $t_0 = (s_0, \$, s_1)$ due to the fact that we marked the beginning of all traces with "$". Both the transition and the token have a purely technical nature and do not bear any meaningful information relevant to the input log behavior. Therefore, we can simply remove the transition t_0 and consider the state s_1 as the initial starting point of the TS. The state s_0 is still present in the TS if other valid transitions connect it to other states. Examples of extracted TSs for $Log1$[1] with two different hyperparameter values are present in Fig. 3.

This method synthesizes FSMs that are perfectly fit (i.e. they can always replay the whole log) *by construction*. It can be easily proved by induction following the same steps as in Theorem 1 in [10]. The described algorithm also gives an opportunity to control the states and transitions created. This allows us to make not perfectly fit models, though give them other features, e.g. being robust to noise. This gives a solid ground for further research, as there are plenty of ways of analysing predictions of token probabilities mentioned above.

The neural network presented above has one *hyperparameter*: the number of neurons in the dense layer. Each additional neuron not only has direct influence on the learning quality, but also doubles the output state space. Technically, the model has more hyperparameters. The number of neurons in the output vector and the embedding method mentioned earlier could also be regarded as such. We believe that they hold less importance and thus excluded them from consideration during our experiments. However, in further experiments they might be considered.

5   Experiments and Discussion

Experiments were done using mixed software and languages. We used Python libraries, namely `lasagne`, `theano` and `agentnet`, for building neural networks. We also used `LDOPA`[2] library to infer TSs with FQR algorithm and calculate metrics.

5.1   Plan of Experiments

All the experiments were conducted according to the following steps:

1. Logs were extracted from the .sq3 database format and processed as described in Sect. 4.
2. Hyperparameters were configured (number of neurons in the dense layer was chosen), the neural network was trained on a log, and the TS was extracted and serialized in the .json format.
3. `LDOPA` was used to convert .json files into internal `LDOPA` object model to calculate the metrics of inferred TSs.

In our experiments we use *Log1, Log2, Log3*. Their key features are presented in Table 3. In conducting the experiments, the "HP Pavillion dv6" PC with Intel Core i7-4510U CPU was used.

---

[1] See Table 3 for the log's details.
[2] The software is avaliable at the website https://pais.hse.ru/research/projects.

Table 3. Log information

Log	Number of traces	Number of unique tokens	Total number of tokens
Log1	8	7	41
Log2	243	143	2444
Log3	774	46	6604

Table 4. Untrained NNs and prefix trees

Log	Model type	Simplicity	Precision
Log2	Untrained net	0.067	0.943
	Prefix tree	0.053	1
Log3	Untrained net 1	0.022	0.731
	Untrained net 2	0.012	0.809
	Prefix tree	0.007	1

5.2 Experiments

Tuning and configuring model parameters influences metrics of inferred TSs. In our experiments we try to determine the nature of the influence. To evaluate the quality of the synthesized TSs, we focus on simplicity and precision metrics and how the model balances between them. In these experiments we do not evaluate the generalization results and consider parameters' influence on it a future work.

In the conducted experiments we compare the described approach to two different algorithms of constructing TSs. The first one is the *k-tail* algorithm of constructing prefix trees [16]. This approach allows to generate models with the maximum precision value of 1. However, such models tend to be very complex and, therefore, have low simplicity. The second one is the *frequency-based reduction* algorithm, which is thoroughly described in [10]. This method allows balancing between simplicity and precision by varying algorithm parameters. We compared the RNN inferred TSs with those inferred by FQR with various parameters in order to have a group of models for comparison.

Our general assumption is that as a neural network trains, it learns to recognize similar behavior fragments in the log even if they are sparsely distributed. An *untrained* neural network possesses a random configuration after initializing, so its reaction to input data is undetermined. As initial weights are assigned random values, random neurons are activated. This results in generating sequences of random states which are still connected by the correctly labeled transitions. Such TSs would be less precise than the prefix tree ones, while being similar in complexity. Table 4 confirms this fact. During the learning period we expect specific neurons to start activating when specific token patterns appear. This leads to the increase of the simplicity metric as same output states are obtained from ϕ.

We still expect very high precision from TSs inferred by this approach, as for each state the chosen loss function forces the model to minimize the chances of predicting wrong (unprobable) tokens. This results in the increase in the number of possible states and reduction of the number of outgoing transitions, which consequently makes the inferred TS more precise.

Comparing RNN Models Depending on Their Parameters. We start by comparing inferred models with varying training times. Figure 4 describes

the change in simplicity and precision values over the training period. Models
in both figures consist of 64 neurons in the dense layer. In the first few training
iterations we observe the expected increase in simplicity, because the network is
able to detect the general features of the log. However, as the number of training
iterations grows, the model starts to distinguish even the slightest differences in
logs, thus creating different states for each pattern and increasing precision of
the TS. Due to the fact that the described architecture is fairly simple, we regard
overfitting as the most probable explanation. This means that the trained model
finds a combination of weights that allows it to be more optimal on the training
sample. Such a combination of weights is not always what we are looking for, as it
harms the ability of a neural network to generalize behaviors. There exist several
methods to overcome this problem, e.g. using *drop-out* techniques or *k-fold cross
validation*. The impact of these methods on the results of the RNN approach
needs further research. Figure 4 illustrates three stages of a model depending on
the training time: underfit, fit[3] and overfit. Both underfit and overfit models are
similar in structure to prefix tree models, that is why metric values return to
the initial figures over time.

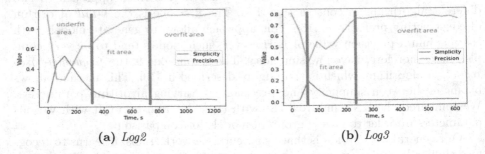

(a) *Log2*   (b) *Log3*

Fig. 4. Simplicity and precision change over time

Next parameter to experiment on is the number of neurons in the dense layer
in neural networks after the training period. The rise of computational power and
classification capability, combined with the exponential growth of the number of
output states and possible overfitting, results in behaviour illustrated on Fig. 5.
As we can see in Fig. 3b, neural networks are capable of learning the prefix tree
models, which also illustrates our assumption about overfitting.

**Comparing the RNN Approach to the Frequency-Based Reduction
Algorithm.** It is necessary to compare models inferred by the neural network
approach with those inferred by the FQR algorithm on close precision and simplicity values. By close values, figures whose residual is less than 0.02 have been
accounted for.

---

[3] Here fit is not used in the context of quality metrics.

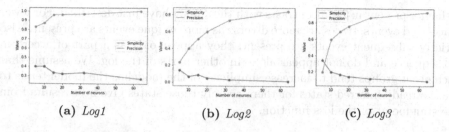

| (a) *Log1* | (b) *Log2* | (c) *Log3* |

Fig. 5. Simplicity and precision change over the number of neurons

Figures 6 and 7 show the figures for *Log2* comparing RNN approach (*red*) and FQR with different parameters [10] (*blue*). For this particular log, RNN proves to be better in both precision and simplicity.

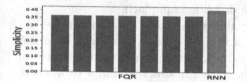

Fig. 6. Simplicity values for close precision, *Log2* (Color figure online)

Fig. 7. Precision values for close simplicity, *Log2* (Color figure online)

However, for *Log3*, the RNN figures are less than those of FQR (Figs. 8 and 9). That is why we need to analyse the structure of both logs to interpret our results and conduct further research.

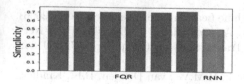
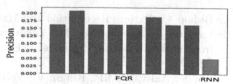

Fig. 8. Simplicity values for close precision, *Log3*

Fig. 9. Precision values for close simplicity, *Log3*

In regard to the most popular sequences of events that occur in the logs, *Log3* has a small amount of very popular sequences of actions. This log also has a relatively small number of unique events. Combined with the a big number of traces and events overall, it is assumed that the neural network is forced to distinguish even relatively similar sets of actions and thus generate new states for these sets as it is the only way of maximizing predictions of tokens appearing.

Behavior present in *Log2* contrasts with the one we have just described. For fewer traces and events, traces are more diverse as more unique events are present. Also strictly subsequent events are present: they appear only as a part of a certain subsequence and do not appear alone in other parts of the log. We assume that such event groups tend to increase simplicity of the model as the model tends to reuse already inferred states for them as only those states give best results from the standpoint of the loss function.

6   Conclusion

In this work a modern neural approach of synthesizing TSs from event logs is developed and implemented. Experiments showed that for some log types, namely for those where similar behaviors appear across the log, this approach gives better results than FQR algorithm. This determines the area of applicability of this method. If one is faced with a log of a structure similar to the structure of *Log2*, it is sensible to use the described approach.

Neural networks are known for their ability to tackle data with noise. As in this work we only built perfectly-fit TSs for the logs, we might have not discovered the full potential of this method yet. Further research in the field of tasks not requiring perfectly fit models is needed in order to find a task where such a model could be of use.

If we regard this approach, different architectures of the inner layers of the net might be considered, as now only a simple neural network with a single hidden dense layer is being experimented on. Moreover, it is concluded that dealing with overfitting using techniques mentioned in 5.2 can potentially provide better results.

References

1. Bishop, C.: Pattern Recognition and Machine Learning. Springer, New York (2006)
2. Lavagno, L., Kishinevsky, M., Yakovlev, A., Cortadella, J.: Deriving Petri Nets from Finite Transition Systems. IEEE Transactions on Computers **47**, 859–882 (1998)
3. van der Aalst, W.M.P., Rubin, V., Verbeek, H.M.W., van Dongen, B.F., Kindler, E., Günther, C.W.: Process mining: a two-step approach to balance between underfitting and overfitting. Softw. Syst. Model. **9**(1), 87 (2008)
4. Solé, M., Carmona, J.: Region-based foldings in process discovery. IEEE Trans. Knowl. Data Eng. **25**(1), 192–205 (2013)
5. Badouel, E., Bernardinello, L., Darondeau, P.: The synthesis problem for elementary net systems is NP-complete. Theor. Comput. Sci. **186**(1), 107–134 (1997)
6. Das, S., Mozer, M.: A unified gradient-descent/clustering architecture for finite state machine induction. In: Morgan Kaufmann Advances in Neural Information Processing Systems, vol. 6, pp. 19–26 (1994)
7. Cook, J., Wolf, A.: Discovering models of software processes from event-based data. ACM Trans. Softw. Eng. Methodol. **7**, 215–249 (1998)

8. Buijs, J.C.A.M., van Dongen, B.F., van der Aalst, W.M.P.: On the role of fitness, precision, generalization and simplicity in process discovery. In: Meersman, R., et al. (eds.) OTM 2012. LNCS, vol. 7565, pp. 305–322. Springer, Heidelberg (2012). https://doi.org/10.1007/978-3-642-33606-5_19

9. Adriansyah, A., Munoz-Gama, J., Carmona, J., van Dongen, B.F., van der Aalst, W.M.P.: Measuring precision of modeled behavior. Inf. Syst. e-Bus. Manag. 13(1), 37–67 (2015)

10. Shershakov, S.A., Kalenkova, A.A., Lomazova, I.A.: Transition systems reduction: balancing between precision and simplicity. In: Koutny, M., Kleijn, J., Penczek, W. (eds.) Transactions on Petri Nets and Other Models of Concurrency XII. LNCS, vol. 10470, pp. 119–139. Springer, Heidelberg (2017). https://doi.org/10.1007/978-3-662-55862-1_6

11. Tax, N., Lu, X., Sidorova, N., Fahland, D., van der Aalst, W.M.P.: The imprecisions of precision measures in process mining. CoRR, abs/1705.03303 (2017)

12. Alpaydin, E.: Introduction to Machine Learning. MIT Press, Cambridge (2010)

13. Weiss, S., Kulikowski, C.: Computer Systems that Learn: Classification and Prediction Methods from Statistics, Neural Nets, Machine Learning, and Expert Systems. Morgan Kaufmann Publishers Inc., Burlington (1991)

14. Mikolov, T., Karafiát, M., Burget, L., Černocký, J., Khudanpur, S.: Recurrent neural network based language model. In: Proceedings of ICSA, pp. 1045–1048 (2010)

15. Jang, E., Gu, S., Poole, B.: Categorical reparameterization with Gumbel-softmax. In: 5th International Conference on Learning Representations (2017)

16. Biermann, A., Feldman, J.: On the synthesis of finite-state machines from samples of their behavior. IEEE Trans. Comput. C-21(6), 592–597 (1972)

Constructing Regular Expressions
from Real-Life Event Logs

Polina D. Tarantsova and Anna A. Kalenkova[✉]

National Research University Higher School of Economics, Moscow, Russia
pdtarantsova@edu.hse.ru, akalenkova@hse.ru

Abstract. Process mining is a new discipline aimed at constructing process models from event logs. Recently several methods for the discovery of transition systems from event logs were introduced. Considering these transition systems as finite state machines classical algorithms for deriving regular expressions can be applied. Regular expressions allow representing sequential process models in a hierarchical way, using sequence, choice, and iterative patterns. The aim of this work is to apply and tune an algorithm deriving regular expressions from transition systems within the process mining domain.

1 Introduction

Process mining is a research field [1] providing methods for the discovery of process models from event logs. *Representational bias*, i.e., a set of restrictions imposed on target process models, plays an important role in process mining. Compact and structured process models clearly represent the underlying processes. One of the aims of the process discovery algorithms is to construct such models from real-life event data. Powerful and widely-used methods for the discovery of process trees (structured models based on sequential, choice, parallel, and iterative patterns) were proposed in [2,3]. Usually, these methods are used under an assumption that not all the traces (sequences of events) of the event log will fit the target process model. In the opposite case, the resulting process model may be too general allowing all the possible behavior. In this work, we assume that the analyst may want to discover fitting and precise structured process models that exactly represent the process behavior. Additionally, we assume that the underlying processes represent sequential behavior.

Regular expressions, which are used to model sequential processes with choice, sequence, and iterative patterns, were selected as a representation bias within this work. The technique proposed in this paper consists of two steps: (1) a transition system is constructed from an event log, using an existing transition system discovery technique [4]; from now on we will call these transitions systems *finite state machines*; (2) a regular expression is synthesized from the given finite state machine. There are several algorithms for constructing regular expressions from finite state machines [5–7]. All these algorithms preserve the language, i.e., finite state machines and resulting regular expressions accept the same set of

© Springer Nature Switzerland AG 2018
W. M. P. van der Aalst et al. (Eds.): AIST 2018, LNCS 11179, pp. 274–280, 2018.
https://doi.org/10.1007/978-3-030-11027-7_26

activity sequences. In this paper we use an algorithm based on the reduction of states [7]. In contrast to the approaches [5,6], this algorithm is iterative and we can iteratively apply heuristics based on the particular properties of the finite state machines constructed from event logs. This algorithm was implemented and tested on real-life event logs showing its capability to construct compact regular expressions in comparison to the regular expressions constructed without the proposed heuristics.

2   Preliminaries

In this section we present basic notions used throughout the paper. Let $\mathcal{A}$ be a set of activities. Let $\mathcal{A}^*$ be a set of all possible sequences of activities. *Language* $\mathcal{L}$ is a set of sequences of activities, i.e., $\mathcal{L} \subseteq \mathcal{A}^*$.

The *regular language* is a language over the set activities $\mathcal{A}$ iff it can be constructed inductively based on the following rules: (1) $\emptyset$ is a regular language; (2) $\{\langle\rangle\}$ is a regular language, denoted as ϵ; (3) $\{\langle a_i \rangle\}$, where $a_i \in \mathcal{A}$, is a regular language, denoted as a_i; (4) let $P, Q \subseteq \mathcal{A}^*$ be regular languages, then (4.1) $P.Q = \{\langle p_1, ..., p_n, q_1, ..., q_m \rangle | \langle p_1, ..., p_n \rangle \in P, \langle q_1, ..., q_m \rangle \in Q\}$ is a regular language, (4.2) $P|Q = \{\sigma | \sigma \in P \cup Q\}$ is a regular language, and (4.3) $P^* = \cup_{i=0}^{\infty} P^n$, where $P^0 = \{\langle\rangle\}$, $P^1 = P$, and $P^n = P^{n-1}.P$, if $n > 1$, is a regular language.

A *finite state machine* is a tuple $FSM = (S, E, T, s_i, S_f)$, where S is a finite set of *states*, E is a finite set of *events* (*activities*), $T \subseteq (S \times E \times S)$ is a set of *transitions*, $s_i \in S$ is an *initial state*, and $S_f \subseteq S$ is a set of *accepting states*. Sequence $t = \langle a_1, ..., a_n \rangle \in \mathcal{A}^*$ is *accepted* by a finite state machine $FSM = (S, \mathcal{A}, T, s_i, S_f)$ iff there exists a sequence of transitions $(s_i, a_1, s_1), (s_1, a_2, s_2), ..., (s_{n-1}, a_n, s_n)$, leading from the initial state s_i to an accepting state $s_n \in S_f$. Language of *FSM*, denoted by $\mathcal{L}(FSM)$, is a set of all accepted sequences of *FSM*.

3   Algorithm Description

This section presents an algorithm for constructing regular expressions from event logs.

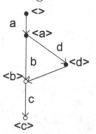

Fig. 1. A finite state machine constructed from the event log E.

The approach presented in this paper consists of two steps: (1) a finite state machine is constructed from an event log, (2) a regular expression is synthesized from the given finite state machine (Fig. 2).

In this work we used a method for constructing finite state machines (transition systems) from event logs [4] based on the prefixes of length 1 together with their subsequent minimization. Let us demonstrate this approach by an example. Since each event log is a set of traces or, in other words, sequences of activities, it represents a language. Consider an event log $E = \{\langle a, b \rangle, \langle a, b, c \rangle, \langle a, d, b \rangle\}$. An example of a finite state machine constructed from this

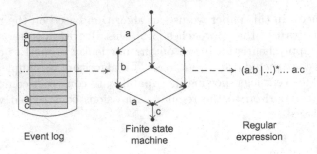

Fig. 2. Discovery approach.

event log is presented in Fig. 1. Accepting states are shown in white and each state corresponds to a sequence of last traversed activities of the length less or equal 1. Although in this concrete example the event log and the language of the corresponding finite state machine coincide, in the general case it is guaranteed that the event log is a subset of the language of the corresponding finite state machine.

In the second step, we apply the approach based on the removal of states to construct regular expressions from finite state machines [7]. The main idea of the state removal approach is to assume that transitions can be labeled not by activities, but by regular expressions. This algorithm starts with the initial finite state machine and then inductively removes the states in order to obtain resulting regular expressions (Fig. 3). At the final steps of the algorithm these resulting expressions label the transitions which lead from the initial state to an accepting state. The order of the processed states significantly affects the size (the length) of the resulting regular expression. We used the strategy proposed in [8]. According to this strategy, the next state which should be removed is a state that produces minimal regular expressions. Thus, the state s with the minimal value $W(s) = (l-1)\sum_{i=1}^{m}|r_{s_is}| + (m-1)\sum_{i=1}^{l}|r_{ss_i}| + (ml-1)|r_{ss}|$, where m (l) is the number of incoming (outgoing) transitions for s, and $r_{s_is_j}$ is a regular expression marking transition from s_i to s_j, will be removed first.

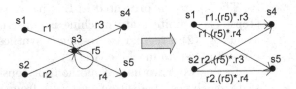

Fig. 3. State removal technique.

Basing on special characteristics of the finite state machines constructed from event logs we propose heuristics aimed to reduce and structure the resulting regular expressions. Let us consider a fragment of a finite state machine presented in Fig. 4.

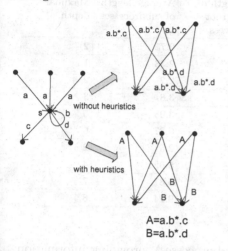

Since this finite state machine was constructed using prefixes of length 1, each state corresponds to a concrete activity and all incoming transitions are marked with this activity. After the removal of state s there will be three regular expressions $a.b^*.c$ and three regular expressions $a.b^*.d$. A repeated regular expression can be replaced by a single symbol denoting a subprocess. This replacement can be applied not only in the first step of the state removal algorithm, but also in the next steps, because (it is confirmed experimentally in the next section) incoming transitions of a state can be labeled by identical regular expressions. Additionally, if incoming transitions are not labeled by identical regular expressions, postfixes, such as $b^*.c$, can also form subprocesses.

Fig. 4. A fragment of a finite state machine.

4   Experimental Results

The state reduction algorithm including the ordering (presented in [8]) and the prefix-based (proposed in this paper) heuristics was implemented [1] and tested on real-life event logs.

Three real-life event logs of a bug tracking system (TS), banking financial system (FS) and flight booking system (BS) were analyzed. First, all these event logs were loaded to ProM (Process Mining Framework) [9] and corresponding finite state machines were constructed. Then these finite state machines were loaded to the tool specially developed by the authors in order to synthesize regular expressions.

Table 1 shows the length of regular expressions constructed without the prefix-based heuristics proposed in this paper, as well as the size (length, length of the root process, average length of subprocesses, depth) of regular expressions constructed with the prefix-based heuristics.

As it follows from the experiment results, the prefix-based heuristics can significantly help to reduce and structure the resulting regular expression. The length of the root regular expression is usually significantly lower than the length of the corresponding regular expression constructed without applying the prefix-based heuristics. Additionally, the proposed approach allows us

---

[1] https://bitbucket.org/akalenkova/mineexpression.

Table 1. The size of the synthesized regular expressions.

Event log	Number of states	Length of RE	Length of RE with replacement	Length of the root process	Average length of subprocesses	Maximal depth
TS	5	48	39	16	2.75	2
FS	7	22	21	14	2.33	1
FS	9	66	54	32	2.2	2
FS	12	397	175	99	3.8	3
FS	13	747	207	97	3.93	5
FS	19	2480	490	214	6.27	6
BS	5	19	16	11	2.5	2
BS	7	569	106	16	5.625	4
BS	9	11000	337	114	5.87	7
BS	11	41953	643	135	9.24	8

to extract repeating regular expressions (subprocesses), providing information about repeating patterns within the event log.

Now let us consider the event log of a bug tracking system (TS) in details. This log is defined over five activities of a bug tracking process: a (*"In Progress"*), b (*"Resolved"*), c (*"Closed"*), d (*"Reopened"*), and e (*"Open"*). The regular expression constructed without the prefix-based heuristics is the following: $c|a.(a)^*|c|b|a.(a)^*.b|c|(e|c.d|a.(a)^*.e|c.d|b|a.(a)^*.b.(d|c.d)).(c.d|a.(a)^*.e|c.d|b|a.(a)^*.b.(d|c.d))^*.(\epsilon|c|a.(a)^*|c|b|a.(a)^*.b|c)$.

The root regular expression constructed using the proposed heuristics is: $c|F|H|(B|\ E|D.(d|C)).(C|E|D.(d|C))^*.(A|F|H)$. The subprocesses are defined in Table 2.

Table 2. Subprocesses of the bug tracking process.

Subprocess	Definition
A	$\epsilon|c$
B	$e|c.d$
C	$c.d$
D	$b|a.(a)^*.b$
E	$a.(a)^*.B$
F	$a.(a)^*.A$
G	$D.(d|C)$
H	$D.A$

This example shows that the regular expression constructed using the proposed heuristics is more structured and can be easy analyzed. For example, one may note there are four main patterns of behavior: c (only *"Closed"* activity), F (repeating *"In Progress"* and then may be *"Closed"*), H (zero or more times *"In Progress"*, then *"Resolved"* and then may be *"Closed"*), and $(B|E|D.(d|C))$. $(C|E|D.(d|C))^*.(A|F|H)$. The last pattern can also be easy analyzed, it contains three explicit parts corresponding to the begging, main execution and termination of the process. All this analysis is not possible with the first regular expression, constructed without the prefix-based heuristics.

5 Conclusion

In this paper we consider an approach for constructing regular expressions from event logs. The main goal of this research is to assess the possibility of applying the regular expression synthesis techniques within process mining field. The proposed approach consists of two steps: first, we construct a finite state machine for an event log using existing techniques, then a regular expression is derived from this finite state machine. For constructing regular expressions the state removal method was selected. Additionally, heuristics aimed at reducing the size of the resulting regular expressions were applied. Besides that, special heuristics applicable for certain type of finite state machines constructed from event logs were proposed. The presented approach was implemented and tested on real-life event logs. It was shown that the proposed heuristics allow us to significantly reduce the size and improve the structure of the resulting regular expressions. In future we plan to tackle other types of finite state machines (not only based on prefixes of size 1) constructed from event logs. Also we plan to investigate the possibility of applying symbolic computations [10] to minimize the result regular expressions.

Acknowledgment. This work was supported by the Basic Research Program at the National Research University Higher School of Economics and funded by the President Grant MK-4188.2018.9.

References

1. van der Aalst, W.: Process Mining: Data Science in Action, 2nd edn. Springer, Heidelberg (2016). https://doi.org/10.1007/978-3-662-49851-4
2. Leemans, S.J.J., Fahland, D., van der Aalst, W.M.P.: Discovering block-structured process models from event logs - a constructive approach. In: Colom, J.-M., Desel, J. (eds.) PETRI NETS 2013. LNCS, vol. 7927, pp. 311–329. Springer, Heidelberg (2013). https://doi.org/10.1007/978-3-642-38697-8_17
3. Leemans, S.J.J., Fahland, D., van der Aalst, W.M.P.: Discovering block-structured process models from incomplete event logs. In: Ciardo, G., Kindler, E. (eds.) PETRI NETS 2014. LNCS, vol. 8489, pp. 91–110. Springer, Cham (2014). https://doi.org/10.1007/978-3-319-07734-5_6
4. van der Aalst, W., Rubin, V., Verbeek, H., van Dongen, B., Kindler, E., Günther, C.: Process mining: a two-step approach to balance between underfitting and overfitting. Softw. Syst. Model. **9**(1), 87 (2008)

5. Kleene, S.: Representation of events in nerve nets and finite automata. In: Automata Studies, pp. 3–41. Princeton University Press, Princeton (1956)
6. Brzozowski, J.: Derivatives of regular expressions. J. ACM **11**(4), 481–494 (1964)
7. Linz, P.: An Introduction to Formal Language and Automata. Jones and Bartlett Publishers Inc., Burlington (2006)
8. Delgado, M., Morais, J.: Approximation to the smallest regular expression for a given regular language. In: Domaratzki, M., Okhotin, A., Salomaa, K., Yu, S. (eds.) CIAA 2004. LNCS, vol. 3317, pp. 312–314. Springer, Heidelberg (2005). https://doi.org/10.1007/978-3-540-30500-2_31
9. van Dongen, B.F., de Medeiros, A.K.A., Verbeek, H.M.W., Weijters, A.J.M.M., van der Aalst, W.M.P.: The ProM framework: a new era in process mining tool support. In: Ciardo, G., Darondeau, P. (eds.) ICATPN 2005. LNCS, vol. 3536, pp. 444–454. Springer, Heidelberg (2005). https://doi.org/10.1007/11494744_25
10. Cardelli, L., Tribastone, M., Tschaikowski, M., Vandin, A.: Symbolic computation of differential equivalences. SIGPLAN Not. **51**(1), 137–150 (2016)

Optimization Problems on Graphs and Network Structures

On Modification of an Asymptotically Optimal Algorithm for the Maximum Euclidean Traveling Salesman Problem

Edward Kh. Gimadi[1,2] and Oxana Yu. Tsidulko[1,2(✉)]

[1] Sobolev Institute of Mathematics SB RAS, Novosibirsk, Russia
[2] Department of Mechanics and Mathematics, Novosibirsk State University,
Novosibirsk, Russia
{gimadi,tsidulko}@math.nsc.ru

Abstract. The known asymptotically optimal algorithm for the Euclidean maximum Traveling Salesman Problem by Serdukov builds approximate solution for the problem around the maximum-weight perfect matching. In this paper we are going to discuss an asymptotically optimal algorithm for the Euclidean maximum TSP with running-time $O(n^3)$, that uses a maximum weight cycle cover of the initial graph as a foundation for constructing the TSP solution. We also prove a number of structural results for the optima of some maximization problems in normed spaces, which follow from the algorithm.

Keywords: Maximum Traveling Salesman Problem · Metric space
Euclidean space · Normed space · Cycle cover
Asymptotically optimal algorithm

1 Introduction

The Traveling Salesman Problem (TSP) is one of the most well-known NP-hard problems. Given an n-vertex undirected complete graph $G(V, E)$ and a weight function $w : E \to \mathbb{R}^+$, the problem is to find a Hamiltonian cycle H of minimum or maximum total weight:

$$W(H) = \sum_{e \in H} w(e) \to \max\,(\min).$$

We refer the reader to the books [10] and [7] for the detailed overview of the known results on the problem. Historically, the most attention was paid to the minimization variant of the TSP, but the maximum TSP is also of great interest [1]. The Max TSP has many applications including the shortest superstring problem [9], matrix reordering in data analysis and clustering [8,11]. Both Max TSP and Min TSP are NP-hard in the general case, so there is no much hope

Supported by Russian Science Foundation (project 16-11-10041).

© Springer Nature Switzerland AG 2018
W. M. P. van der Aalst et al. (Eds.): AIST 2018, LNCS 11179, pp. 283–293, 2018.
https://doi.org/10.1007/978-3-030-11027-7_27

to obtain fast (polynomial time) exact algorithms for these problems. Thus, the study of special subclasses of the problem is an important aspect in the field of operations research.

One of the most studied sub-classes is the Metric Max TSP, where the weights of edges satisfy the triangle inequality. Another important variant of the problem is the Euclidean Max TSP (Max ETSP). In this case the vertices of a given graph correspond to points in $\mathbb{R}^k$ for some $k \geq 1$ and the weight of each edge is determined by the Euclidean distance between its endpoints. It is known that Max ETSP is NP-hard, if the dimension of the space $k \geq 3$ [2].

In 1987 Serdukov [12] presented the first polynomial-time asymptotically optimal algorithm for the Max ETSP. Later the algorithm was modified and simplified in [5,6].

Definition 1. *An approximation algorithm A for a maximization problem is said to have a guaranteed relative error ε_A, if*

$$\frac{OPT(X) - F_A(X)}{OPT(X)} \leq \varepsilon_A$$

for any input X, where $F_A(X)$ is the value of the approximate solution obtained by algorithm A, and $OPT(X)$ is the optimum.

Definition 2. *For an optimization problem on an n-vertex graph, an approximation algorithm is called asymptotically optimal if the relative error is a function of the number of vertices n, and $\varepsilon_A(n) \to 0$ as $n \to \infty$.*

Asymptotically optimal algorithms for the Max ETSP from papers [5,6,12] start with constructing the maximum weight matching M^*, which can be represented as a set $\{e_1, \ldots, e_\mu\} \subset \mathbb{R}^k$ of (closed) straight intervals in Euclidean space, $\mu = \lfloor n/2 \rfloor$. The intervals are sorted in non-decreasing order of their weights (length). Choosing parameter t in a spacial way, the last t intervals are declared light, while the remaining intervals are declared heavy. The following lemma establishes a fact that was crucial for proving the asymptotic optimality of these algorithms.

Lemma 1 ([12]). *Suppose that an arbitrary set of t straight line segments is given in Euclidean space $\mathbb{R}^k$ of fixed dimension k. Then, the smallest angle between two segments from this set is upper bounded by a constant $\alpha(k,t)$ such that $\alpha(k,t) \to 0$ as $t \to \infty$. Moreover*

$$\sin^2 \frac{\alpha(k,t)}{2} \leq \frac{\gamma_k}{t^{2/(k-1)}}, \tag{1}$$

where the constant $\gamma_k \leq (k\sqrt{\pi}/2)^{\frac{2}{k-1}}$.

The heavy and light edges in the algorithm A_E from [12] are used to form a multigraph $\overline{G}$ (see Fig. 1), that consists of four Hamiltonian cycles H_1, H_2, H_3, H_4, where each H_i contains all the edges of M^*. The structure of these Hamiltonian cycles differs in cases of odd and even μ. The output of the algorithm A_E is the heaviest of the four Hamiltonian cycles. In case of odd n, at the final step of the algorithm vertex $v \notin M^*$ is arbitrary inserted into the constructed cycle.

The key points used in [12] to prove the asymptotic optimality of the algorithm A_E when $t = \lceil n^{\frac{k-1}{k+1}} \rceil$ are Lemma 1 and the fact that the total weight of the heavy edges is at least $(1 - \frac{t}{\mu})W(M^*)$.

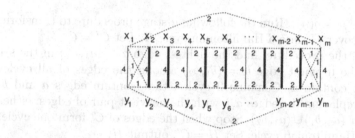

Fig. 1. Multigraph $\overline{G}$ from [12]. Heavy edges of M^* are represented by bold vertical lines, and light edges are represented by thin vertical lines. The numbers refer to the number of copies of the corresponding edges in the multigraph $\overline{G}$.

Later a new more simple version A' of algorithm A_E was presented in [5, 6]. In contrast to A_E, an asymptotically optimal solution obtained by algorithm A' does not contain the edges of the maximum weight matching. This circumstance later played an essential role in implementing the asymptotically optimal approach to such an important generalization of the Traveling Salesman Problem as the Euclidean Maximum m-Peripatetic Salesman Problem [3]. In the latter work, the edges of maximal matching are used as the foundation in constructing each of m edge-disjoint traveling salesmen routes.

In this paper we discuss a new modification of the Serdukov's approach for the Euclidean Max TSP, that starts with constructing an optimal cycle cover. In Sect. 2 we describe the algorithm. In Sect. 3.1 we prove that the algorithm is asymptotically optimal in the Euclidean space, and as a structural corollary we show that in the Euclidean space the weight of optimal solution to the maximum cycle cover problem and the double weight of the maximum weight matching are asymptotically equal. In Sect. 3.2 using the approach from [13] we show that the results obtained in the previous section can be extended to the case of an arbitrary normed space.

2   Modified Algorithm $\widetilde{A}$ Based on Maximum Cycle Cover

In this section we construct the main algorithm $\widetilde{A}$ for the Max TSP with running-time $O(n^3)$, that uses maximum weight cycle cover $C^* = (C_1, C_2, \ldots, C_\mu)$ of the initial graph as a foundation for constructing the solution to TSP. The analysis of algorithm $\widetilde{A}$ will be given in Sect. 3.

2.1 Description of Algorithm $\widetilde{A}$

Stage 1. Given a complete n-vertex complete undirected graph G, construct a cycle cover $C^* = (C_1, C_2, \ldots, C_\mu)$ of maximum total weight. Let μ be the number of cycles in this cover. If $\mu = o(n)$, go to Stage 2. Otherwise, perform Stage 3.

Stage 2. (The case $\mu = o(n)$). Run the following μ-stage procedure to transform the optimal cycle cover C^* into a Hamiltonian cycle H. Let $C' = C^*$.

At each step of the procedure, find the lightest edge $a = (a_1, a_2)$ in the set of edges C' and the heaviest edge $b = (b_1, b_2)$ among the edges of all cycles in C' that do not contain a. Combine 2 cycles, that contain edges a and b, into one cycle by replacing the edges a, b with the heaviest pair of edges: either $(a_1, b_1) \cup (a_2, b_2)$ or $(a_2, b_1) \cup (a_1, b_2)$. Stop when the edges of C' form one cycle, which is clearly a Hamiltonian cycle. Set $H = C'$, output H.

Stage 3. (The case of $\mu = O(n)$).

3.1. Select an edge e_i of minimum weight in each cycle C_i of the cycle cover C^*, $i = 1, \ldots, \mu$. Let $C = \{C_1 \setminus e_1, \ldots, C_\mu \setminus e_\mu\}$ be the family of remaining chains.

3.2. Let V' be the set of vertices of the selected edges, $|V'| = 2\mu$. Let $G' = G[V']$ be the corresponding induced subgraph of G. Note that $M' = \{e_1, \ldots, e_\mu\}$ is a maximum weight perfect matching in G'.

3.3. In subgraph G' construct a Hamiltonian cycle H' that doesn't contain edges $\{e_1, \ldots, e_\mu\}$. The detailed implementation of this step will be given in Subsect. 2.2 (Algorithm A' [5]).

3.4. The obtained Hamiltonian cycle H' consists of two perfect matchings M'_1 and M'_2. Denote by $\widetilde{M}$ the matching with a larger total weight.

3.5. As a result of Stage 3 return the Hamiltonian cycle $H = C \cup \widetilde{M}$ (see Fig. 2).

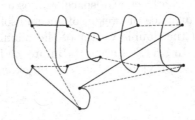

Fig. 2. The Hamiltonian cycle $H = C \cup \widetilde{M}$. The bold curves are the chains of C. The bold lines correspond to the edges of $\widetilde{M}$. The dashed lines together with $\widetilde{M}$ form Hamiltonian cycle H' in subgraph G'.

2.2   Algorithm A': Constructing Hamiltonian Cycle H' in G'

Preliminary Step

Recall that the edges $\{e_1, \ldots, e_\mu\}$ in the 2μ-vertex graph G' form the perfect matching M' of maximum weight. Fix parameter $t \leq \mu/2$. Sort the edges of M' in non-decreasing order with respect to their weights: We will refer to the first $(\mu - t)$ edges of M' as heavy edges, and the last t edges as light.

Algorithm A' constructs a Hamiltonian cycle H', which does not include edges of $M' = \{e_1, \ldots, e_\mu\}$, in G'.

Stage 1. Let angle $\alpha = \alpha(k, t)$ satisfy inequality (1).

Definition 3. *An α-chain is a sequence of edges (intervals in $\mathbb{R}^k$), where the angle between any two neighboring edges of the sequence is at most α. The first edge in the sequence is called the master edge, and the last one is the inferior edge.*

We are going to build a set $\mathcal{I}$ of α-chains. Each α-chain will consist only of heavy edges. Note that an edge is a one-element α-chain.

Start with $\mathcal{I} = \{e_1, e_2, \ldots, e_t\}$ consisting of the first t heaviest edges of M'. Set $j = t$. In the current t-set $\mathcal{I}$, find a pair of α-chains such that the angle between their master edges is at most α. Join these chains into one α-chain by setting their master edges to be neighbors in the new sequence and assign one of the end edges of the joined chain (one of the former inferior edges) to be the new master edge. Set $j := j + 1$. If $j < \mu - t$, then append one more heavy edge e_j to the current set $\mathcal{I}$ and repeat this block.

Otherwise, we have obtained a sequence $\mathcal{C}' = \{C_1', \ldots, C_t'\}$ of t α-chains such that each chain consists of a sequence of heavy edges, where an angle between any neighboring (consecutive) pair of edges in this chain is at most $\alpha = \alpha(k, t)$.

Stage 2. Let's regard the sequence $\mathcal{C}'$ as a cycle, i.e. the α-chain C_t' is followed by the α-chain C_1'. Let the edges of the α-chains $C_1', \ldots, C_t'$ be enumerated so that $C_r' = \{e_{\nu_{r-1}+1}, \ldots, e_{\nu_r - 1}\}$, $1 \leq r \leq t$, where $\nu_1 < \nu_2 < \ldots < \nu_t$ are the numbers reserved for the remaining light edges of the maximum weight matching M' ($\nu_0 = 0$, $\nu_t = \mu$).

Arbitrary place t light edges of the maximum weight matching M' to the positions $\nu_1, \nu_2, \ldots, \nu_t = \mu$. Finally, we have a sequence $\mathcal{S} = \{C_1', e_{\nu_1}, C_2', e_{\nu_2}, \ldots, C_t', e_{\nu_t}\}$ of t α-chains consisting of heavy edges, which alternate with the t light edges (Fig. 3).

Stage 3. Construct Hamiltonian cycle H' in the following way.

We assume that the sequence of edges of the maximum weight matching $M' = \{e_1, e_2, \ldots, e_\mu\}$ is given according to their order in the sequence $\mathcal{S}$, $e_j = (x_j, y_j)$, $j = 1, \ldots, \mu$. Now we are going to construct a partial tour T consisting of the end vertices of the light edge $e_\mu = (x_\mu, y_\mu)$ and of two $(\mu - 1)$-vertex paths.

Step 1. Set $T = x_\mu \cup y_\mu$; $u_1 := x_1$; $v_1 := y_1$ and $j = 2$.

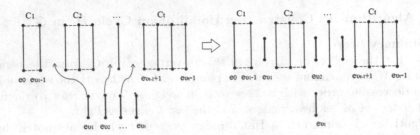

Fig. 3. The bold lines correspond to the edges of M', while the dashed lines indicates the α-chains. The t light edges of M' were placed to the positions between the α-chains. The last light edge e_{ν_t} is placed between the first and the t-th α-chains.

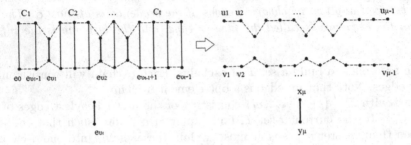

Fig. 4. The bold lines correspond to the edges of M', the dashed lines correspond to the partial tour. We connect each two neighboring edges of M' with the edges of the partial tour in a best possible way. Note that the edges of the perfect matching are not included to the partial tour.

Step 2. While $1 < j < \mu$ do: if

$$w(u_{j-1}, x_j) + w(v_{j-1}, y_j) \geq w(u_{j-1}, y_j) + w(v_{j-1}, x_j),$$

then set $u_j = x_j$; $v_j = y_j$; otherwise, set $u_j = y_j$ and $v_j = x_j$. Supplement the partial tour T with a pair of new edges (Fig. 4) and increase j by 1:

$$T := T \cup (u_{j-1}, u_j) \cup (v_{j-1}, v_j).$$

Step 3. At the previous step we have obtained a partial tour consisting of the end vertices of the light edge $e_\mu = (x_\mu, y_\mu)$ and of the two non-intersecting paths $(u_1, u_2, \ldots, u_{\mu-1})$ and $(v_1, v_2, \ldots, v_{\mu-1})$:

$$T = (u_1, u_2, \ldots, u_{\mu-1}) \cup (v_1, v_2, \ldots, v_{\mu-1}) \cup \{x_\mu\} \cup \{y_\mu\}.$$

Close T into a 2μ-vertex cycle (Fig. 5) by adding a pair of two-edge chains $(u_{\mu-1}, y_\mu, v_1) \cup (v_{\mu-1}, x_\mu, u_1)$ or $(u_{\mu-1}, x_\mu, v_1) \cup (v_{\mu-1}, y_\mu, u_1)$ with the greatest total weight.

The description of the algorithm A' is complete.

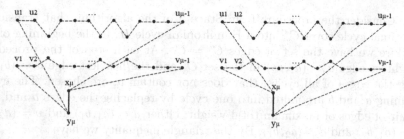

Fig. 5. The partial tour is closed into a Hamiltonian cycle H', according to the best of two possible variants

Theorem 1 ([5]). *Let G' be an undirected 2μ-vertex complete graph in Euclidean space $\mathbb{R}^k$ of a fixed dimension. Set parameter $t = \lceil (2\mu)^{\frac{k-1}{k+1}} \rceil$. Algorithm A' builds an asymptotically optimal solution H' in G' for the Max ETSP. The relative error $\varepsilon_{A'}(n)$ of algorithm A' satisfies:*

$$\varepsilon_{A'}(n') \le \frac{\beta'}{(n')^{\frac{2}{k+1}}}, \tag{2}$$

where $n' = 2\mu$, $\beta' \le (k\sqrt{\pi}/2)^{\frac{2}{k-1}}$. The total weight of the approximate solution H' satisfies:

$$2W(M')(1 - \varepsilon_{A'}(n')) \le W(H') \le 2W(M'), \tag{3}$$

where M' is the maximum weight perfect matching in G'.

Remark 1. The obtained Hamiltonian cycle H' consists of 2 perfect matchings M_1 and M_2. Note that by construction of H', $M' \cup M_1$ and $M' \cup M_2$ are also Hamiltonian cycles. Thus, adding M_1 or M_2 to the set of chains $C = C^* \setminus M'$ at Step 3.5 of the main algorithm $\widetilde{A}$ gives a Hamiltonian cycle in the initial graph G (see Fig. 2).

3   Asymptotic Optimality of Algorithm $\widetilde{A}$

3.1   The Case of Euclidean Space

Next statement shows the asymptotic optimality of algorithm $\widetilde{A}$ in the case of $\mu = o(n)$ for the Metric Max TSP and the Euclidean Max TSP.

Theorem 2. *In the case of $\mu = o(n)$, algorithm $\widetilde{A}$ gives asymptotically optimal solutions for the Metric Max TSP and the Euclidean Max TSP with guaranteed relative error*

$$\varepsilon_{\widetilde{A}}(n) \le \frac{\mu}{n}. \tag{4}$$

Proof. Consider the procedure from Stage 2 of the algorithm, that transforms the optimal cycle cover C^* into a Hamiltonian cycle $\widetilde{H}$. In the beginning of the procedure we have the set of edges $C' = C^*$. At each step of the procedure, the lightest edge $a = (a_1, a_2)$ in the set C' and the heaviest edge $b = (b_1, b_2)$ among the edges of all cycles, that does not contain a, are chosen. The cycles containing a and b are united into one cycle by replacing the edges a and b by the pair of edges of maximum total weight: either $u = (a_1, b_1)$ and $v = (a_2, b_2)$ or $c = (a_1, b_2)$ and $d = (a_2, b_1)$. By the triangle inequality we have

$$w(b) \leq w(u) + w(c) \quad \text{and} \quad w(b) \leq w(v) + w(d).$$

Summing these inequalities, we obtain

$$2w(b) \leq w(u) + w(v) + w(c) + w(d)$$

and, therefore,

$$w(b) \leq \max\big(w(u) + w(v), w(c) + w(d)\big).$$

Thus, the new pair of edges compensate the weight of the heaviest edge b, and the loss of weight does not exceed the weight $w(a)$ of the lightest edge.

After μ steps, we obtain a Hamiltonian cycle H, and the total loss of weight does not exceed the weight of μ lightest edges of C^*, which is at most $(\mu/n)W(C^*)$. Thus,

$$(1 - \mu/n)W(C^*) \leq W(H) \leq W(H^*) \leq W(C^*), \tag{5}$$

where H^* is the maximum weight Hamiltonian cycle. Finally, for the relative error we have:

$$\varepsilon_{\widetilde{A}}(n) = \frac{W(H^*) - W(H)}{W(H^*)} = 1 - \frac{W(H)}{W(H^*)} \leq 1 - \frac{(1 - \frac{\mu}{n})W(C^*)}{W(H^*)} \leq \frac{\mu}{n}.$$

Next claim shows the asymptotic optimality of algorithm $\widetilde{A}$ in the case of $\mu = O(n)$ for the Euclidean Max TSP.

Theorem 3. *In the case of* $\mu = O(n)$, *algorithm* $\widetilde{A}$ *gives an asymptotically optimal solution for the Euclidean Max TSP, if the dimension k of the Euclidean space $\mathbb{R}^k$ is fixed. The relative error satisfies*

$$\varepsilon_{\widetilde{A}}(n) = \frac{\varepsilon_{A'}(n')}{3} \leq \frac{\beta}{(n')^{\frac{2}{k+1}}}, \tag{6}$$

$n' = 2\mu$, $\mu = O(n)$, $\beta \leq (k\sqrt{\pi}/2)^{\frac{2}{k-1}}/3$.

Proof. Let M^* be the maximum weight perfect matching, C^* be the maximum weight cycle cover, and H^* be the maximum weight Hamiltonian cycle in the given graph.

At Step 3.1 algorithm $\widetilde{A}$ deletes the lightest edge e_i from each cycle C_i in the cycle cover C^* and sets $M' = \{e_1, \ldots e_\mu\}$. Since each cycle in the cycle cover consists of at least 3 edges, from $W(C_i) \geq 3w(e_i)$ for all $i = 1, \ldots, \mu$, it follows that:

$$W(C^*) \geq 3W(M').$$

Using (3), for the matching $\widetilde{M}$, which is the heaviest matching in the Hamiltonian cycle H' formed at Step 3.3 of algorithm $\widetilde{A}$, we have:

$$W(\widetilde{M}) \geq \frac{W(H')}{2} \geq \frac{2\left(1 - \varepsilon_{A'}(n')\right)W(M')}{2} = \left(1 - \varepsilon_{A'}(n')\right)W(M').$$

Finally, due to the construction of the desired Hamiltonian cycle H, the relative error of algorithm $\widetilde{A}$ is

$$\varepsilon_{\widetilde{A}}(n) = 1 - \frac{W(H)}{W(H^*)} \leq 1 - \frac{W(H)}{W(C^*)} = 1 - \frac{W(C^* \backslash M') + W(\widetilde{M})}{W(C^*)}$$

$$\leq 1 - \frac{W(C^* \backslash M') + \left(1 - \varepsilon_{A'}(n')\right)W(M')}{W(C^*)} = 1 - \frac{W(C^*) - \varepsilon_{A'}(n')W(M')}{W(C^*)}$$

$$= \frac{\varepsilon_{A'}(n')W(M')}{W(C^*)} \leq \frac{\varepsilon_{A'}(n')}{3} \leq \frac{\beta}{(n')^{\frac{2}{k+1}}}.$$

Remark 2. The running-time of algorithm $\widetilde{A}$ is determined by the time one needs to construct a maximum weight cycle cover C^*, which is $O(n^3)$ [4].

Remark 3. For the practical use of algorithm $\widetilde{A}$ one can set the following threshold in choosing between Stages 2 and 3. If the number of cycles μ in optimal the cycle cover at most $n^{\frac{k+1}{k+3}}\left(\beta^{k+1}/4\right)^{\frac{1}{k+3}}$, perform Stage 2, otherwise perform Stage 3. In this case for the total relation error of the algorithm we have:

$$\varepsilon_{\widetilde{A}}(n) \leq \left(\frac{\beta^{k+1}}{4n^2}\right)^{\frac{1}{k+3}}.$$

Corollary 1. *Let M^* be the maximum weight perfect matching, C^* be the maximum weight cycle cover, and H^* be the maximum weight Hamiltonian cycle of an n-vertex complete graph $G(V, E)$ in Euclidean space of fixed dimension. Then*

$$2W(M^*) = W(H^*)(1 + o(1)) = W(C^*)(1 + o(1)).$$

In other words, in Euclidean space of fixed dimension the weights of the optimal C^*, H^* and the doubled weight of M^* are asymptotically equal.

Proof. The first equality follows from (3) for an n-vertex graph. The second equality follows from (5) and from the proof of Theorem 3, where having:

$$1 - \frac{W(H)}{W(C^*)} \leq \frac{\beta}{(n')^{\frac{2}{k+1}}} \quad \text{and} \quad W(C^*) \geq W(H^*) \geq W(H)$$

we obtain $W(C^*) = W(H^*)(1 + o(1))$.

3.2   The Case of an Arbitrary Normed Space

In 2014 Shenmaier [13] introduced the notation of a remote angle $\alpha(x, y)$ between two vectors x and y in an arbitrary normed space:

$$\alpha(x, y) = \min\{\|x/\|x\| - y/\|y\|\|, \|x/\|x\| + y/\|y\|\|\}.$$

He proved that substituting the angle, defined by the dot product, with the remote angle $\alpha(x, y)$ in Serdukov's algorithm [12] (or its modification [5]), one can obtain an asymptotically optimal algorithm A_N for the Max TSP in an arbitrary normed space. The relative error of algorithm A_N is $\varepsilon_{A_N} \leq (k+1)/\lfloor n^{\frac{1}{k+1}} \rfloor$, where k is the dimension of the space.

It is easy to see that due to the same modifications our algorithm $\widetilde{A}$ would also give asymptotically optimal solutions to the Max TSP in an arbitrary normed space. First note, that in the case of small number of cycles in the optimal cycle cover ($\mu = o(n)$) according to Theorem 2 algorithm $\widetilde{A}$ is asymptotically optimal in any metric space. For the case of large number of cycles in the optimal cycle cover ($\mu = O(n)$) the asymptotic optimality follows strictly from Theorem 3 and the result by Shenmaier [13].

Corollary 2. *Using the remote angle instead of an angle, defined by the dot product, at Step 3.3, our algorithm $\widetilde{A}$ also gives asymptotically optimal solutions for the Max TSP in an arbitrary normed space. The relative error of the algorithm in the case of executing Stage 3 is $\varepsilon_{\widetilde{A}}(n) = \varepsilon_{A_N}(n')/3$, where $n' = 2\mu$, $\mu = O(n)$ is the number of cycles in the optimal cycle cover C^*.*

Now we can extend the structural results from the Corollary 1 to the case of an arbitrary normed space.

Corollary 3. *Let M^* be the maximum weight perfect matching, C^* be the maximum weight cycle cover, and H^* be the maximum weight Hamiltonian cycle of a complete graph $G(V, E)$ in an arbitrary normed space of fixed dimension. Then the weights of the optimal C^*, H^* and the doubled weight of M^* are asymptotically equal: $2W(M^*) = W(H^*)(1 + o(1)) = W(C^*)(1 + o(1))$.*

4   Conclusion

This paper present a new modification $\widetilde{A}$ of an approximation algorithm for the Max TSP with $O(n^3)$ running-time. Maximum weight cycle cover C^* of the initial graph is used as a foundation for constructing the approximate solutions. It is shown that if the number of cycles in C^* is $\mu = o(n)$, algorithm $\widetilde{A}$ is asymptotically optimal for the Metric Max TSP (and, thus, also for the Euclidean Max TSP). In the case of $\mu = O(n)$ algorithm $\widetilde{A}$ is asymptotically optimal for the Euclidean Max TSP and due to [13] can be easily modified to give asymptotically optimal solutions for the Max TSP in an arbitrary normed space. The performance and analysis of algorithm $\widetilde{A}$ do not depend neither on the parity of the number of vertices in the initial graph, nor on the parity of the number of edges μ in the maximum weight matching.

As a corollary, we've showed that, given a complete undirected weighted graph in an arbitrary normed space, the weights of the maximum Hamiltonian cycle, the maximum weight cycle cover and the doubled weight of the maximum weight matching are asymptotically equal.

For the future work it would be interesting to extend the ideas of the algorithm, so it could find approximate solutions with improved approximation ratio for the Metric Max TSP, when $\mu = O(n)$.

References

1. Barvinok, A.I., Gimadi, E.Kh., Serdyukov, A.I.: The maximum TSP. In: Gutin, G., Punnen, A.P. (eds.) The Traveling Salesman Problem and Its Variations, pp. 585–608. Kluwer Academic Publishers, Dordrecht (2002)
2. Barvinok, A., Fekete, S.P., Johnson, D.S., Tamir, A., Woeginger, G.J., Woodroofe, R.: The geometric maximum traveling salesman problem. J. ACM **50**(5), 641–664 (2003)
3. Baburin, A.E., Gimadi, E.Kh.: On the asymptotic optimality of an algorithm for solving the maximum m-PSP in a multidimensional Euclidean space. Proc. Steklov Inst. Math. **272**(1), 1–13 (2011)
4. Gabow, H.N.: An efficient reduction technique for degree-restricted subgraph and bidirected network flow problems. In: Proceedings of the 15th Annual ACM Symposium on Theory of Computing, Boston, USA, 25–27 April 1983, pp. 448–456. ACM, New York (1983)
5. Gimadi, E.Kh.: A new version of the asymptotically optimal algorithm for solving the Euclidean maximum traveling salesman problem. In: Proceedings of the 12th Baykal International Conference 2001, Irkutsk, vol. 1, pp. 117–123 (2001). (in Russian)
6. Gimadi, E.Kh.: Asymptotically optimal algorithm for finding one and two edge-disjoint traveling salesman routes of maximal weight in Euclidean space. Proc. Steklov Inst. Math. **263**(2), 56–67 (2008)
7. Gutin, G., Punnen, A.P. (eds.): The Traveling Salesman Problem and ITS Variations. Kluver Academic Publishers, Dordrecht/Boston/London (2002)
8. Johnson, O., Liu, J.: A traveling salesman approach for predicting protein functions. Source Code Biol. Med. **1**(3), 9–16 (2006)
9. Kaplan, H., Lewenstein, M., Shafrir, N., Sviridenko, M.: Approximation algorithms for asymmetric TSP by decomposing directed regular multigraphs. J. ACM **52**(4), 602–626 (2005)
10. Lawler, E.L., Lenstra, J.K., Rinnoy Kan, A.H.G., Shmoys, D.B. (eds.): The Traveling Salesman Problem: A Guided Tour of Combinatorial Optimization. Wiley, Chichester (1985)
11. Ray, S.S., Bandyopadhyay, S., Pal, S.K.: Gene ordering in partitive clustering using microarray expressions. J. Biosci. **32**(5), 1019–1025 (2007)
12. Serdyukov, A.I.: An asymptotically optimal algorithm for the maximum traveling salesman problem in Euclidean space. Upravlyaemye sistemy, Novosibirsk, vol. 27, pp. 79–87 (1987). (in Russian)
13. Shenmaier, V.V.: Asymptotically optimal algorithms for geometric Max TSP and Max m-PSP. Discret. Appl. Math. **163**(2), 214–219 (2014)

Exact Algorithms for the Special Cases of Two Hard to Solve Problems of Searching for the Largest Subset

Alexander Kel'manov[1,2], Vladimir Khandeev[1,2(⊠)], and Anna Panasenko[1,2]

[1] Sobolev Institute of Mathematics, 4 Koptyug Ave., 630090 Novosibirsk, Russia
[2] Novosibirsk State University, 2 Pirogova St., 630090 Novosibirsk, Russia
{kelm,khandeev,a.v.panasenko}@math.nsc.ru

Abstract. We consider two problems of searching for the largest subset in a finite set of points in Euclidean space. Both problems have applications, in particular, in data analysis and data approximation. In each problem, an input set and a positive real number are given and it is required to find a subset (i.e., a cluster) of the largest size under a constraint on the value of a quadratic function. The points in the input set which are outside the sought subset are treated as a second cluster. In the first problem, the function in the constraint is the sum over both clusters of the intracluster sums of the squared distances between the elements of the clusters and their centers. The center of the first (i.e., sought) cluster is unknown and determined as the centroid, while the center of the second one is fixed in a given point in Euclidean space (without loss of generality in the origin). In the second problem, the function in the constraint is the sum over both clusters of the weighted intracluster sums of the squared distances between the elements of the clusters and their centers. As in the first problem, the center of the first cluster is unknown and determined as the centroid, while the center of the second one is fixed in the origin. In this paper, we show that both problems are strongly NP-hard. Also, we present exact algorithms for the cases of these problems in which the input points have integer components. If the space dimension is bounded by some constant, the algorithms are pseudopolynomial.

Keywords: Euclidean space · 2-clustering · Largest subset
NP-hardness · Exact algorithm · Pseudopolynomial-time solvability

1 Introduction

In the paper, we study two optimization problems. Our goal is to analyze the computational complexity of the problems and construct for these problems effective algorithms with guaranteed accuracy.

The research is motivated by insufficient study of the problems, by the absence of algorithmic results for them, and by importance of these problems in some applications (see next section).

© Springer Nature Switzerland AG 2018
W. M. P. van der Aalst et al. (Eds.): AIST 2018, LNCS 11179, pp. 294–304, 2018.
https://doi.org/10.1007/978-3-030-11027-7_28

The paper has the following structure. Section 2 contains the problems for-mulation, their interpretation, related problems, their distinctive features, and known algorithmic results. In the same section we announce the results obtained. In Sect. 3, we analyze the computational complexity of the problems considered. Section 4 contains some auxiliary problems and algorithms for them. Finally, the step-by-step description of algorithms for the considered problems and jus-tification of their properties (accuracy and time complexity) are presented in Sect. 5.

2 Problems Formulation and Related Problems, Known and Obtained Results

Everywhere below $\mathbb{R}$ denotes the set of real numbers, $\|\cdot\|$ denotes the Euclidean norm, and $\langle\cdot,\cdot\rangle$ denotes the scalar product; *centroid* (geometric center) of a set is the arithmetic mean of the points from this set; *center* of a cluster $\mathcal{C}$ is a point $x \in \mathbb{R}^d$ in the sum $\sum_{y \in \mathcal{C}} \|y - x\|^2$.

The problems under consideration are as follows.

Problem 1. Given: an N-element set $\mathcal{Y}$ of points in d-dimensional Euclidean space and a number $\alpha \in (0,1)$. *Find:* a subset $\mathcal{C} \subset \mathcal{Y}$ of the largest cardinality such that

$$f(\mathcal{C}) = \sum_{y \in \mathcal{C}} \|y - \overline{y}(\mathcal{C})\|^2 + \sum_{y \in \mathcal{Y} \setminus \mathcal{C}} \|y\|^2 \leq \alpha \sum_{y \in \mathcal{Y}} \|y - \overline{y}(\mathcal{Y})\|^2 , \qquad (1)$$

where $\overline{y}(\mathcal{C}) = \frac{1}{|\mathcal{C}|} \sum_{y \in \mathcal{C}} y$ and $\overline{y}(\mathcal{Y}) = \frac{1}{|\mathcal{Y}|} \sum_{y \in \mathcal{Y}} y$ are the centroids of the subset $\mathcal{C}$ and the given set $\mathcal{Y}$, respectively.

Problem 2. Given: a set $\mathcal{Y} = \{y_1, \ldots, y_N\}$ of points in $\mathbb{R}^d$ and a number $\alpha \in (0,1)$. *Find:* a subset $\mathcal{C} \subset \mathcal{Y}$ of the largest cardinality such that

$$g(\mathcal{C}) = |\mathcal{C}| \sum_{y \in \mathcal{C}} \|y - \overline{y}(\mathcal{C})\|^2 + |\mathcal{Y} \setminus \mathcal{C}| \sum_{y \in \mathcal{Y} \setminus \mathcal{C}} \|y\|^2 \leq \alpha N \sum_{y \in \mathcal{Y}} \|y - \overline{y}(\mathcal{Y})\|^2 . \qquad (2)$$

There are some clusterization problems related to the considered ones. The closest related problem to Problem 1 is the following

Problem 3 (2-MSSC with given center). Given: an N-element set $\mathcal{Y}$ of points in Euclidean space of dimension d. *Find:* a 2-partition of $\mathcal{Y}$ into clusters $\mathcal{C}$ and $\mathcal{Y} \setminus \mathcal{C}$ minimizing the value of $f(\mathcal{C})$.

Abbreviation *MSSC* corresponds to *Minimum Sum-of-Squares Clustering*. In Problem 3, one needs to minimize $f(\mathcal{C})$, i.e., the sum over both clusters of the intracluster sums of the squared distances between the elements of the clusters and their centers. The center of the cluster $\mathcal{C}$ is unknown and determined as the centroid, while the center of the cluster $\mathcal{Y} \setminus \mathcal{C}$ is fixed in a given point in Euclidean space (without loss of generality in the origin). In Problem 1, one needs to find a cluster C of the largest cardinality under upper constraint on the value of $f(C)$. This constraint is defined by the right-hand part of (1), i.e., by the fraction of the total quadratic variation of the input set $\mathcal{Y}$ with respect to its centroid $\overline{y}(\mathcal{Y})$.

Problem 3 is close to the *MSSC* problem, which is also well-known since the last century as the *k-Means* problem [1–6]. The complexity of this problem was studied in [7–10]. In [7], it was shown that its basic two-cluster variant, i.e., 2-*MSSC* (or 2-*Means*) problem, is strongly NP-hard. Recall that in 2-*MSSC* it is required to find 2-partition minimizing the value of

$$\sum_{y \in C} \|y - \overline{y}(C)\|^2 + \sum_{y \in \mathcal{Y} \setminus C} \|y - \overline{y}(\mathcal{Y} \setminus C)\|^2 \, .$$

In 2-*MSSC*, in contrast to Problem 3, the both cluster centers of C and $\mathcal{Y} \setminus C$ are unknown (in Problem 3, the center of only one cluster is unknown). Strong NP-hardness of Problem 3 was proved in [11,12].

The closest related problem to Problem 2 is the following

Problem 4 (Cardinality-weighted 2-MSSC with given center). Given: an N-element set $\mathcal{Y}$ of points in d-dimensional Euclidean space. *Find:* a 2-partition of $\mathcal{Y}$ into clusters C and $\mathcal{Y} \setminus C$ minimizing the value of $g(C)$.

In this problem, it is required to minimize $g(C)$, i.e., the sum over both clusters of the cardinality-weighted intracluster sums of the squared distances between the elements of the clusters and their centers. As in Problem 3, the center of the cluster C is unknown and determined as the centroid, while the center of the cluster $\mathcal{Y} \setminus C$ is fixed in the origin. In Problem 2, it is required to find a cluster C of the largest cardinality under upper constraint on the value of $g(C)$. This constraint is defined by the right-hand part of (2), i.e., by the fraction of the total cardinality-weighted quadratic variation of the input set $\mathcal{Y}$ with respect to its centroid $\overline{y}(\mathcal{Y})$.

Problem 4 is close to the known [13–20] *Quadratic Euclidean Min-Sum All-Pairs 2-clustering problem*. In this problem, it is required to find a 2-partition minimizing the value of

$$|C| \sum_{y \in C} \|y - \overline{y}(C)\|^2 + |\mathcal{Y} \setminus C| \sum_{y \in \mathcal{Y} \setminus C} \|y - \overline{y}(\mathcal{Y} \setminus C)\|^2 \, ,$$

where the both centers are unknown (in Problem 4, only one center is unknown). The strong NP-hardness of this problem was proved in [21,22] and the strong NP-hardness of Problem 4 was shown in [23,24].

All mentioned clusterization problems have a simple geometric interpretation and are closely related to Data science, Data approximation, Data mining, Machine learning, Statistics, Combinatorial geometry. In these research areas, clusterization algorithms are the key tools (see, for example, [25–31] and papers cited therein). Developing new tools for clusterization is of great interest for these areas.

By this time, there are no available algorithmic results for considered Problems 1 and 2, so below we present known results only for the closest problems, namely, for Problems 3 and 4.

Firstly, let us recall known results for Problem 3. The only known result for the general case of Problem 3 is a 2-approximation polynomial algorithm with time complexity $\mathcal{O}(dN^2)$ which was proposed in [32]. The following results were obtained for variant of Problem 3 in which cluster cardinalities are given. Below this variant is referred to as 2-*MSSC with given center and cluster cardinalities* (Problem 9). We present algorithmic results for this variant since we use one of them to construct our algorithm for considered Problem 1.

Strong NP-hardness of this variant of Problem 3 follows from [33–35].

In [36], a 2-approximation polynomial algorithm with time complexity $\mathcal{O}(dN^2)$ was proposed for the problem.

A polynomial-time approximation scheme (PTAS) having $\mathcal{O}(dN^{2/\varepsilon+1}(9/\varepsilon)^{3/\varepsilon})$ time complexity, where ε is a relative error, was proposed in [37].

A randomized algorithm was proposed in [38]. Under assumption $M \geq \beta N$, where $\beta \in (0, 1)$ is some constant, and given $\varepsilon > 0$ and $\gamma \in (0, 1)$, the algorithm finds a $(1 + \varepsilon)$-approximate solution of the problem with probability not less than $1 - \gamma$ in $\mathcal{O}(dN)$ time. In the same paper, the conditions were found under which the algorithm finds a $(1 + \varepsilon_N)$-approximate solution of the problem in $\mathcal{O}(dN^2)$ time with probability not less than $1 - \gamma_N$, where $\varepsilon_N \to 0$ and $\gamma_N \to 0$ as $N \to \infty$, i.e., the conditions under which the algorithm is asymptotically exact.

It follows from [39] that the problem is solvable in $\mathcal{O}(d^2 N^{2d})$ time, which is polynomial when the space dimension d is fixed. A faster exact algorithm with $\mathcal{O}(dN^{d+1})$ running time was proposed in [40]. Additionally, an exact algorithm for the case of integer inputs was presented in [41]. The time complexity of the algorithm is $\mathcal{O}(dN(2MD + 1)^d)$, where D is the maximum absolute coordinate value in the input set. If the space dimension is fixed, the algorithm is pseudopolynomial and its time complexity is $\mathcal{O}(N(MD)^d)$.

In [42], it was shown that there is no FPTAS for the problem unless $P = NP$. In the same paper, the algorithm was proposed which allows finding a $(1 + \varepsilon)$-approximate solution in $\mathcal{O}(dN^2(\sqrt{2q/\varepsilon} + 2)^d)$ time for given $\varepsilon \in (0, 1)$. For the fixed space dimension d, the algorithm runs in $\mathcal{O}(N^2(1/\varepsilon)^{q/2})$ time and implements an FPTAS.

Secondly, let us recall known results for Problem 4. In [23,24] it was shown that there are no FPTAS for this problem unless $P = NP$. The following results were obtained for the variant of Problem 4 in which cluster cardinalities are given. In Sect. 4, this variant is referred to as *Cardinality-weighted 2-MSSC with given center and cluster cardinalities* (Problem 10). Below we only list algorithmic results for this variant since their characteristics are similar to the characteristics of the algorithms for Problem 3 (although the algorithms for these problems are not identical).

A 2-approximation algorithm was constructed in [43]. An exact algorithm for the integer case of the problem was proposed in [44]. A $(1 + \varepsilon)$-approximation algorithm, which implements an FPTAS in the case of the fixed space dimension, was constructed in [45]. The modification of this algorithm with improved running time was proposed in [46]. This modification implements a PTAS for instances of dimension $d = \mathcal{O}(\log N)$. A randomized algorithm was proposed in [47].

In this paper, we present the first results for Problems 1 and 2. Namely, we prove the strong NP-hardness of Problems 1 and 2 and present exact algorithms for their special cases in which the input points have integer-valued coordinates; if the space dimension is fixed, both algorithms are pseudopolynomial.

3   Computational Complexity

Before analyzing the computational complexity of Problems 1 and 2, note that the right-hand sides of (1) and (2) do not depend on $\mathcal{C}$, and for a given input are constants. Therefore, Problems 1 and 2 in a property verification form can be formulated in the following way.

Problem 5. Given: an N-element set $\mathcal{Y}$ of points in d-dimensional Euclidean space, a positive integer M, real numbers $A > 0$ and $\alpha \in (0,1)$. *Question:* is there in $\mathcal{Y}$ a subset $\mathcal{C}$ of cardinality at least M such that

$$f(\mathcal{C}) \leq A?$$

Problem 6. Given: a set $\mathcal{Y} = \{y_1, \ldots, y_N\}$ of points in $\mathbb{R}^d$, a positive integer M, numbers $\alpha \in (0,1)$ and $B > 0$. *Question:* is there in $\mathcal{Y}$ a subset $\mathcal{C}$ of cardinality at least M such that

$$g(\mathcal{C}) \leq B?$$

Property verification form of Problems 3 and 4 is as follows.

Problem 7. Given: an N-element set $\mathcal{Y}$ of points in d-dimensional Euclidean space, real numbers $A > 0$ and $\alpha \in (0,1)$. *Question:* is there in $\mathcal{Y}$ a subset $\mathcal{C}$ such that

$$f(\mathcal{C}) \leq A?$$

Problem 8. Given: a set $\mathcal{Y} = \{y_1, \ldots, y_N\}$ of points in $\mathbb{R}^d$, numbers $\alpha \in (0,1)$ and $B > 0$. *Question:* is there in $\mathcal{Y}$ a subset $\mathcal{C}$ such that

$$g(\mathcal{C}) \leq B?$$

Proposition 1. *Problems 5 and 6 are strongly NP-complete.*

Proof. Obviously, both problems are in NP. Therefore, the proposition follows from the fact that strongly NP-complete Problems 7 and 8 are the special cases of Problems 5 and 6 with $M = 1$.   □

Thus, optimization Problems 1 and 2 are strongly NP-hard since Problems 5 and 6 are their reformulation in a property verification form.

4   Algorithms Foundations

In order to substantiate our algorithms, we need two auxiliary problems with exact algorithms for their special cases.

First, we need the following auxiliary

Problem 9 (2-MSSC with given center and cluster cardinalities). *Given:* a set $\mathcal{Y} = \{y_1, \ldots, y_N\}$ in Euclidean space of dimension d and a positive integer M. *Find:* a subset $\mathcal{C} \subseteq \mathcal{Y}$ of the size M minimizing the value of $f(\mathcal{C})$.

The algorithm for special case of Problem 9 in which components of all points in the input set are integer is presented below.

Algorithm $\mathcal{A}_1$.
Input: a set $\mathcal{Y}$ and a positive integer M.
Step 1. Find the value of D by formula

$$D = \max_{y \in \mathcal{Y}} \max_{j \in \{1, \ldots, d\}} |(y)^j|, \tag{3}$$

where $(y)^j$ denotes the j-th coordinate of a point y. Construct the set $\mathcal{D}$ by formula

$$\mathcal{D} = \{x \mid (x)^j = \frac{1}{M}(v)^j, \ (v)^j \in \mathbb{Z}, \ |(v)^j| \leq MD, \ j = 1, \ldots, q\}. \tag{4}$$

Step 2. For each $x \in \mathcal{D}$, construct the set $\mathcal{Y}(x)$ of M points $y \in \mathcal{Y}$ having the largest values of $\langle y, x \rangle$. Calculate

$$f^x(\mathcal{Y}(x)) = \sum_{y \in \mathcal{Y}(x)} \|y - x\|^2 + \sum_{y \in \mathcal{Y} \setminus \mathcal{Y}(x)} \|y\|^2.$$

Step 3. Find the point $x_A = \arg \min_{x \in \mathcal{D}} f^x(\mathcal{Y}(x))$ and corresponding subset $\mathcal{Y}(x_A)$. As a solution to the problem, take $\mathcal{C}_{\mathcal{A}_1}^M = \mathcal{Y}(x_A)$. If there are several solutions then take any of them.
Output: the set $\mathcal{C}_{\mathcal{A}_1}^M$.

Remark 1. It was proved in [41] that if components of all points in $\mathcal{Y}$ are integers in the interval $[-D, D]$ then algorithm $\mathcal{A}_1$ finds an optimal solution of Problem 9 in $\mathcal{O}(dN(2MD + 1)^d)$ time. This running time is caused by the equality $|\mathcal{D}| = (2MD + 1)^d$.

Next, we need the second auxiliary

Problem 10 (Cardinality-weighted 2-MSSC with given center and cluster cardinalities). *Given:* an N-element set $\mathcal{Y}$ of points in d-dimensional Euclidean space and a positive integer number $M \leq N$. *Find:* a partition of $\mathcal{Y}$ into two non-empty clusters $\mathcal{C}$ and $\mathcal{Y} \setminus \mathcal{C}$ such that

$$g(\mathcal{C}) \to \min,$$

subject to constraint $|\mathcal{C}| = M$.

Finally, we need the following algorithm which allows one to find a solution of Problem 10 in the case of integer coordinates of the input points of $\mathcal{Y}$.

Algorithm $\mathcal{A}_2$.
Input: a set $\mathcal{Y}$ and a positive integer M.
Step 1. Find the value of D by formula (3). Construct the set $\mathcal{D}$ by formula (4).
Step 2. For each $x \in \mathcal{D}$, construct the set $\mathcal{Y}(x)$ of M points in $y \in \mathcal{Y}$ having the smallest values of

$$h^x(y) = (2M - N) \|y\|^2 - 2M \langle y, x \rangle .$$

Calculate

$$g^x(\mathcal{Y}(x)) = M \sum_{y \in \mathcal{Y}(x)} \|y - x\|^2 + (N - M) \sum_{y \in \mathcal{Y} \setminus \mathcal{Y}(x)} \|y\|^2 .$$

Step 3. Find the point $x_A = \arg\min_{x \in \mathcal{D}} g^x(\mathcal{Y}(x))$ and corresponding subset $\mathcal{Y}(x_A)$. As a solution to the problem, take $\mathcal{C}_{A_2}^M = \mathcal{Y}(x_A)$. If there are several solutions then take any of them.
Output: the set $\mathcal{C}_{A_2}^M$.

Remark 2. It was proved in [44] that if components of all points in $\mathcal{Y}$ are integers in the interval $[-D, D]$ then the algorithm $\mathcal{A}_2$ finds an optimal solution of Problem 10 in $\mathcal{O}(dN(2MD + 1)^d)$ time.

5   Algorithms

The algorithms formulated in Sect. 4 are the computational basis for algorithms presented below for Problems 1 and 2. Let us formulate the following algorithm for solving Problem 1.

Algorithm $\mathcal{A}_3$.
Input: a set $\mathcal{Y}$ and a number α.
Step 1. Compute $A = \alpha \sum_{y \in \mathcal{Y}} \|y - \overline{y}(\mathcal{Y})\|^2$.
Step 2. For every $M = 1, \ldots, N$ using algorithm $\mathcal{A}_1$ find an exact solution $\mathcal{C}_{A_1}^M$ of Problem 9 and calculate for this solution the value of the objective function $f(\mathcal{C}_{A_1}^M)$.
Step 3. In the family $\{\mathcal{C}_{A_1}^M, M = 1, \ldots, N\}$ of the sets obtained in Step 2 find a set $\mathcal{C}_{A_1}$ of maximum cardinality for which $f(\mathcal{C}_{A_1}) \leq A$.
Output: the set $\mathcal{C}_{A_1}$.
The algorithm for solving Problem 2 is similar; the main difference between the algorithms is in constructing a feasible solution of the problem at Step 2.

Algorithm $\mathcal{A}_4$.
Input: a set $\mathcal{Y}$ and a number α.
Step 1. Compute $A = \alpha N \sum_{y \in \mathcal{Y}} \|y - \overline{y}(\mathcal{Y})\|^2$.

Step 2. For every $M = 1, \ldots, N$ using algorithm $\mathcal{A}_2$ find an exact solution $\mathcal{C}_{A_2}^M$ of Problem 10 and calculate for this solution the value of the objective function $g(\mathcal{C}_{A_2}^M)$.

Step 3. In the family $\{\mathcal{C}_{A_2}^M, M = 1, \ldots, N\}$ of the sets obtained in Step 2 find a set $\mathcal{C}_{A_2}$ of maximum cardinality for which $g(\mathcal{C}_{A_2}) \leq A$.

Output: the set $\mathcal{C}_{A_2}$.

The following theorem is true.

Theorem 1. *Suppose that the points of $\mathcal{Y}$ have integer components lying in $[-D, D]$. Then algorithms $\mathcal{A}_3$ and $\mathcal{A}_4$ find exact solutions of Problems 1 and 2 in $\mathcal{O}(dN^2(2ND + 1)^d)$ time.*

Proof. Let us prove the accuracy bound of algorithm $\mathcal{A}_3$. Let $\mathcal{C}_1^*$ be the optimal solution of Problem 1, $M_1^* = |\mathcal{C}_1^*|$. Note that algorithm $\mathcal{A}_3$ finds an admissible solution of Problem 9 with $M = M_1^*$. Since $\mathcal{C}_{A_1}^{M^*}$ is the optimal solution of the same problem,

$$f(\mathcal{C}_{A_1}^{M^*}) \leq f(\mathcal{C}_1^*) \leq A.$$

Therefore, $\mathcal{C}_{A_1}^{M^*}$ was considered at Step 3 of the algorithm, and the family $\{\mathcal{C}_{A_2}^M, M = 1, \ldots, N \mid f(\mathcal{C}_{A_1}) \leq A\}$ is not empty. Moreover, from the definition of Step 3 we get

$$|\mathcal{C}_{A_1}| \geq |\mathcal{C}_{A_1}^{M^*}| = M_1^* = |\mathcal{C}_1^*|.$$

On the other hand, since $\mathcal{C}_{A_1}$ is an admissible solution of Problem 1, we have $|\mathcal{C}_{A_1}| \leq |\mathcal{C}_1^*|$; therefore, $|\mathcal{C}_{A_1}| = |\mathcal{C}_1^*|$.

The proof of the optimality of algorithm $\mathcal{A}_4$ is similar to the proof of the optimality of algorithm $\mathcal{A}_3$.

Let us estimate the time complexity of algorithm $\mathcal{A}_3$ (the time complexity of algorithm $\mathcal{A}_4$ is the same). Step 1 is done in $\mathcal{O}(dN)$ operations. The most time consuming Step 2 requires $\mathcal{O}(dN^2(2ND + 1)^d)$ operations since for every $M = 1, \ldots, N$ algorithm $\mathcal{A}_1$ requires $\mathcal{O}(dN(2MD + 1)^d)$ operations. Finally, Step 3 can be done in $\mathcal{O}(N)$ operations. Summing up the costs required at all steps yields the time complexity bound for algorithm $\mathcal{A}_3$.

Thus, the time complexity of both algorithms is equal to $\mathcal{O}(dN^2(2ND + 1)^d)$. $\qquad\square$

Remark 3. In the case of fixed space dimension, algorithms $\mathcal{A}_3$ and $\mathcal{A}_4$ are pseudopolynomial, since in that case their running time is equal to $\mathcal{O}(N^2(ND)^d)$.

6   Conclusion

In this paper, we prove the strong NP-hardness of two different quadratic Euclidean clusterization problems. Also we present the first algorithms for these problems. Our algorithms allow finding exact solutions in the case of integer problems instances. If the space dimension is bounded by some constant, the algorithms are pseudopolynomial. Thus, we show that both problems are solvable in a pseudopolynomial time in the case of fixed space dimension and integer problems instances.

It is clear that the proposed algorithms are suitable for solving practical problems of only small dimension. However, one can consider these algorithms as a starting point for obtaining improved algorithmic solutions.

An important direction of further studies is the substantiation of the effective approximation algorithms with guaranteed accuracy for the considered problems.

In our opinion, these algorithms will be useful, in particular, in Data mining, Pattern recognition, and Machine learning.

Acknowledgments. The study of Problem 1 was supported by the Russian Science Foundation, project 16-11-10041. The study of Problem 2 was supported by the Russian Foundation for Basic Research, projects 16-07-00168 and 18-31-00398, by the Russian Academy of Science (the Program of Basic Research), project 0314-2016-0015, and by the Russian Ministry of Science and Education under the 5-100 Excellence Programme.

References

1. MacQueen, J.B.: Some methods for classification and analysis of multivariate observations. In: Proceedings of the 5th Berkeley Symposium on Mathematical Statistics and Probability, University of California Press, Berkeley, vol. 1, pp. 281–297 (1967)
2. Rao, M.: Cluster analysis and mathematical programming. J. Am. Stat. Assoc. **66**, 622–626 (1971)
3. Hansen, P., Jaumard, B., Mladenovich, N.: Minimum sum of squares clustering in a low dimensional space. J. Classif. **15**, 37–55 (1998)
4. Hansen, P., Jaumard, B.: Cluster analysis and mathematical programming. Math. Program. **79**, 191–215 (1997)
5. Fisher, R.A.: Statistical Methods and Scientific Inference. Hafner, New York (1956)
6. Jain, A.K.: Data clustering: 50 years beyond k-means. Pattern Recogn. Lett. **31**(8), 651–666 (2010)
7. Aloise, D., Deshpande, A., Hansen, P., Popat, P.: NP-hardness of Euclidean sum-of-squares clustering. Mach. Learn. **75**(2), 245–248 (2009)
8. Drineas, P., Frieze, A., Kannan, R., Vempala, S., Vinay, V.: Clustering large graphs via the singular value decomposition. Mach. Learn. **56**, 9–33 (2004)
9. Dolgushev, A.V., Kel'manov, A.V.: On the algorithmic complexity of a problem in cluster analysis. J. Appl. Ind. Math. **5**(2), 191–194 (2011)
10. Mahajan, M., Nimbhorkar, P., Varadarajan, K.: The planar k-means problem is NP-hard. Theoret. Comput. Sci. **442**, 13–21 (2012)
11. Kel'manov, A.V., Pyatkin, A.V.: On the complexity of a search for a subset of "similar" vectors. Doklady Math. **78**(1), 574–575 (2008)
12. Kel'manov, A.V., Pyatkin, A.V.: On a version of the problem of choosing a vector subset. J. Appl. Ind. Math. **3**(4), 447–455 (2009)
13. Brucker, P.: On the complexity of clustering problems. In: Henn, R., Korte, B., Oettli, W. (eds.) Optimization and Operations Research. LNEMS, vol. 157, pp. 45–54. Springer, Heidelberg (1978). https://doi.org/10.1007/978-3-642-95322-4_5
14. Bern, M., Eppstein, D.: Approximation algorithms for geometric problems. In: Approximation Algorithms for NP-Hard Problems, pp. 296–345. PWS Publishing, Boston (1997)

15. Indyk, P.: A sublinear time approximation scheme for clustering in metric space. In: Proceedings of the 40th Annual IEEE Symposium on Foundations of Computer Science (FOCS), pp. 154–159 (1999)
16. de la Vega, F., Kenyon, C.: A randomized approximation scheme for metric max-cut. J. Comput. Syst. Sci. **63**, 531–541 (2001)
17. de la Vega, F., Karpinski, M., Kenyon, C., Rabani, Y.: Polynomial time approximation schemes for metric min-sum clustering. Electronic Colloquium on Computational Complexity (ECCC). Report No. 25 (2002)
18. Hasegawa, S., Imai, H., Inaba, M., Katoh, N., Nakano, J.: Efficient algorithms for variance-based k-clustering. In: Proceedings of the 1st Pacific Conference on Computer Graphics and Applications, Pacific Graphics 1993, Seoul, Korea, vol. 1. pp. 75–89. World Scientific, River Edge (1993)
19. Inaba, M., Katoh, N., Imai, H.: Applications of weighted Voronoi diagrams and randomization to variance-based k-clustering: (extended abstract), 6–8 June 1994, Stony Brook, NY, USA, pp. 332–339. ACM, New York (1994)
20. Sahni, S., Gonzalez, T.: P-complete approximation problems. J. ACM **23**, 555–566 (1976)
21. Ageev, A.A., Kel'manov, A.V., Pyatkin, A.V.: NP-hardness of the Euclidean max-cut problem. Doklady Math. **89**(3), 343–345 (2014)
22. Ageev, A.A., Kel'manov, A.V., Pyatkin, A.V.: Complexity of the weighted max-cut in Euclidean space. J. Appl. Ind. Math. **8**(4), 453–457 (2014)
23. Kel'manov, A.V., Pyatkin, A.V.: NP-hardness of some quadratic Euclidean 2-clustering problems. Doklady Math. **92**(2), 634–637 (2015)
24. Kel'manov, A.V., Pyatkin, A.V.: On the complexity of some quadratic Euclidean 2-clustering problems. Comput. Math. Math. Phys. **56**(3), 491–497 (2016)
25. Bishop, C.M.: Pattern Recognition and Machine Learning. Springer Science+Business Media, LLC, New York (2006)
26. James, G., Witten, D., Hastie, T., Tibshirani, R.: An Introduction to Statistical Learning. Springer Science+Business Media, LLC, New York (2013). https://doi.org/10.1007/978-1-4614-7138-7
27. Hastie, T., Tibshirani, R., Friedman, J.: The Elements of Statistical Learning, 2nd edn. Springer, Heidelberg (2009). https://doi.org/10.1007/978-0-387-84858-7
28. Aggarwal, C.C.: Data Mining: The Textbook. Springer, Heidelberg (2015). https://doi.org/10.1007/978-3-319-14142-8
29. Goodfellow, I., Bengio, Y., Courville, A.: Deep Learning. Adaptive Computation and Machine Learning Series. The MIT Press, Cambridge (2017)
30. Shirkhorshidi, A.S., Aghabozorgi, S., Wah, T.Y., Herawan, T.: Big data clustering: a review. In: Murgante, B., et al. (eds.) ICCSA 2014. LNCS, vol. 8583, pp. 707–720. Springer, Cham (2014). https://doi.org/10.1007/978-3-319-09156-3_49
31. Pach, J., Agarwal, P.K.: Combinatorial Geometry. Wiley, New York (1995)
32. Kel'manov, A.V., Khandeev, V.I.: A 2-approximation polynomial algorithm for a clustering problem. J. Appl. Ind. Math. **7**(4), 515–521 (2013)
33. Gimadi, E.Kh., Kel'manov, A.V., Kel'manova, M.A., Khamidullin, S.A.: A posteriori detection of a quasi periodic fragment in numerical sequences with given number of recurrences. Siberian J. Ind. Math. **9**(1(25)), 55–74 (2006). (in Russian)
34. Gimadi, E.K., Kel'manov, A.V., Kel'manova, M.A., Khamidullin, S.A.: A posteriori detecting a quasiperiodic fragment in a numerical sequence. Pattern Recogn. Image Anal. **18**(1), 30–42 (2008)
35. Baburin, A.E., Gimadi, E.K., Glebov, N.I., Pyatkin, A.V.: The problem of finding a subset of vectors with the maximum total weight. J. Appl. Ind. Math. **2**(1), 32–38 (2008)

36. Dolgushev, A.V., Kel'manov, A.V.: An approximation algorithm for solving a problem of cluster analysis. J. Appl. Ind. Math. **5**(4), 551–558 (2011)
37. Dolgushev, A.V., Kel'manov, A.V., Shenmaier, V.V.: Polynomial-time approximation scheme for a problem of partitioning a finite set into two clusters. Proc. Steklov Inst. Math. **295**(Suppl. 1), 47–56 (2016)
38. Kel'manov, A.V., Khandeev, V.I.: A randomized algorithm for two-cluster partition of a set of vectors. Comput. Math. Math. Phys. **55**(2), 330–339 (2015)
39. Gimadi, E.K., Pyatkin, A.V., Rykov, I.A.: On polynomial solvability of some problems of a vector subset choice in a Euclidean space of fixed dimension. J. Appl. Ind. Math. **4**(1), 48–53 (2010)
40. Shenmaier, V.V.: Solving some vector subset problems by Voronoi diagrams. J. Appl. Ind. Math. **10**(4), 560–566 (2016)
41. Kel'manov, A.V., Khandeev, V.I.: An exact pseudopolynomial algorithm for a problem of the two-cluster partitioning of a set of vectors. J. Appl. Ind. Math. **9**(4), 497–502 (2015)
42. Kel'manov, A.V., Khandeev, V.I.: Fully polynomial-time approximation scheme for a special case of a quadratic euclidean 2-clustering problem. J. Appl. Ind. Math. **56**(2), 334–341 (2016)
43. Kel'manov, A.V., Motkova, A.V.: Polynomial-time approximation algorithm for the problem of cardinality-weighted variance-based 2-clustering with a given center. Comput. Math. Math. Phys. **58**(1), 130–136 (2018)
44. Kel'manov, A.V., Motkova, A.V.: Exact pseudopolynomial algorithms for a balanced 2-clustering problem. J. Appl. Ind. Math. **10**(3), 349–355 (2016)
45. Kel'manov, A., Motkova, A.: A fully polynomial-time approximation scheme for a special case of a balanced 2-clustering problem. In: Kochetov, Y., Khachay, M., Beresnev, V., Nurminski, E., Pardalos, P. (eds.) DOOR 2016. LNCS, vol. 9869, pp. 182–192. Springer, Cham (2016). https://doi.org/10.1007/978-3-319-44914-2_15
46. Kel'manov, A., Motkova, A., Shenmaier, V.: An approximation scheme for a weighted two-cluster partition problem. In: van der Aalst, W.M.P., et al. (eds.) AIST 2017. LNCS, vol. 10716, pp. 323–333. Springer, Cham (2018). https://doi.org/10.1007/978-3-319-73013-4_30
47. Kel'manov, A., Khandeev, V., Panasenko, A.: Randomized algorithms for some clustering problems. In: Eremeev, A., Khachay, M., Kochetov, Y., Pardalos, P. (eds.) OPTA 2018. CCIS, vol. 871, pp. 109–119. Springer, Cham (2018). https://doi.org/10.1007/978-3-319-93800-4_9

On a Problem of Summing Elements Chosen from the Family of Finite Numerical Sequences

Alexander Kel'manov[1,2], Ludmila Mikhailova[1(✉)], and Semyon Romanchenko[1]

[1] Sobolev Institute of Mathematics, Koptyug Avenue, 630090 Novosibirsk, Russia
{kelm,mikh,rsm}@math.nsc.ru
[2] Novosibirsk State University, 2 Pirogova Street, 630090 Novosibirsk, Russia

Abstract. The tuple of permutations and the tuple of indices are required to be found in the problem considered in order to minimize the sum of elements chosen from the given family of finite numerical sequences subject to some constraints on the elements choice. Namely, given the family of L numerical nonnegative N-element sequences and a positive integer J, it is required to minimize the sum of J intra-sums. Each element corresponds to one element in one of L input sequences, and all possible L-permutations are admissible in this one-to-one correspondence in each intra-sum of L elements. In addition, there are some constraints on the indices of the summed sequence elements. The problem solution is a pair of tuples, namely, (1) a tuple of J permutations on L elements, and (2) a tuple of JL increasing indices. The paper presents an exact polynomial-time algorithm with $\mathcal{O}(N^5)$ running time for this problem. In particular, the problem is induced by an applied problem of noiseproof searching for repetitions of the given tuple of elements with their possible permutations at each tuple repeat, and finding the positions of these elements in the numerical sequence distorted by noise under some constraints on unknown positions of elements. The applied problem noted is related, for example, to the remote monitoring of several moving objects with possible arbitrary displacements (permutations) of these objects.

Keywords: Optimal summing · Finite numerical sequences
Permutations · Exact polynomial-time algorithm

1 Introduction

The paper aims at studying one problem of optimal summing of elements chosen from the family of finite numerical sequences. The subject of this research is algorithmic approximability of the problem. Our goal is to construct a polynomial-time algorithm with the guaranteed accuracy.

The research is motivated by the lack of available algorithms for the problem under consideration despite its importance, in particular, in the remote monitoring of some objects (see next section). Moreover, it is the generalization of one

© Springer Nature Switzerland AG 2018
W. M. P. van der Aalst et al. (Eds.): AIST 2018, LNCS 11179, pp. 305–317, 2018.
https://doi.org/10.1007/978-3-030-11027-7_29

optimization problem studied before (see next section). Therefore, this study is of interest, as well.

The paper has the following structure. The problem formulation, its interpretation, and known results for the particular case of the problem are presented in Sect. 2. Also, our result obtained is announced in this section. Section 3 contains two auxiliary problems and exact polynomial-time algorithms for their solution which are necessary to justify the proposed algorithm. Finally, our algorithm for the main problem considered is given and justified in Sect. 4.

2   Problem Formulation and Related Problems

Hereinafter we denote the set of real numbers by $\mathbb{R}$, the set of non-negative real numbers by $\mathbb{R}_{\geq}$, and the set of positive integer numbers by $\mathbb{N}$. The permutation on L elements and the set of all possible permutations on L elements are designated as π and Π respectively. Finally, the j-th permutation, for every $j = 1, \ldots, J$, is denoted by π_j, and its i-th element, for every $i = 1, \ldots, L$, is denoted by $\pi_j(i)$.

We consider the following

Problem 1. Given the family $\mathcal{S} = \Big\{ \{s_1(n), \ldots, s_L(n)\} \Big| s_i(n) \in \mathbb{R}_{\geq}, i = 1, \ldots, L,$

$n = 1, \ldots, N \Big\}$ of numerical sequences and a positive integer J such that $JL \leq N$. *Find* a tuple $(n_1, \ldots, n_{JL})$ of indices, where $n_k \in \{1, \ldots, N\}$, $k = 1, \ldots, JL$, and a tuple $(\pi_1, \ldots, \pi_J)$ of J permutations on L elements such that

$$S(n_1, \ldots, n_{JL}, \pi_1, \ldots, \pi_J) = \sum_{j=1}^{J} \sum_{i=1}^{L} s_{\pi_j(i)}(n_{(j-1)L+i}) \longrightarrow \min \qquad (1)$$

subject to constraints

$$1 \leq n_1 < n_2 < \ldots < n_{JL} \leq N. \qquad (2)$$

In the particular case of Problem 1, when $\pi_j(i) = i$ for all $j = 1, \ldots, J$, $i = 1, \ldots, L$, i.e., when permutations are prohibited, it is required to find an index tuple $(n_1, \ldots, n_{JL})$ minimizing the value of

$$\sum_{j=1}^{J} \sum_{i=1}^{L} s_i(n_{(j-1)L+i})$$

under the same constraints (2). This particular case was studied in [1]. The authors presented an exact polynomial algorithm with running time $\mathcal{O}(JLN^2)$.

This particular case [1] is induced by the optimization model where each element $s_i(n)$, $n = 1, \ldots, N$, $i = 1, \ldots, L$, is a squared distance between the point $x_i \in \mathbb{R}^d$ in the given collection $\{x_1, \ldots, x_L\}$ of L points and the element $y_n \in \mathbb{R}^d$ in a sample sequence $(y_1, \ldots, y_N)$. In other words, the i-th element in the family S is the result of comparisons (by distance) between the i-th point in the collection $\{x_1, \ldots, x_L\}$ and each element in the sequence $(y_1, \ldots, y_N)$. The inequalities (2) define the tuple $(n_1, \ldots, n_{JL})$ of indices in the sequence $(y_1, \ldots, y_N)$. The indices in this tuple are admissible for summing in (1).

The correspondence between the applied problem inducing this particular case and its model is as follows. The point x_i and the collection $\{x_1, \ldots, x_L\}$ of points in the model correspond to the i-th object and the collection of L objects of some interest respectively. The sample sequence $(y_1, \ldots, y_N)$ corresponds to a d-dimensional time series obtained as a measurement result. There are some measurement results corresponding to the objects of interest in this time series. These results are indexed by the tuple $(n_1, \ldots, n_{JL})$. Every sequence s_i in the family S corresponds to the result of comparing the i-th object in the collection $\{x_1, \ldots, x_L\}$ and each element of time ordered measurement results. The applied problem can be interpreted as a noiseproof detection of the given pattern collection $\{x_1, \ldots, x_L\}$ repeated J times in the signal $(y_1, \ldots, y_N)$. There are no pattern permutations in each pattern collection repetition in the applied problem. One can see that the interval $n_m - n_{m-1}$, $m = 2, \ldots, JL$, between two consecutive patterns is not constant, but it is bounded in accordance with the inequalities (2).

Problem 1 is induced by the similar but more complex applied problem. Namely, it is allowed that the patterns (objects) are arbitrarily reordered with each repetition of the pattern collection. This rearrangement of objects is represented by permutations in the objective function (1).

We present a polynomial algorithm with running time $\mathcal{O}(N^5)$ which allows us to find an optimal solution for Problem 1.

3   Algorithm Foundation

To solve Problem 1 two auxiliary problems and exact polynomial algorithms for their solution are necessary. In addition, exact algorithms are needed for two well-known problems, namely, (1) *Minimum weight perfect matching in a complete bipartite graph* [2], (2) *Nearest neighbor search* [3].

The first auxiliary problem has the following formulation.

Problem 2. Given the family $\mathcal{F} = \left\{ \{f_1(n), \ldots, f_L(n)\} \,\middle|\, f_i(n) \in \mathbb{R}_{\geq}, \, i = 1, \ldots, L, \, n = a, \ldots, b - 1; \, L \leq b - a, \, a, b \in \mathbb{N} \right\}$ *of sequences. Find an index tuple* $(n_1, \ldots, n_L)$, *where* $n_i \in \{a, \ldots, b - 1\}$, $i = 1, \ldots, L$, *and a permutation* $\pi \in \Pi$ *such that*

$$F(n_1, \ldots, n_L, \pi) = \sum_{i=1}^{L} f_{\pi(i)}(n_i) \to \min, \qquad (3)$$

while the elements of $(n_1, \ldots, n_L)$ *satisfy the following constraints*

$$a \leq n_1 < \ldots < n_L < b.$$

To solve this problem, it will be transformed (reduced) into *Minimum weight perfect matching* problem.

Let $\mathcal{G} = (\mathcal{V} \cup \mathcal{U}, \mathcal{V} \times \mathcal{U}, w)$ be a complete bipartite graph with vertex parts $\mathcal{V} = \{v_a, \ldots, v_{b-1}\}$ and $\mathcal{U} = \{u_1, \ldots, u_{b-a}\}$. Each vertex v_n, $n = a, \ldots, b-1$, in $\mathcal{V}$ corresponds to the index n in $\{a, \ldots, b-1\}$. The vertices $u_1, \ldots, u_L$ in $\mathcal{U}$ correspond to the indices in $\{1, \ldots, L\}$ of the sequences in the input family $\mathcal{F}$. The vertices $u_{L+1}, \ldots, u_{b-a}$ are auxiliary ones and necessary only to make vertex parts equal by size.

Let the edge weights be given as

$$w_{v_n u_j} = \begin{cases} f_j(n), & j = 1, \ldots, L, \\ 0, & j = L+1, \ldots, b-a, \end{cases} \tag{4}$$

for each $n = a, \ldots, b-1$. The weight of each edge (v, u), where $v \in \mathcal{V}$ and u is an auxiliary vertex in $\mathcal{U}$, is zero. While the weight of each edge (v_n, u_l), $n = a, \ldots, b-1$, $l = 1, \ldots, L$, is equal to the value of the n-th element in the l-th sequence in $\mathcal{F}$.

Define a bijection $r : \{a, \ldots, b-1\} \longrightarrow \{1, \ldots, b-a\}$ between the sets of indices of $\mathcal{V}$ and $\mathcal{U}$ for every perfect matching $\mathcal{P}$ in $\mathcal{G}$. Hereinafter the weight of the perfect matching $\mathcal{P} = \{(v_n, u_{r(n)}), n = a, \ldots, b-1\}$ is designated as $W(\mathcal{P})$.

The following lemma is true.

Lemma 1. *Let* $\mathcal{P}^* = \{(v_n, u_{r^*(n)}), n = a, \ldots, b-1\}$ *be the minimum weight perfect matching in* $\mathcal{G}$, *where* r^* *is the optimal bijection. Then in Problem 2 the optimal tuple* $\eta^*(a, b) = (n_1^*, \ldots, n_L^*)$ *and the optimal permutation* $\pi^*(a, b)$ *can be found by the following formulas:*

$$n_1^* = \min\left\{n \mid r^*(n) \in \{1, \ldots, L\}, n \in \{a, \ldots, b-1\}\right\}, \tag{5}$$

$$n_i^* = \min\left\{n \mid r^*(n) \in \{1, \ldots, L\}, n \in \{n_{i-1}^* + 1, \ldots, b-1\}\right\}, \quad i = 2, \ldots, L, \tag{6}$$

and

$$\pi^*(a, b) = (r^*(n_1^*), \ldots, r^*(n_L^*)). \tag{7}$$

The optimal value $F^*(a, b)$ *of the objective function* (3) *can be found as*

$$F^*(a, b) = W(\mathcal{P}^*). \tag{8}$$

Proof. Consider an arbitrary perfect matching $\mathcal{P}$ in $\mathcal{G}$. Let

$$\mathcal{M}(\mathcal{P}) = \left\{n \mid r(n) \in \{1, \ldots, L\}, n \in \{a, \ldots, b-1\}\right\} \tag{9}$$

be the set of indices of the vertices in $\mathcal{V}$ which are adjacent to the vertices $u_1, \ldots, u_L$ in $\mathcal{U}$.

Let $\eta(\mathcal{P}) = (n_1, \ldots, n_L)$ be the tuple with the following components:

$$n_1 = \min\{n \mid n \in \mathcal{M}(\mathcal{P})\}, \tag{10}$$

$$n_j = \min\Big\{n \mid n \in \mathcal{M}(\mathcal{P}), \ n \in \{n_{j-1}+1,\ldots,b-1\}\Big\}, \ j = 2,\ldots,L, \qquad (11)$$

and let

$$\pi(\mathcal{P}) = (r(n_1),\ldots,r(n_L)). \qquad (12)$$

It follows from (9), (10), and (11) that the sequence $n_1,\ldots,n_L$ is an increasing one. Moreover, the tuple $\eta(\mathcal{P})$ and the permutation $\pi(\mathcal{P})$ are admissible in Problem 2.

Denote by $\mathcal{P}_1$ a set of edges in the matching $\mathcal{P}$ which are incident to vertices $u_1,\ldots,u_L$. Then from (9) we have $\mathcal{P}_1 = \{(v_n, u_{r(n)}), \ n \in \mathcal{M}(\mathcal{P})\}$.

According to (4) $w(e) = 0$ for each edge $e \in \mathcal{P} \setminus \mathcal{P}_1$. Consequently, for the weight of the matching $\mathcal{P}$, we have

$$W(\mathcal{P}) = W(\mathcal{P}_1) + W(\mathcal{P} \setminus \mathcal{P}_1) = W(\mathcal{P}_1) = \sum_{n \in \mathcal{M}(\mathcal{P})} w(v_n, u_{r(n)})$$

$$= \sum_{i=1}^{L} w(v_{n_i}, u_{r(n_i)}) = \sum_{i=1}^{L} f_{r(n_i)}(n_i) = F(\eta(\mathcal{P}), \pi(\mathcal{P})). \qquad (13)$$

Now, let us prove (8). Consider an arbitrary permutation $\pi = (\pi(1),\ldots,\pi(L))$ and an arbitrary tuple $\eta = (n_1,\ldots,n_L)$ admissible in Problem 2. For this tuple and this permutation we show the existence of a perfect matching $\mathcal{P}$ in $\mathcal{G}$ such that

$$\eta = \eta(\mathcal{P}), \quad \pi = \pi(\mathcal{P}), \qquad (14)$$

where $\eta(P)$ and $\pi(P)$ are defined by (10), (11), and (12).

To do so, we construct an example of $\mathcal{P}$ in the following way. Take the set $\{(v_{n_k}, u_{\pi(k)}), \ k = 1,\ldots,L\}$ of edges. This set is a matching in $\mathcal{G}$. Add $b - a - L$ edges to this matching to obtain a perfect one. All these added edges have zero weight, since they are incident to the vertices $u_{L+1},\ldots,u_{b-a}$ in $\mathcal{U}$. Then it is easy to see that (14) is true for the constructed perfect matching $\mathcal{P}$. Consequently, in accordance with (13), we have

$$F(\eta, \pi) = F(\eta(\mathcal{P}), \ \pi(\mathcal{P})) = W(\mathcal{P}). \qquad (15)$$

Further, since $\mathcal{P}^*$ is the minimum weight perfect matching, for the weight of the constructed matching $\mathcal{P}$ we have

$$W(\mathcal{P}) \geq W(\mathcal{P}^*) = F(\eta(\mathcal{P}^*), \pi(\mathcal{P}^*)). \qquad (16)$$

From (15) and (16) we obtain $F(\eta, \pi) \geq F(\eta(\mathcal{P}^*), \pi(\mathcal{P}^*))$. This inequality means that $F^*(a,b) = F(\pi(\mathcal{P}^*), \eta(\mathcal{P}^*))$. Finally, combining this equality with (13), we prove (8). Formulas (5), (6), and (7) follow immediately from (10), (11), and (12). $\qquad \square$

Recall that the minimum weight perfect matching can be found by a well-known algorithm (see, for example [2]) in polynomial time. This time is equal to $\mathcal{O}((b-a)^3)$ for the graph $\mathcal{G}$.

Thus, we have the following algorithm for auxiliary Problem 2.

Algorithm $\mathcal{A}_1$.

INPUT: the family $\mathcal{F}$ of sequences.

STEP 1. Construct $\mathcal{G}$ using (4) and find the minimum weight perfect matching $\mathcal{P}^*$ in $\mathcal{G}$ using the well-known algorithm [2].

STEP 2. Find the optimal permutation $\pi^*(a, b)$ and the optimal tuple $\eta^*(a, b)$ by (5), (6), and (7).

STEP 3. Compute $W(\mathcal{P}^*)$ and $F^*(a, b)$ by (8).

OUTPUT: the permutation $\pi^*(a, b)$, the tuple $\eta^*(a, b) = (n_1^*, \ldots, n_L^*)$, and the value $F^*(a, b)$.

The following proposition is true.

Proposition 1. *Algorithm $\mathcal{A}_1$ finds the optimal solution of Problem 2 in $\mathcal{O}((b-a)^3)$ time.*

Proof. The solution optimality follows immediately from Lemma 1. Evaluate the time complexity of the algorithm. We need $\mathcal{O}((b - a)^2)$-time to construct $\mathcal{G}$ at Step 1. At the same step finding the perfect matching $\mathcal{P}^*$ requires $\mathcal{O}((b - a)^3)$ time. Steps 2 and 3 are executed in $\mathcal{O}(b - a)$ time. So, the running time of Algorithm $\mathcal{A}_1$ is $\mathcal{O}((b - a)^3)$.   □

The second auxiliary problem has the following formulation.

Problem 3. *Given* positive integers L, J, and N such that $JL \leq N$, the family $\mathcal{Q} = \{q(n, m) \in \mathbb{R}_\geq \mid 1 \leq n < m \leq N+1;\ L \leq m-n \leq N-(J-1)L,\ n, m \in \mathbb{N}\}$. *Find* an index tuple $(\mu_1, \ldots, \mu_{J+1})$, where $\mu_j \in \{1, \ldots, N+1\}$, $j = 1, \ldots, J+1$, such that

$$Q(\mu_1, \ldots, \mu_{J+1}) = \sum_{j=1}^{J} q(\mu_j, \mu_{j+1}) \to \min, \qquad (17)$$

subject to constraints

$$1 = \mu_1 < \ldots < \mu_{J+1} = N + 1, \qquad (18)$$

$$\mu_{j+1} - \mu_j \geq L, \ j = 1, \ldots, J. \qquad (19)$$

Remark 1. It follows from (18) and (19) that

$$\mu_{j+1} - \mu_j \leq N - (J - 1)L, \ j = 1, \ldots, J. \qquad (20)$$

To solve Problem 3 we use an exact algorithm for the well-known *Nearest neighbor search* problem (see, for example, [3]). *Given* the family $\mathcal{H} = \{h(n, m) \in \mathbb{R}_\geq \mid 1 \leq n \leq m \leq N + 1;\ n, m \in \mathbb{N}\}$ and a positive integer J. *Find* an index tuple $(\nu_1, \ldots, \nu_{J+1})$, where $\nu_j \in \{1, \ldots, N+1\}$, $j = 1, \ldots, J+1$, such that

$$H(\nu_1, \ldots, \nu_{J+1}) = \sum_{j=1}^{J} h(\nu_j, \nu_{j+1}) \to \min, \qquad (21)$$

subject to constraints

$$1 = \nu_1 \leq \nu_2 \leq \ldots \leq \nu_{J+1} = N + 1.$$

As it is known, the solution of the *Nearest neighbor search* problem can be found in $\mathcal{O}(JN^2)$ time.

Lemma 2. *Let* $(\nu_1^*, \ldots, \nu_{J+1}^*)$ *be the optimal solution of the Nearest neighbor search problem with*

$$h(n, m) = \begin{cases} q(n, m), & n = 1, \ldots, N - L + 1, \\ & \quad m = n + L, \ldots, \min\{N + 1, n + N - (J-1)L\}, \\ +\infty, & n = 1, \ldots, N + 1, m = n, \ldots, \min\{n + L - 1, N + 1\}, \quad (22) \\ +\infty, & n = 1, \ldots, 1 + (J-1)L, \\ & \quad m = n + N - (J-1)L + 1, \ldots, N + 1, \end{cases}$$

and H^* *be the optimal value of the objective function* (21). *Then the components of the optimal tuple* $(\mu_1^*, \ldots, \mu_{J+1}^*)$ *and the optimal value* Q^* *of the function* (17) *can be found by the following formulas*

$$\mu_i^* = \nu_i^*, \quad i = 1, \ldots, J + 1, \tag{23}$$

$$Q^* = H^*. \tag{24}$$

Proof. First, let us show that the tuple $(\nu_1^*, \ldots, \nu_{J+1}^*)$ is an admissible one in Problem 3. Assume the contrary, i.e., there exists $i \in \{1, \ldots J\}$ such that $\nu_{i+1}^* - \nu_i^* < L$ or $\nu_{i+1}^* - \nu_i^* > N - (J-1)L$. Then, we have

$$H^* = H(\nu_1^*, \ldots, \nu_{J+1}^*) = \sum_{j=1}^{J} h(\nu_j^*, \nu_{j+1}^*) =$$

$$\sum_{j=1}^{i} h(\nu_j^*, \nu_{j+1}^*) + h(\nu_i^*, \nu_{i+1}^*) + \sum_{j=i+1}^{J} h(\nu_j^*, \nu_{j+1}^*) = +\infty,$$

since the first and the third terms in the penultimate expression are non-negative and the second term is $+\infty$ in accordance with (22).

On the other hand, the value of the objective function (21) is finite for every tuple admissible in Problem 3. This is contradictory to the optimality of $(\nu_1^*, \ldots, \nu_{J+1}^*)$ in the *Nearest neighbor search* problem. So, $(\nu_1^*, \ldots, \nu_{J+1}^*)$ satisfies (19) and (20). In addition, it satisfies (18) and so, it is an admissible tuple in Problem 3.

Let us show that the tuple $(\nu_1^*, \ldots, \nu_{J+1}^*)$ is optimal in Problem 3. Let $(\mu_1, \ldots, \mu_{J+1})$ be an arbitrary tuple such that it satisfies the restrictions (18) and (19). Then, using (17), (21), and (22), we obtain

$$Q(\mu_1, \ldots, \mu_{J+1}) = \sum_{j=1}^{J} q(\mu_j, \mu_{j+1}) = \sum_{j=1}^{J} h(\mu_j, \mu_{j+1}) = H(\mu_1, \ldots, \mu_{J+1}). \tag{25}$$

Thus, for an arbitrary tuple $(\mu_1, \ldots, \mu_{J+1})$, admissible in Problem 3, we have

$$Q(\mu_1, \ldots, \mu_{J+1}) = H(\mu_1, \ldots, \mu_{J+1}) \geq H^*$$
$$= H(\nu_1^*, \ldots, \nu_{J+1}^*) = Q(\nu_1^*, \ldots, \nu_{J+1}^*).$$

Consequently, $Q^* = Q(\nu_1^*, \ldots, \nu_{J+1}^*)$ and by (25) we have

$$Q^* = Q(\nu_1^*, \ldots, \nu_{J+1}^*) = H(\nu_1^*, \ldots, \nu_{J+1}^*) = H^*.$$

In such a way, the formulas (23) and (24) are proved.   □

Thus, we have the following algorithm for auxiliary Problem 3.

Algorithm $\mathcal{A}_2$.

INPUT: positive integers L, J, and the family $\mathcal{Q}$.

STEP 1. Find the family $\mathcal{H}$ by (22). Compute the optimal value H^* and find the optimal tuple $(\nu_1^*, \ldots, \nu_{J+1}^*)$ of the *Nearest neighbor search* problem.

STEP 2. Compute the optimal value Q^* of the objective function of Problem 3 using (24) and find the optimal tuple $(\mu_1^*, \ldots, \mu_{J+1}^*)$ of indices using (23).

OUTPUT: the tuple $(\mu_1^*, \ldots, \mu_{J+1}^*)$ and the value Q^*.

The following proposition is true.

Proposition 2. *Algorithm $\mathcal{A}_2$ finds an optimal solution for Problem 3 in $\mathcal{O}(JN^2)$ time.*

Proof. The solution optimality follows from Lemma 2. Evaluate the running time of Algorithm $\mathcal{A}_2$. Step 1 implements the algorithm for the *Nearest neighbor search* problem, so it is executed in $\mathcal{O}(JN^2)$ time. Step 2 requires $\mathcal{O}(N)$ time. Consequently, the running time of Algorithm $\mathcal{A}_2$ is $\mathcal{O}(JN^2)$.   □

4   Exact Algorithm

Let, as above $a, b \in \mathbb{N}$, in addition, let $M, T \in \mathbb{N}$. Define two sets of integer tuples:

$$\Omega_M(a, b, T) = \{(n_1, \ldots, n_M) |\ n_i \in \mathbb{N},\ i = 1, \ldots, M,$$
$$a \leq n_1 < \ldots < n_M \leq b,\ T \leq n_j - n_{j-1},\ j = 2, \ldots, M\},\quad (26)$$

and

$$\Phi_L(a, b) = \{(n_1, \ldots, n_L) |\ n_i \in \mathbb{N},\ i = 1, \ldots, M, a \leq n_1 < \ldots < n_L < b\}.\quad (27)$$

Remark 2. If $a = 1$, $b = N$, $M = JL$, and $T = 1$, where N, L, and J are instances of Problem 1, then it is easy to see that $\Omega_{JL}(1, N, 1)$ is the set of all admissible tuples in Problem 1. If $1 \leq a < b \leq N$ and $b - a \geq L$, then $\Phi_L(a, b)$ is the set of all tuples admissible in Problem 2.

Next, we need the following

Property 1. If $s,t \in \mathbb{N}$, $1 \leq s < t \leq M$, and $(n_1,\ldots,n_{JL}) \in \Omega_{JL}(1,N,1)$, where N, L, and J are instances of Problem 1, then

$$t - s \leq n_t - n_s. \tag{28}$$

To prove Property 1, it is sufficient to sum up the inequalities $T \leq n_i - n_{i-1}$ included in the definition (26) over all $i = s+1,\ldots,t$ with $T = 1$.

To justify our algorithm, we need the following

Proposition 3. *For the set of the tuples admissible in Problem 1 we have*

$$\Omega_{JL}(1,N,1) = \bigcup_{\substack{(m_2,\ldots,m_J)\in\Omega_{J-1}(L+1,N-L+1,L), \\ m_1=1,\, m_{J+1}=N+1}} \prod_{j=1}^{J} \Phi_L(m_j, m_{j+1}). \tag{29}$$

Proof. To prove this proposition, we show the mutual inclusion of the sets in the left and right parts of (29).

Direct inclusion of the sets. Consider a tuple $(n_1,\ldots,n_{JL}) \in \Omega_{JL}(1,N,1)$. Let us show that its subtuple $(n_{L+1}, n_{2L+1}, \ldots, n_{(J-1)L+1}) \in \Omega_{J-1}(L+1, N - L+1, L)$. Indeed, for the subtuple components, according to (28), we have

$$n_{L+1} - n_1 \geq L + 1 - 1 = L,$$

$$n_{JL} - n_{(J-1)L+1} \geq JL - ((J-1)L+1) = L - 1.$$

Consequently,

$$n_{L+1} \geq L + n_1 \geq L + 1,$$

$$n_{(J-1)L+1} \leq n_{JL} - L + 1 \leq N - L + 1.$$

From this and the definition (26) we obtain

$$L + 1 \leq n_{L+1} < \ldots < n_{(J-1)L+1} \leq N - L + 1.$$

Moreover, in accordance with (28) for every $j = 2, \ldots, J - 1$ we have

$$n_{jL+1} - n_{(j-1)L+1} \geq jL + 1 - ((j-1)L+1) = L.$$

In such a way, with $a = L + 1$, $b = N - L + 1$, $M = J - 1$, and $T = L$ all inequalities in the definition (26) are satisfied. Consequently,

$$(n_{L+1}, n_{2L+1}, \ldots, n_{(J-1)L+1}) \in \Omega_{J-1}(L+1, N - L + 1, L).$$

Finally, since $(n_1,\ldots,n_{JL}) \in \Omega_{JL}(1,N,1)$, for every $j = 1,\ldots,J$ we have

$$(n_{(j-1)L+1}, \ldots, n_{jL}) \in \Phi_L(m_j, m_{j+1}),$$

where $m_1 = 1$, $m_j = n_{(j-1)L+1}$, $j = 2, \ldots, J$, and $m_{J+1} = N+1$. This completes the proof of the direct inclusion.

Inverse inclusion of the sets. Let

$$m_1 = 1, \ (m_2, \ldots, m_J) \in \Omega_{J-1}(L+1, N-L+1, L), \ m_{J+1} = N+1.$$

Consider the tuple

$$(n_1^j, \ldots, n_L^j) \in \Phi_L(m_j, m_{j+1}), \ j = 1, \ldots, J. \tag{30}$$

Let us show that the composite tuple

$$(n_1^1, \ldots, n_L^1, n_1^2, \ldots, n_L^2, \ldots, n_1^J, \ldots, n_L^J) \in \Omega_{JL}(1, N, 1).$$

Indeed, it follows from (27) and (30) that $n_L^{j-1} < m_j \leq n_1^j$, for every $j = 2, \ldots, J$. The inequality $n_{i-1}^j < n_i^j$, $i = 2, \ldots, L$, holds for every $j = 1, \ldots, J$, since (30). Finally, $n_1^1 \geq 1$ and $n_L^J < N+1$.

So, if $a = 1, b = N, M = JL$, and $T = 1$, then the components of the composite tuple $(n_1^1, \ldots, n_L^1, n_1^2, \ldots, n_L^2, \ldots, n_1^J, \ldots, n_L^J)$ satisfy all conditions included in the definition (26). Thus, the inverse inclusion of the sets is proved. $\square$

Let us formulate the following main

Lemma 3. *Let N, J, and L be the instances of Problem 1. Let $\eta^*(a, b)$, $\pi^*(a, b)$ be the optimal solution of Problem 2 for each $a = 1, \ldots, N - L + 1$, and $b = a + L, \ldots, \min\{N+1, a+N-(J-1)L\}$ with*

$$f_i(j) = s_i(j), \ j = a, \ldots, b-1, \ i = 1, \ldots, L,$$

where $s_i(j)$ is the sequence in the input family S of Problem 1, and $F^*(a, b)$ be the optimal value of the objective function (3).

Let $(\mu_1^*, \ldots, \mu_{J+1}^*)$ be the optimal solution of Problem 3 with

$$q(n, m) = F^*(n, m), \ n = 1, \ldots, N - L + 1,$$
$$m = n + L, \ldots, \min\{N+1, n+N-(J-1)L\}, \tag{31}$$

in the family $\mathcal{Q}$, and Q^* be the optimal value of the function (17).

Then the optimal tuples $(n_1^*, \ldots, n_{JL}^*)$ and $(\pi_1^*, \ldots, \pi_L^*)$ of Problem 1 can be found by the following formulas:

$$(n_{(j-1)L+1}^*, \ldots, n_{jL}^*) = \eta^*(\mu_j^*, \mu_{j+1}^*), \ j = 1, \ldots, J, \tag{32}$$
$$\pi_j^* = \pi^*(\mu_j^*, \mu_{j+1}^*), \ j = 1, \ldots, J, \tag{33}$$

and the optimal value S^* of the function (1) can be found as

$$S^* = Q^*. \tag{34}$$

Proof. In accordance with Remark 2 $\Omega_{JL}(1, N, 1)$ is the set of tuples admissible in Problem 1, and $\Phi_L(n, m)$ is the set of tuples admissible in Problem 2 for every $n, m \in \{1, \ldots, N+1\}$ such that $m - n \geq L$. So,

$$S^* = \min_{(n_1, \ldots, n_{JL}) \in \Omega_{JL}(1, N, 1)} \ \min_{(\pi_1, \ldots, \pi_J) \in \Pi^J} \sum_{j=1}^{J} \sum_{i=1}^{L} s_{\pi_j(i)}(n_{i+(j-1)L}), \tag{35}$$

$$F^*(n,m) = \min_{(n_1,\ldots,n_L)\in\Phi_L(n,m)} \min_{\pi\in\Pi} \sum_{i=1}^{L} s_{\pi(i)}(n_i). \qquad (36)$$

Then, using (29), (35), and (36), for the right-hand part of (35) we obtain

$$S^* =$$

$$\min_{\substack{(m_2,\ldots,m_J)\in\Omega_{J-1}(L+1,N-L+1,L)\\ m_1=1,\ m_{J+1}=N+1}} \sum_{j=1}^{J} \min_{\substack{(n_{(j-1)L+1},\ldots,n_{jL})\in\\ \Phi_L(m_j,m_{j+1})}} \min_{\pi_j\in\Pi} \sum_{i=1}^{L} s_{\pi_j(i)}\big(n_{i+(j-1)L}\big)$$

$$\min_{\substack{(m_2,\ldots,m_J)\in\Omega_{J-1}(L+1,N-L+1,L)\\ m_1=1,\ m_{J+1}=N+1}} \sum_{j=1}^{J} \min_{(\nu_1,\ldots,\nu_L)\in\Phi_L(m_j,m_{j+1})} \min_{\pi\in\Pi} \sum_{i=1}^{L} s_{\pi(i)}(\nu_i) =$$

$$\min_{\substack{(m_2,\ldots,m_J)\in\Omega_{J-1}(1,N-L+1,L)\\ m_1=1,\ m_{J+1}=N+1}} \sum_{j=1}^{J} F^*(m_j,m_{j+1}) = Q^*, \qquad (37)$$

since $\{(m_1,\ldots,m_{J+1})|\ m_1 = 1,\ (m_2,\ldots,m_J) \in \Omega_{J-1}(L+1,N-L+1),\ m_{J+1} = N+1\}$ is the set of tuples admissible in Problem 3 in accordance with (18), (19), and (26). Thus, (34) is proved.

Let us show that (32) and (33) hold. Since $(\mu_1^*,\ldots,\mu_{J+1}^*)$ is the optimal tuple of Problem 3, and Q^* is the optimal value of the objective function of this problem in accordance with (37), we have

$$S^* = \sum_{j=1}^{J} F^*(\mu_j^*,\mu_{j+1}^*), \qquad (38)$$

where $F^*(\mu_j^*,\mu_{j+1}^*)$, at every $j = 1,\ldots,J$, is the optimal value of the objective function of Problem 2.

By definitions (3), (32), and (33) we have

$$F^*(\mu_j^*,\mu_{j+1}^*) = F(n_{(j-1)L+1}^*,\ldots,n_{jL}^*,\pi_j^*) = \sum_{i=1}^{L} s_{\pi_j^*(i)}\big(n_{(j-1)L+i}^*\big) \qquad (39)$$

for every $j = 1,\ldots,J$.

Consider the tuple

$$(n_1^*,\ldots,n_{JL}^*) = ((n_1^*,\ldots,n_L^*),\ldots,(n_{(J-1)L+1}^*,\ldots,n_{JL}^*))$$

of indices and the sequence $(\pi_1^*,\ldots,\pi_J^*)$ of permutations defined by (32) and (33). For this tuple and these permutations, from (1), (38), and (39) we obtain

$$S(n_1^*,\ldots,n_{JL}^*,\pi_1^*,\ldots,\pi_J^*) =$$

$$\sum_{j=1}^{J}\sum_{i=1}^{L} s_{\pi_j^*(i)}\big(n_{(j-1)L+i}^*\big) = \sum_{j=1}^{J} F(\pi_j^*,n_{(j-1)L+1}^*,\ldots,n_{jL}^*) = S^*.$$

Consequently, the tuple $(n_1^*,\ldots,n_{JL}^*)$ and the sequence of permutations $(\pi_1^*,\ldots,\pi_L^*)$ defined by (32) and (33) are optimal in Problem 1. $\qquad\square$

Now, present the algorithm for Problem 1.

Algorithm $\mathcal{A}$.

INPUT: the family $\mathcal{S}$ of sequences and a positive integer J.

STEP 1 (solving the family of Problems 2). For each $a = 1, \ldots, N-L+1$, and $b = a+L, \ldots, \min\{N+1, a+N-(J-1)L\}$, put $f_i(j) = s_i(j)$, $j = a, \ldots, b-1$, $i = 1, \ldots, L$. Using Algorithm $\mathcal{A}_1$, find the optimal solution $\pi^*(a,b)$, $\eta^*(a,b)$ and compute the optimal value $F^*(a,b)$ of the objective function (3).

STEP 2 (solving Problem 3). Find the family $\mathcal{Q}$ in accordance with (31). Find the optimal solution $(\mu_1^*, \ldots, \mu_{J+1}^*)$ and compute the optimal value Q^* of the objective function (17), using Algorithm $\mathcal{A}_2$.

STEP 3. Compute the value S^* of the objective function (1) by (34). Construct $(n_1^*, \ldots, n_{JL}^*)$ and $(\pi_1^*, \ldots, \pi_L^*)$, using (32) and (33).

OUTPUT: the tuples $(n_1^*, \ldots, n_{JL}^*)$, $(\pi_1^*, \ldots, \pi_L^*)$ and the value S^*.

The following theorem is true.

Theorem 1. *Algorithm $\mathcal{A}$ finds an optimal solution of Problem 1 in $\mathcal{O}(N^5)$ time.*

Proof. The optimality of the solution follows from Proposition 1, Proposition 2, and Lemma 3. At Step 1 we have to solve Problem 2 not more then $(N-L+1)^2$ times. Every solution requires $\mathcal{O}((b-a)^3)$ time. Since $b - a \leq (N-(J-1)L)$, the running time of Step 1 is bounded by $\mathcal{O}((N-L+1)^2(N-(J-1)L)^3)$. Step 2 is running in $\mathcal{O}(JN^2)$ time, since this step implements the algorithm for Problem 3. The running time of Step 3 is $\mathcal{O}(JL)$. Since $JL \leq N$, the total running time is $\mathcal{O}(N^5)$. □

Remark 3. If J and L are bounded by some constants, then the time complexity of Algorithm $\mathcal{A}$ is $\mathcal{O}(N^5)$. In a particular case, when $L = N$, we have $J = 1$ and the running time of the algorithm is $\mathcal{O}(N^3)$.

5   Conclusion

An exact polynomial time algorithm for one optimization problem arising in applications connected with the analysis of time series has been presented in this paper. Although this algorithm is a polynomial, it might be difficult to use it on large data. Therefore, one of the main lines for the further research on the problem is the construction of faster approximation algorithms with a guaranteed accuracy. In connection with the known Big data problem, constructing of approximation algorithms with linear or sublinear time complexity for the considered problem is of a specific interest.

Acknowledgments. The study presented in Sect. 3 was supported by the Russian Science Foundation, project 16-11-10041. The study presented in Sects. 2 and 4 was supported by the Russian Foundation for Basic Research, projects 16-07-00168, and by the Russian Academy of Science (the Program of Basic Research), project 0314-2016-0015, and by the Russian Ministry of Science and Education under the 5–100 Excellence Programme.

References

1. Kel'manov, A.V., Mikhailova, L.V., Khamidullin, S.A.: A posteriori joint detection of a recurring tuple of reference fragments in a quasi-periodic sequence. Comput. Math. Math. Phys. **48**(12), 2276–2288 (2008)
2. Papadimitrou, C., Steiglitz, K.: Combinatorial Optimization: Algorithms and Complexity. Prentice-Hall, Upper Saddle River (1982)
3. Bellman, R.: Dynamic Programming. Princeton University Press, Princeton (1957)

Efficient PTAS for the Euclidean CVRP with Time Windows

Michael Khachay[1,2,3](✉)(iD) and Yuri Ogorodnikov[1,2](✉)(iD)

[1] Krasovsky Institute of Mathematics and Mechanics, Ekaterinburg, Russia
{mkhachay,yogorodnikov}@imm.uran.ru
[2] Ural Federal University, Ekaterinburg, Russia
[3] Omsk State Technical University, Omsk, Russia

Abstract. The Capacitated Vehicle Routing Problem (CVRP) is the well-known combinatorial optimization problem having a wide range of practical applications in operations research. It is known that the problem is NP-hard and remains intractable even in the Euclidean plane. Although the problem is hardly approximable in the general case, some of its geometric settings can be approximated efficiently. Unlike other versions of CVRP, approximability of the Capacitated Vehicle Routing Problem with Time Windows (CVRPTW) by the algorithms with performance guarantees seems to be weakly studied so far. To the best of our knowledge, the recent Quasi-Polynomial Time Approximation Scheme (QPTAS) proposed by L. Song et al. appears to be the only one known result in this field. In this paper, we propose the first Efficient Polynomial Time Approximation Scheme (EPTAS) for CVRPTW extending the classic approach of M. Haimovich and A. Rinnooy Kan.

Keywords: Capacitated Vehicle Routing Problem · Time windows
Efficient Polynomial Time Approximation Scheme

1 Introduction

The Capacitated Vehicle Routing Problem (CVRP) is the widely known combinatorial optimization problem introduced by Dantzig and Ramser in their seminal paper [5]. This problem has a wide range of applications in operations research (see, e.g. [4] and references within). In addition, CVRP settings stem from specific models of unsupervised learning [1]. For instance, in [7], CVRP is considered as a clustering problem, where the clusters are formed by customers serviced by the same route and intra-cluster costs are defined by lengths of the corresponding routes.

As the well-known *k-means* clustering problem [2,10], CVRP remains NP-hard even in finite dimensional Euclidean spaces (see, e.g. [17]). Although the problem is hardly approximable in general, its geometric settings can be approximated rather well. Most of the known results in this field date back to the

This research was supported by Russian Science Foundation, grant no. 14-11-00109.

W. M. P. van der Aalst et al. (Eds.): AIST 2018, LNCS 11179, pp. 318–328, 2018.
https://doi.org/10.1007/978-3-030-11027-7_30

famous papers by Haimovich and Rinnooy Kan [8] and Arora [3]. To the best of our knowledge, the most recent among them are the Quasi-Polynomial Time Approximation Scheme (QPTAS) proposed in [6] for the Euclidean plane and extended in [15,16] to the case of finite number of non-intersecting time windows, and the Efficient Polynomial Time Approximation Scheme (EPTAS) introduced in [12] for the CVRP in the Euclidean space of an arbitrary dimension $d > 1$. Although, for the Capacitated Vehicle Routing Problem with Time Windows (CVRPTW) there is a significant success in development of branch-and-cut exact methods and heuristics that can solve its numerous instances coming from the practice efficiently (see surveys in [14,17]), the known results concerning the algorithms with performance guarantees remain still very rare.

In this paper, we try to bridge this gap and propose the first EPTAS for CVRPTW extending the famous approach proposed by Haimovich and Rinnooy Kan [8] and some other previous results [11,12,15,16].

The rest of the paper is structured as follows. In Sect. 2 we provide a mathematical statement of CVRPTW. In Sect. 3, we discuss the main idea of the scheme proposed, provide its rigorous description, and claim our main result. Its proof for the simplest non-trivial case of the problem defined in the Euclidean plane is presented in Sect. 4. Finally, in Sect. 5 we come to conclusions and list some open questions.

2   Problem Statement

In the simplest setting of the Capacitated Vehicle Routing Problem with Time Windows (CVRPTW), we are given by a set $X = \{x_1, \ldots, x_n\}$ of *customers* and a set $T = \{t_1, \ldots, t_p\}$ of consecutive *time windows*. It is assumed that, for any $1 \leq j < p$, the time window t_j *precedes* time window t_{j+1}, i.e. the set T is ordered. In the sequel, we denote this order by $\preceq$. Each customer x_i has the unit non-splittable demand, which should be serviced in a given time window $t(x_i) \in T$. Service is carried out by a fleet of *vehicles* arranged at a given *depot* x_0. Each vehicle has the same *capacity* q and visits customers assigned to it in a cyclic route starting and finishing at the depot x_0. The goal is to visit all the customers minimizing the total transportation cost subject to capacity and time windows constraints.

The mathematical statement of the CVRPTW can be defined as follows. The instance is given by the weighted complete digraph $G = (X \cup \{x_0\}, E, w)$, a partition

$$X_1 \cup \ldots \cup X_p = X, \tag{1}$$

and the capacity bound $q \in \mathbb{N}$. Here,

(i) the non-negative weighting function $w \colon E \to \mathbb{R}_+$ defines transportation costs for location pairs (x_i, x_j), such that, for any route $R = x_{i_1}, \ldots, x_{i_s}$, its cost $w(R) = \sum_{j=1}^{s-1} w(x_{i_j}, x_{i_{j+1}})$

(ii) any subset X_j of partition (1) consists of customers x_i, which should be serviced in time window $t(x_i) = t_j \in T$

(iii) for any feasible vehicle route $R = x_0, x_{i_1}, \ldots, x_{i_s}, x_0$, the following constraints

$$s \leq q \quad \text{(capacity)} \tag{2}$$
$$t(x_{i_j}) \preceq t(x_{i_{j+1}}) \quad \text{(time windows)} \tag{3}$$

are valid.

It is required to find a collection of feasible routes $S = \{R_1, \ldots, R_l\}$ that visits each customer once and has the minimum total transportation cost $w(S) = \sum_{i=1}^{l} w(R_i)$.

If the weighting function w satisfies the triangle inequality, then CVRPTW is called *metric* and transportation costs $w(x_i, x_j)$ are called distances between locations x_i and x_j. In this paper, we consider the Euclidean CVRPTW, where $X \cup \{x_0\} \subset \mathbb{R}^d$ and $w(x_i, x_j) = \|x_i - x_j\|_2$.

3 Approximation Scheme

Our scheme (Algorithm 1) extends the approach proposed in the seminal paper by Haimovich and Rinnooy-Kan [8].

Its main idea is quite simple and consists of the following points

(i) decomposition of the initial CVRPTW instance with p time windows to a single CVRPTW subinstance induced by the *outer* customers and p independent subinstances of the classic CVRP that describe servicing of the *inner* ones
(ii) the famous Iterated Tour Partition (ITP) heuristic (Algorithm 2) reducing the CVRP to an appropriate instance of the Traveling Salesman Problem (TSP) induced by a subset of the inner customers, which should be serviced in the fixed time window t_j, $j \in \{1, \ldots, p\}$
(iii) an upper bound for the optimum TSP* of the TSP instance enclosed in the Euclidean sphere of a given radius.

At first glance, the worst case time complexity of Algorithm 1 is determined by running time of the dynamic programming procedure applied to finding the exact solution of CVRPTW (due to Step 2), i.e. the proposed scheme should be extremely inefficient.

Actually, it is not so. As we show in Sect. 4, for any $\varepsilon > 0$, $p \geq 1$, $\rho \geq 1$, and $q \in \mathbb{N}$ there exist a bound $\tilde{K} = \tilde{K}(\varepsilon, p, \rho, q)$ such that Eq. (4) is satisfied by at least one $1 \leq k \leq \tilde{K}$. Therefore, for $n > \tilde{K}$, the solution S_0 can be obtained by dynamic programming with time complexity bound $O(qk^2 2^k)$ (see, e.g. [9]), which does not depend on n. Further, for finding ρ-approximate solutions for the inner Euclidean TSP problem, one can employ an arbitrary approximation algorithm. Since running time of the ITP is $O(n^2)$, the overall time complexity is mainly determined by the complexity TIME(TSP, ρ, n) of such an algorithm. The main result is claimed in Theorem 1. For the sake of simplicity, we present it for the case of Euclidean plane postponing more general result to the forthcoming paper.

Algorithm 1. Approximation Scheme for the Euclidean CVRPTW

Input: an instance of the Euclidean CVRPTW defined by a complete graph $G(X \cup \{x_0\}, E, w)$, a capacity q, and partition $X_1 \cup \ldots \cup X_p = X$

Parameters: $\varepsilon > 0$ and $\rho \geq 1$

Output: an $(1 + \varepsilon)$-approximate solution S_{APP} of the given CVRPTW instance

1: relabel the customers in the order $r_1 \geq \ldots \geq r_n$, where $r_i = w(x_0, x_i)$ is a distance between the customer x_i and the depot x_0
2: for the given $\varepsilon > 0$, find the smallest number $k = k(\varepsilon, p, q)$ such that

$$q\left(2k + p(1 + \rho)\right)\frac{r_k}{\sum_{i=1}^{n} r_i} + 2q\sqrt{\pi}p\rho\sqrt{\frac{r_k}{\sum_{i=1}^{n} r_i}} < \varepsilon \qquad (4)$$

or set $k = n$, if equation (4) is violated by any $k < n$.
3: split the set X to subsets $X(k) = \{x_1, \ldots, x_{k-1}\}$ and $X'(k) = X \setminus X(k)$ of *outer* and *inner* customers, respectively
4: find an optimal solution S_0 for the CVRPTW subinstance defined by the subgraph $G\langle X(k) \cup x_0\rangle$, partition $(X_1 \cap X(k)) \cup \ldots \cup (X_p \cap X(k))$, and capacity q by dynamic programming
5: find a ρ-approximate solution H of the auxiliary TSP instance defined by the subgraph $G\langle X'(k)\rangle$. Split H into p cycles $H_1, \ldots, H_p$ such that each cycle H_j spans $X'(k) \cap X_j$ (Fig. 1)
6: **for all** $j \in \{1, \ldots, p\}$ **do**
7:     obtain a partial CVRP solution S_j by employing the ITP heuristic (Algorithm 2) to the cycle H_j and CVRP subinstance defined by $G\langle X'(k) \cap X_j \cup \{x_0\}\rangle$ and capacity q
8: **end for**
9: output the solution $S_{APP} = S_0 \cup S_1 \cup \ldots \cup S_p$

Theorem 1. *For any $\varepsilon > 0$ an $(1 + \varepsilon)$-approximate solution for the Euclidean CVRPTW can be obtained in time* $\text{TIME}(\text{TSP}, \rho, n) + O(n^2) + O(qk^2 2^k)$, *where* $k = k(\varepsilon, p, q, \rho) = O\left(p(\rho + 1)\exp(O(q/\varepsilon))\exp\left(O(\sqrt{pq/\varepsilon})\right)\right)$.

Remark 1. As it follows from Theorem 1, the scheme proposed is EPTAS for any fixed p and q and remains PTAS for $q = O(\log \log n)$ and $p = O(\log \log n)$.

4   Proof Sketch

We start with description of the well-known ITP heuristic (Algorithm 2). As it is shown in [8,13], for the cost $w(S_{ITP})$ of the ITP-produced solution S_{ITP}, there exists the following upper bound, which remains valid for an arbitrary non-negative weighting function w.

Lemma 1. *For $\bar{r} = 1/|X| \sum_{i=1}^{|X|} r_i$ and $l = \lceil |X|/q \rceil$,*

$$w(S_{ITP}) \leq 2l\bar{r} + \left(1 - \frac{l}{|X|}\right)w(H) \leq 2l\bar{r} + (1 - 1/q)w(H). \qquad (5)$$

Algorithm 2. ITP heuristic

Input: an instance of CVRP defined by a complete weighted digraph $G(X \cup \{x_0\}, E, w)$ and capacity q, and an arbitrary Hamiltonian circuit H in the subgraph $G\langle X \rangle$
Output: an approximate solution S_{ITP} of the given CVRP instance
1: **for all** $x \in X$ **do**
2:     starting from the vertex x, split the circuit H into $l = \lceil |X|/q \rceil$ chains, such that each of them, except maybe one, spans q vertices
3:     connecting endpoints of each chain with the depot x_0 directly, construct a set $S(x)$ of l routes
4: **end for**
5: output the solution $S_{\text{ITP}} = \arg \min \{ w(S(x)) : x \in X \}$

Lemma 1 can be easily extended to the case of metric CVRPTW. Indeed, consider an instance of the metric CVRPTW defined by the graph $G(X \cup \{x_0\}, E, w)$, partition $X_1 \cup \ldots \cup X_p = X$, and capacity q. Let H be a Hamiltonian cycle spanning all the customers X. For any j, obtain a Hamiltonian cycle H_j for the subset X_j by shortcutting the cycle H (Fig. 1). Then, for each CVRP subinstance defined by the subgraph $G\langle X_j \cup \{x_0\} \rangle$ and capacity q and for the appropriate cycle H_j, apply Algorithm 2 to construct the partial solution S_j. For the combined solution $S = S_1 \cup \ldots \cup S_p$, the following upper bound is valid.

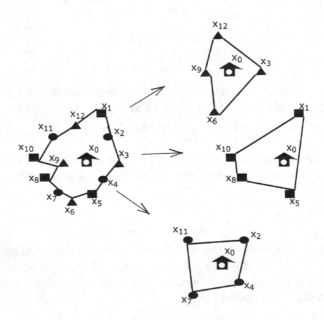

Fig. 1. Shortcutting of a Hamiltonian cycle. Triangles, squares and circles denote customers, which should be serviced in different time windows.

Lemma 2.

$$w(S) \leq p\left(1 - \frac{1}{q}\right)w(H) + \frac{2}{q}\sum_{i=1}^{|X|}r_i + 2pr_{\max}, \tag{6}$$

where $r_{\max} = \max\{r_1, \ldots, r_{|X|}\}$.

Proof. by Lemma 1, for each S_j, we have

$$w(S_i) \leq 2\left\lceil\frac{n_j}{q}\right\rceil\bar{r}_j + \left(1 - \frac{1}{q}\right)w(H_j), \tag{7}$$

where $n_j = |X_j|$ and $\bar{r}_j = 1/n_j\sum\{r_i : x_i \in X_j\}$. Since, by construction, $w(H_j) \leq w(H)$ and $\lceil n_j/q\rceil \leq n_j/q + 1$,

$$w(S) = \sum_{i=1}^{p}w(S_j) \leq \left(1 - \frac{1}{q}\right)\sum_{j=1}^{p}w(H_j) + 2\sum_{j=1}^{p}\left\lceil\frac{n_j}{q}\right\rceil\bar{r}_i \leq$$

$$\leq p\left(1 - \frac{1}{q}\right)w(H) + \frac{2}{q}\sum_{j=1}^{p}n_j\bar{r}_j + 2\sum_{j=1}^{p}\bar{r}_j \leq p\left(1 - \frac{1}{q}\right)w(H) + \frac{2}{q}\sum_{i=1}^{|X|}r_i + 2pr_{\max}.$$

Lemma 2 is proved.

In particular, in the case when the cycle H is found by ρ-approximation algorithm applied to the auxiliary TSP instance (with optimum $\text{TSP}^*(X)$) defined by the subgraph $G\langle X\rangle$,

$$w(S) \leq p\left(1 - \frac{1}{q}\right)\rho\text{TSP}^*(X) + \frac{2}{q}\sum_{i=1}^{|X|}r_i + 2pr_{\max}.$$

Notice that, for an arbitrary feasible solution of the metric CVRP, the following simple lower bound

$$w(S) \geq \frac{2}{q}\sum_{i=1}^{|X|}r_i \tag{8}$$

is also valid.

Further, let the customers $x_1, \ldots, x_n$ (elements of the set X) be ordered by decreasing their distances $r_1 \geq r_2 \geq \ldots \geq r_n$ from the depot x_0. Given some $k \in \{1, \ldots, n\}$, consider a decomposition of the initial CVRP instance to two independent subinstances induced by the subsets $X(k) = \{x_1, \ldots, x_{k-1}\}$ and $X'(k) = X \setminus X(k)$ of outer and inner customers and the same depot x_0. Denote by $\text{OPT}(X)$, $\text{OPT}(X(k))$, and $\text{OPT}(X'(k))$ optimum values of the initial instance and the produced subinstances, respectively. Then, the following lemma holds the metric CVRP.

Lemma 3. *For any $1 \leq k \leq n$,*

$$\text{OPT}(X(k)) + \text{OPT}(X'(k)) \leq \text{OPT}(X) + 4(k - 1)r_k. \tag{9}$$

It is easy to verify that the claim of Lemma 3 remains valid for the case of metric CVRPTW as well.

The next step to our proof of Theorem 1 is concerned with an upper bound the optimum value of the Euclidean TSP instance enclosed in a sphere of a given radius R. As we mentioned above, in this paper we restrict ourselves to the simplest non-trivial case of the Euclidean plane described in the following lemma. As it was shown in [11,12], the similar bounds can be obtained for any fixed-dimensional Euclidean space.

Lemma 4. *Let an instance of the Euclidean TSP be given by a set $X = \{x_1, \ldots, x_n\}$ that be enclosed in a circle of radius $R = \max\{r_i \colon i \in \{1, \ldots, n\}\}$ centered at x_0. Then, for the optimum $\mathrm{TSP}^*(X)$,*

$$\mathrm{TSP}^*(X) \le 2R + 4\sqrt{\pi R \sum_{i=1}^{n} r_i}. \tag{10}$$

Proof. For some $h < \pi$, partition the enclosing circle to $\lceil 2\pi/h \rceil$ sectors such that all of them, except maybe one, have h as a value of their central angle. Consider the cyclic route walking back and forth all the obtained radii and augmented by the double connection of each point x_i to the closest radius. Transform this 'strange' route to a Hamiltonian cycle with shortcutting by the triangle inequality. Denote the length of the tour obtained by $L(h)$. Since the distance between x_i and the nearest radius is at most $r_i \sin(h/2) \le r_i h/2$ and the total length of all the radii does not exceed $R(2\pi/h + 1)$, we have

$$L(h) \le 2R + \frac{4\pi R}{h} + h \sum_{i=1}^{n} r_i. \tag{11}$$

The rhs of (11) is minimized by

$$h^* = 2\sqrt{\frac{\pi R}{\sum_{i=1}^{n} r_i}} \in (0, 2\sqrt{\pi}].$$

Therefore,

$$\mathrm{TSP}^*(X) \le L(h^*) \le 2R + 4\sqrt{\pi R \sum_{i=1}^{n} r_i}.$$

Lemma is proved.

Consider a solution $S = S_0 \cup S_1 \cup \ldots \cup S_p$ provided by Algorithm 1 for some k. By construction, $w(S_0) = \mathrm{OPT}(X_k)$. At this point, we are ready to estimate the relative approximation error of the solution S

$$e(k) = \frac{w(S) - \mathrm{OPT}(X)}{\mathrm{OPT}(X)} = \frac{\mathrm{OPT}(X(k)) + \mathrm{APP}(X'(k)) - \mathrm{OPT}(X)}{\mathrm{OPT}(X)}, \tag{12}$$

where $\mathrm{APP}(X'(k)) = \sum_{i=1}^{p} w(S_i)$.

Lemma 5.

$$e(k) \leq q\left(2k + p(1+\rho)\right) \frac{r_k}{\sum_{i=1}^{n} r_i} + 2q\sqrt{\pi}p\rho\sqrt{\frac{r_k}{\sum_{i=1}^{n} r_i}} \tag{13}$$

Proof. Indeed,

$$e(k) = \frac{\mathrm{OPT}(X(k)) + \mathrm{APP}(X'(k)) - \mathrm{OPT}(X)}{\mathrm{OPT}(X)}$$

$$= \frac{\left(\mathrm{OPT}(X(k)) + \mathrm{OPT}(X'(k)) - \mathrm{OPT}(X)\right) + \mathrm{APP}(X'(k)) - \mathrm{OPT}(X'(k))}{\mathrm{OPT}(X)}.$$

Since,

$$\mathrm{OPT}(X(k)) + \mathrm{OPT}(X'(k)) - \mathrm{OPT}(X) \leq 4(k-1)r_k, \text{ by Lemma 3,}$$

$$\mathrm{OPT}(X) \geq 2/q\sum_{i=1}^{n} r_i \text{ and } \mathrm{OPT}(X'(k)) \geq 2/q\sum_{i=k}^{n} r_i, \text{ by Eq. (8),}$$

$$\mathrm{APP}(X'(k)) \leq p\left(1 - \frac{1}{q}\right)\rho\mathrm{TSP}^*(X'(k)) + \frac{2}{q}\sum_{i=k}^{n} r_i + 2pr_k, \text{ by Lemma 2, and}$$

$$\mathrm{TSP}^*(X'(k)) \leq 2r_k + 4\sqrt{\pi r_k \sum_{i=k}^{n} r_i}, \text{ by Lemma 4,}$$

we obtain

$$e(k) \leq \left(2q(k-1) + \rho pq + pq\right) \frac{r_k}{\sum_{i=1}^{n} r_i} + 2\sqrt{\pi}p\rho q\sqrt{\frac{r_k}{\sum_{i=1}^{n} r_i}}$$

$$\leq q\left(2k + p(1+\rho)\right) \frac{r_k}{\sum_{i=1}^{n} r_i} + 2q\sqrt{\pi}p\rho\sqrt{\frac{r_k}{\sum_{i=1}^{n} r_i}}.$$

Lemma 5 is proved.

The following technical lemma is the last stair to the proof of our main result.

Lemma 6. *For any $\varepsilon > 0$, $p \geq 1$, $\rho \geq 1$, and $q \in \mathbb{N}$, there exists $\tilde{K} = \tilde{K}(\varepsilon, p, \rho, q)$, such that the equation*

$$q\left(2k + p(1+\rho)\right) \frac{r_k}{\sum_{i=1}^{n} r_i} + 2q\sqrt{\pi}p\rho\sqrt{\frac{r_k}{\sum_{i=1}^{n} r_i}} < \varepsilon \tag{14}$$

holds at least for one $1 \leq k \leq \tilde{K}$.

Proof. Suppose, for some $\tilde{K}$, Eq. (14) is violated by any natural k from the interval $[1, \tilde{K}]$, i.e., for any $1 \leq k \leq \tilde{K}$,

$$s_k^2 + \frac{2\sqrt{\pi}p\rho}{2k + p(1+\rho)}s_k - \frac{\varepsilon}{q(2k + p(1+\rho))} \geq 0 \tag{15}$$

for $s_k = \sqrt{\frac{r_k}{\sum_{i=1}^{n} r_i}}$. Since the lhs of Eq. (15) has the roots of different sign, Eq. (15) implies

$$s_k \geq -\frac{\sqrt{\pi}p\rho}{2k + p(1 + \rho)} + \sqrt{\frac{\pi p^2 \rho^2}{(2k + p(1 + \rho))^2} + \frac{\varepsilon}{q(2k + p(1 + \rho))}}$$

or

$$s_k \geq -\frac{B}{2k + A} + \sqrt{\left(\frac{B}{2k + A}\right)^2 + \frac{C}{2k + A}} \geq -\frac{B}{2k + A} + \frac{\sqrt{C}}{\sqrt{2k + A}},$$

where $A = p(1 + \rho)$, $B = \sqrt{\pi}p\rho$, and $C = \varepsilon/q$.

Hence,

$$s_k^2 \geq \frac{C}{2k + A} + \frac{B^2}{(2k + A)^2} - \frac{2B\sqrt{C}}{(2k + A)^{3/2}} \geq \frac{C}{2k + A} - \frac{2B\sqrt{C}}{(2k + A)^{3/2}} \tag{16}$$

Further, we suppose that n is sufficiently large, i.e. $\tilde{K} \leq n$. Therefore,

$$\sum_{k=1}^{\tilde{K}} s_k^2 = \sum_{k=1}^{\tilde{K}} \frac{r_k}{\sum_{i=1}^{n} r_i} \leq 1$$

and

$$1 \geq \sum_{k=1}^{\tilde{K}} s_k^2 \geq \sum_{k=1}^{\tilde{K}} \frac{C}{2k + A} - \sum_{k=1}^{\tilde{K}} \frac{2B\sqrt{C}}{(2k + A)^{3/2}}$$

$$\geq \int_1^{\tilde{K}} \frac{C dx}{2x + A} - \frac{2B\sqrt{C}}{(2 + A)^{3/2}} - \int_1^{\tilde{K}} \frac{2B\sqrt{C} dx}{(2x + A)^{3/2}}$$

$$\geq \frac{C}{2} \ln \frac{2\tilde{K} + A}{2 + A} - \frac{2B\sqrt{C}}{(2 + A)^{3/2}} - \frac{1}{2} \frac{B\sqrt{C}}{\sqrt{2 + A}}$$

$$\geq \frac{C}{2} \ln \frac{2\tilde{K} + A}{2 + A} - \frac{5}{2} \frac{B\sqrt{C}}{\sqrt{2 + A}} \geq \frac{C}{2} \ln \frac{2\tilde{K} + A}{2 + A} - \frac{5B}{2} \sqrt{\frac{C}{A}},$$

since $B\sqrt{C} > 0$. Thus,

$$\ln \frac{2\tilde{K} + A}{2 + A} \leq \frac{2}{C}\left(1 + \frac{5B}{2}\sqrt{\frac{C}{A}}\right) = \frac{2}{C} + \frac{5B}{\sqrt{AC}}$$

and

$$\tilde{K} \leq A \cdot \exp\left(\frac{2}{C} + \frac{5B}{\sqrt{AC}}\right) = p(\rho + 1) \cdot \exp\left(2 \cdot \frac{q}{\varepsilon}\right) \cdot \exp\left(5\sqrt{\pi}p\rho\sqrt{\frac{1}{p(1 + \rho)}}\sqrt{\frac{q}{\varepsilon}}\right)$$

$$\leq p(\rho + 1) \cdot (\exp(q/\varepsilon))^2 \cdot \left(exp(\sqrt{(pq)/\varepsilon})\right)^{5\sqrt{\pi\rho}},$$

since $A = p(\rho + 1) \geq 2$. Lemma 6 is proved.

The proof of Theorem 1 follows straightforward from the lemmas above.

5   Conclusion

In this paper, we proposed new approximation scheme for the Capacitated Vehicle Routing Problem with Time Windows. To the best of our knowledge, the scheme proposed is the first EPTAS for this problem for any fixed capacity q and number of time windows p. Our scheme remains PTAS with respect to n even for $q = \log\log n$ and $p = \log\log n$. Although the proof presented in this paper concerns the simplest case of the problem (Euclidean plane, single depot), the scheme proposed seems to be applicable to the more general setting of CVRPTW. In forthcoming paper, we are going to present the full proof for multiple depots and Euclidean spaces of an arbitrary fixed dimension $d > 1$.

References

1. Aggarwal, C.C., Reddy, C.K.: Data Clustering: Algorithms and Applications. Chapman & Hall/CRC Data Mining and Knowledge Discovery Series. Chapman and Hall/CRC, Boca Raton (2013)
2. Aloise, D., Deshpande, A., Hansen, P., Popat, P.: NP-hardness of Euclidean sum-of-squares clustering. Mach. Learn. **75**(2), 245–248 (2009). https://doi.org/10.1007/s10994-009-5103-0
3. Arora, S.: Polynomial time approximation schemes for Euclidean traveling salesman and other geometric problems. J. ACM **45**, 753–782 (1998)
4. Braekers, K., Ramaekers, K., van Nieuwenhuyse, I.: The vehicle routing problem: state of the art classification and review. Comput. Ind. Eng. **99**, 300–313 (2016). http://www.sciencedirect.com/science/article/pii/S0360835215004775
5. Dantzig, G.B., Ramser, J.H.: The truck dispatching problem. Manag. Sci. **6**(1), 80–91 (1959)
6. Das, A., Mathieu, C.: A quasipolynomial time approximation scheme for Euclidean capacitated vehicle routing. Algorithmica **73**, 115–142 (2015)
7. Expósito-Izquierdo, C., Rossi, A., Sevaux, M.: A two-level solution approach to solve the clustered capacitated vehicle routing problem. Comput. Ind. Eng. **91**, 274–289 (2016). http://www.sciencedirect.com/science/article/pii/S0360835215004684
8. Haimovich, M., Rinnooy Kan, A.H.G.: Bounds and heuristics for capacitated routing problems. Math. Oper. Res. **10**(4), 527–542 (1985)
9. van Hoorn, J.J.: A note on the worst case complexity for the capacitated vehicle routing problem. In: Research Memorandum, vol. 5. Faculteit der Economische Wetenschappen en Bedrijfskunde (2010)
10. Kel'manov, A.V., Pyatkin, A.V.: Complexity of certain problems of searching for subsets of vectors and cluster analysis. Comput. Math. Math. Phys. **49**(11), 1966–1971 (2009). https://doi.org/10.1134/S0965542509110128
11. Khachai, M.Y., Dubinin, R.D.: Approximability of the vehicle routing problem in finite-dimensional Euclidean spaces. Proc. Steklov Inst. Math. **297**(1), 117–128 (2017). https://doi.org/10.1134/S0081543817050133
12. Khachay, M., Dubinin, R.: PTAS for the Euclidean capacitated vehicle routing problem in R^d. In: Kochetov, Y., Khachay, M., Beresnev, V., Nurminski, E., Pardalos, P. (eds.) DOOR 2016. LNCS, vol. 9869, pp. 193–205. Springer, Cham (2016). https://doi.org/10.1007/978-3-319-44914-2_16

13. Khachay, M., Zaytseva, H.: Polynomial time approximation scheme for single-depot Euclidean capacitated vehicle routing problem. In: Lu, Z., Kim, D., Wu, W., Li, W., Du, D.-Z. (eds.) COCOA 2015. LNCS, vol. 9486, pp. 178–190. Springer, Cham (2015). https://doi.org/10.1007/978-3-319-26626-8_14
14. Kumar, S., Panneerselvam, R.: A survey on the vehicle routing problem and its variants. Intell. Inf. Manag. **4**, 66–74 (2012)
15. Song, L., Huang, H.: The Euclidean vehicle routing problem with multiple depots and time windows. In: Gao, X., Du, H., Han, M. (eds.) COCOA 2017. LNCS, vol. 10628, pp. 449–456. Springer, Cham (2017). https://doi.org/10.1007/978-3-319-71147-8_31
16. Song, L., Huang, H., Du, H.: Approximation schemes for Euclidean vehicle routing problems with time windows. J. Comb. Optim. **32**(4), 1217–1231 (2016). https://doi.org/10.1007/s10878-015-9931-5
17. Toth, P., Vigo, D.: Vehicle Routing: Problems, Methods, and Applications, 2nd edn. MOS-SIAM Series on Optimization. SIAM (2014)

Innovative Systems

Generic Distributed In Situ Aggregation for Earth Remote Sensing Imagery

Ramon Antonio Rodriges Zalipynis[(✉)]

National Research University Higher School of Economics, Moscow, Russia
rodriges@wikience.org

Abstract. Earth remote sensing imagery come from satellites, unmanned aerial vehicles, airplanes, and other sources. National agencies, commercial companies, and individuals across the globe collect enormous amounts of such imagery daily. Array DBMS are one of the prominent tools to manage and process large volumes of geospatial imagery. The core data model of an array DBMS is an N-dimensional array. Recently we presented a geospatial array DBMS – ChronosDB – which outperforms SciDB by up to 75× on average. We are about to launch a Cloud service running our DBMS. SciDB is the only freely available distributed array DBMS to date. Remote sensing imagery are traditionally stored in files of sophisticated formats, not in databases. Unlike SciDB, ChronosDB does not require importing files into an internal DBMS format and works with imagery "in situ": directly in their native file formats. This is one of the many virtues of ChronosDB. It has now certain aggregation capabilities, but this paper focuses on more advanced aggregation queries which still constitute a large portion of a typical workload applied to remote sensing imagery. We integrate the aggregation types into the data model, present the respective algorithms to perform aggregations in a distributed fashion, and thoroughly compare the performance of our technique with SciDB. We carried out experiments on real-world data on 8- and 16-node clusters in Microsoft Azure Cloud.

Keywords: ChronosDB · SciDB · Cloud computing · Array DBMS · Satellite imagery · In Situ · Big data · Landsat

1 Introduction

Earth remote sensing imagery is attractive both for computer science researchers and practitioners across a wide variety of disciplines. From the research perspective, the imagery carry a wealth of analytic opportunities and are truly "Big Data". For example, Planet Labs, a new startup company aiming to "index" objects on our planet, downloads and processes over 5 TB of data daily [12]. From the practical perspective, the imagery are crucial in decision-making for urban planning, agriculture, emergency monitoring, and many other important applications [7]. Finally, these imagery form an ever-growing commercial sector

© Springer Nature Switzerland AG 2018
W. M. P. van der Aalst et al. (Eds.): AIST 2018, LNCS 11179, pp. 331–342, 2018.
https://doi.org/10.1007/978-3-030-11027-7_31

and reveal an unprecedented increase in popularity: Amazon and Google allocate petabytes of space to provide satellite imagery via commercial clouds [5].

An array DBMS is one of the options to process geospatial imagery keeping pace with the imagery volumes and the analytical needs. To date, SciDB [4] is the only freely available distributed array DBMS. ChronosDB is a new distributed array DBMS radically different from SciDB. First, it works on top of files in diverse raster file formats and does not require complex, error-prone, and slow importing of remote sensing imagery into an internal DBMS format. The import is a major bottleneck as it may sometimes take more time than the imagery analysis itself. Second, ChronosDB leverages a new delegation approach: the algorithms are designed such that they always delegate a significant portion of the required work to external, powerful command line tools. Finally, an efficient execution engine coordinates the work of tools and the movement of files between the tools to enable distributed execution of queries on a computer cluster.

We started from a high-performance raster data dissemination server and its applications [9,15–18,23]. Than we enabled raster data processing on just a single machine reaching 3× to 193× faster performance than SciDB [19]. After that, we targeted certain N-d array operations on a computer cluster (Azure Cloud) achieving over 1000× acceleration over SciDB [22]. We than added specific 2-d operations on satellite imagery of diverse formats and types [20,25]. Finally, we significantly rethought, reworked, redesigned, and extended our previous approaches and built ChronosDB – a new array DBMS [21]. ChronosDB is at least 3× faster than SciDB for all queries. ChronosDB can be over 1000× faster than SciDB. On average, ChronosDB is by up to 75× faster than SciDB [21].

This paper considers additional types of aggregation the need for which frequently occurs during the analysis of Earth remote sensing imagery. We integrate the new types into ChronosServer, as they are experimental. Our previous work included the aggregation of an N-d array only over the first axis. However, it is important to be able to obtain aggregations over an arbitrary set of array axes.

For example, one may have an N-d array consisting of 4 dimensions: (*time*, *elevation*, *latitude*, *longitude*). In this case, it might be desirable to get an aggregate array (*time*, *latitude*, *longitude*) or (*time*, *elevation*, *latitude*). The former aggregate shows the average over all available vertical levels while the latter is a zonal average. Other aggregation functions can be applied such as *min* or *max*. In addition, a user may wish to aggregate over several dimensions to get a resulting array like (*elevation*, *longitude*).

The rest of the paper is organized as follows. Section 2 briefly overviews the architecture and the data model of ChronosServer. Section 3 formally defines the aggregation operation we tackle and presents the respective algorithms. Section 4 gives performance evaluation on real-world satellite imagery in Microsoft Azure Cloud. Section 5 surveys the related work. Section 6 concludes the paper.

2   ChronosServer

2.1   Multidimensional Array Model

In this paper, an N-dimensional array (N-d array) is the mapping $A : D_1 \times D_2 \times \cdots \times D_N \mapsto \mathbb{T}$, where $N > 0$, $D_i = [0, l_i) \subset \mathbb{Z}$, $0 < l_i$ is a finite integer which is said to be the *size* or *length* of ith dimension, and $\mathbb{T}$ is a standard numeric type. In this paper, $i \in [1, N] \subset \mathbb{Z}$. Let us denote the N-d array A by

$$A\langle l_1, l_2, \ldots, l_N \rangle : \mathbb{T} \tag{1}$$

By $l_1 \times l_2 \times \cdots \times l_N$ denote the *shape* of A, by $|A|$ denote the *size* of A such that $|A| = \prod_i l_i$. A *cell* value of A with index $(x_1, x_2, \ldots, x_N)$ is referred to as $A[x_1, x_2, \ldots, x_N]$, where $x_i \in D_i$. Each cell value of A is of type $\mathbb{T}$. An array may be initialized after its definition by listing its cell values: $A\langle 2, 2 \rangle$: int $= \{\{1, 2\}, \{3, 4\}\}$, where A is 2-d array of integers, $A[1, 0] = 3$, $|A| = 4$, and the shape of A is 2×2. A graphical array illustration is in Fig. 2.

Indexes x_i are optionally mapped to specific values of ith dimension by *coordinate* arrays $A.d_i \langle l_i \rangle : \mathbb{T}_i$, where $\mathbb{T}_i$ is a totally ordered set, and $d_i[j] < d_i[j+1]$ for all $j \in D_i$. In this case, A is defined as

$$A(d_1, d_2, \ldots, d_N) : \mathbb{T} \tag{2}$$

A *hyperslab* $A' \sqsubseteq A$ is an N-d subarray of A defined by the notation

$$A[b_1 : e_1, \ldots, b_N : e_N] = A'(d'_1, \ldots, d'_N) \tag{3}$$

where $b_i, e_i \in \mathbb{Z}$, $0 \leqslant b_i \leqslant e_i < l_i$, $d'_i = d_i[b_i : e_i]$, $|d'_i| = e_i - b_i + 1$, and for all $y_i \in [0, e_i - b_i]$ the following holds

$$A'[y_1, \ldots, y_N] = A[y_1 + b_1, \ldots, y_N + b_N] \tag{4a}$$
$$d'_i[y_i] = d_i[y_i + b_i] \tag{4b}$$

Equations (4a) and (4b) state that A and A' have a common coordinate subspace over which cell values of A and A' coincide. Note that the original dimensionality is preserved even if some $b_i = e_i$ (in this case, "$: e_i$" may be omitted in (3)). Also, "$b_i : e_i$" may be omitted in (3) if $b_i = 0$ and $e_i = |d_i| - 1$.

2.2   Datasets

A *dataset* $\mathbb{D} = (A, H, M, P)$ contains a *user-level* array $A(d_1, \ldots, d_N) : \mathbb{T}$ and the set of *system-level* arrays $P = \{(A_{key}, key, M_{key}, node_{key})\}$, where $A_{key} \sqsubseteq A$, $key = \langle N \rangle$: int is the key for A_{key}, $node_{key}$ is the ID of the cluster node storing A_{key}, M_{key} is metadata for A_{key}, $H\langle t_1, \ldots, t_N \rangle$: int such that $A_{key} = A[h_1 : h'_1, \ldots, h_N : h'_N]$, where $h_i = H[k_1, \ldots, k_i, \ldots, k_N]$, and $h'_i = H[k_1, \ldots, k_i + 1, \ldots, k_N]$. This means that array A is divided by $(N - 1)$-d hyperplanes on subarrays: this is a common practice, see top of Fig. 3. We ingest a time series of 2-d satellite scenes into a set of 3-d cubes using our retiling algorithm, Sect. 2.4.

A user-level array is never stored explicitly: operations with A are mapped to a sequence of operations with respective arrays from P. Let us call a user-level array and a system-level array an array and a subarray respectively for short. Dataset metadata $M = \{(key, val)\}$ includes general dataset properties (name, description, contacts, etc.) and metadata valid for all $p \in P$ (data type $\mathbb{T}$, storage format, date, cartographic projection, etc.). For example, $M = \{(name = \text{"Landsat 8 Band 1"}), (type = \text{int16}), (format = \text{GeoTIFF})\}$. Let us refer to an element in a tuple $p = (A_{key}, key, \ldots) \in P$ as $p.A$ for A_{key}, $p.key$ for key, etc.

2.3 ChronosServer Architecture

ChronosServer runs on a computer cluster of commodity hardware. Files of diverse raster file formats are distributed between cluster nodes without changing their formats. A file is always stored entirely on a node in contrast to parallel or distributed file systems. Workers are launched at each node and are responsible for data processing. A single Gate at a dedicated node receives client queries and coordinates workers.

Gate stores metadata for all datasets and subarrays. Consider a dataset $\mathbb{D} = (A, H, M, P)$. Arrays $A.d_i$ and elements of $\forall p \in P$ except $p.A$ are stored on Gate. In practice, array axes usually have coordinates such that $A.d_i[j] = start + j \times step$, where $j \in [0, |A.d_i|) \subset \mathbb{N}$, $start, step \in \mathbb{R}$. Only $|A.d_i|$, $start$ and $step$ values have to be usually stored. The array model merit is that it has been designed to be generic as much as possible but allowing the establishment of 1:1 mapping of a $p \in P$ to a real dataset file at the same time.

Upon startup workers connect to Gate and receive the list of all available datasets and file naming rules. Workers scan their local file systems to discover datasets and create $p \in P$ by parsing file names or reading file metadata. Workers transmit to Gate the described information.

2.4 Generic N-d Retiling

Data providers disseminate satellite imagery in very diverse forms. For example, Landsat scenes overlap irregularly and sides of a scene bounding polygon are not parallel to the coordinate axes. A satellite usually does not capture exactly the same area during each pass due to the orbit drift: scenes for different dates may overlap for 99% and be shifted relatively to each other. Even basic raster operations with raw scenes from the same dataset are complex.

We ingest raw satellite imagery with a retiling algorithm presented earlier [24]. It performs tiling of subarrays from their initial diverse forms to a fixed one: N-d arrays with shape $s_1 \times s_2 \times \cdots \times s_N$ and edges parallel to the coordinate axes. The algorithm is used on the "data cooking" (data preprocessing) stage. Uniformity of subarrays greatly simplifies raster algorithms. More importantly, often this kind of data transformation makes it possible to leverage existing command line tools. For example, some tools refuse to perform operations on rasters that do not cover exactly the same area. The data cooking time is negligible compared to the SciDB data import time (Sect. 4).

The basic idea is to cut each $p \in P$ onto smaller pieces $P' = \{p' : p' \sqsubseteq p\}$, assign each piece a key, and merge all pieces with the same key into a single, new subarray. We skip detailed description of the algorithm due to space constraints.

Thick blue lines in Fig. 1 are borders of the initial subarrays, dashed red lines are borders of the resulting subarrays (an overlap is not shown). Resulting subarrays will share border cells if overlap is given: e.g., if overlap $= 1$, $A[5 : 8, 1 : 4]$ and $A[5 : 8, 3 : 6]$ will share $A[5 : 8, 3 : 4]$. This allows performing many raster operations in parallel without network exchange of border cell values (e.g. calculate convolution for cells $A[2 : 6, 4 : 4]$).

This ingestion step allows us to transform a time series of 2-d satellite scenes into a set of 3-d subarrays separated by hyperplanes.

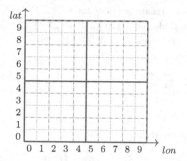

Fig. 1. An example of system-level arrays retiling: $s_1 = s_2 = 2$. (Color figure online)

3 Generic N-d Array Aggregation

3.1 Formal Definition

Let us define the array aggregation operation over an arbitrary set of array axes. The definition below fits the data model of ChronosServer and provides a formal framework for further algorithmic development.

The aggregate of an N-d array $A(d_1, d_2, \ldots, d_N) : \mathbb{T}$ over a set of axes $d_{aggr} \subset \{d_1, d_2, \ldots, d_N\}$ is the K-d array $A_{aggr}(d_{y_1}, \ldots, d_{y_K}) : \mathbb{T}$ such that $K = N - |d_{aggr}|$, $y_k < y_{k+1}$, $y_k \in [1, N]$, $d_{y_k} \notin d_{aggr}$, $A_{aggr}[x_{y_1}, \ldots, x_{y_K}] = f_{aggr}(cells(A[g(1), g(2), \ldots, g(N)]))$, where $\forall x_{y_k} \in [0, |d_{y_k}|)$, $g(i) = 0 : |d_i| - 1$ if $d_i \in d_{aggr}$, otherwise $g(i) = x_i$, $cells : A' \mapsto \mathbb{T}$ is a multiset of all cell values of an array $A' \sqsubseteq A$, $f_{aggr} : \mathbb{T} \mapsto w \in \mathbb{T}$ is an aggregation function. Note that the condition $K > 0$ always holds, otherwise the result is not an array but a scalar.

Algorithm 1 Computing a new key for a subarray

Input: $d_{aggr}, key\langle N \rangle = \{k_1, k_2, \ldots, k_N\}$ ▷ *key* is an input key
Output: key_{aggr} ▷ key_{aggr} is a new key
1: **function** GET-NEW-KEY
2:     $key_{aggr} \leftarrow \langle N - |d_{aggr}| \rangle = \{\}$
3:     $k \leftarrow 0$
4:     **for each** $j = 0, j < N, j{+}{+}$ **do**
5:        **if** $d_j \notin d_{aggr}$ **then**
6:           $key_{aggr}[k] \leftarrow key[j]$
7:           $k \leftarrow k + 1$

Consider an example in Fig. 2. The source 3-d array is aggregated over one of its axis resulting in a 2-d array, or over all of its axes except the first resulting in a time series.

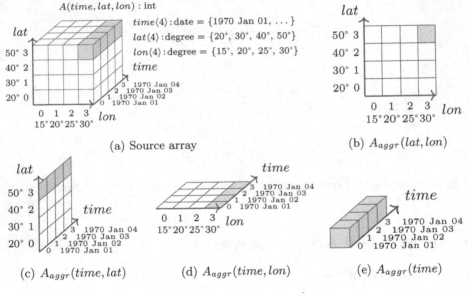

(a) Source array

(b) $A_{aggr}(lat, lon)$

(c) $A_{aggr}(time, lat)$

(d) $A_{aggr}(time, lon)$

(e) $A_{aggr}(time)$

Fig. 2. Aggregation examples.

Recall that we reference subarrays by keys which are essentially 1-d arrays. The subarrays of the newly aggregated array have different keys, but the most part of a key is inherited from the source subarrays. Let us define a function that will help us to compute the keys of new subarrays. It is easier to express it as a pseudocode, Algorithm 1.

3.2 Distributed In Situ Aggregation Algorithm

Recall that an array is partitioned into a set of subarrays which are distributed among cluster nodes. Algorithm 2 performs the aggregation over an arbitrary set of array axes. This algorithm is executed on the Gate and uses ChronosServer operation execution facilities [22], not the new execution engine [21]. Lines prefixed by worker(*node*) contain messages sent to the worker №*node*. The messages contain instructions APPLY, STREAM, and CONSUME. Several such messages can be packed into a single network message. The algorithm is outlined as if the aggregation is the only operation that is running in ChronosServer.

APPLY instruction launches a command on a local cluster node. STREAM [node] *dataset*$_1$ and CONSUME [count] *dataset*$_2$ instructions are used to exchange intermediate data. STREAM instruction sends all subarrays of *dataset*$_1$ to the node with identifier *node*. CONSUME accepts *count* subarrays from other nodes into the dataset named *dataset*$_2$.

Intermediate datasets are prefixed with "$" and are reused in subsequent commands. Intermediate datasets are not registered as regular datasets on the Gate to save time and network traffic. Dataset quantiles are specified after a colon ":". They are designed to fully leverage the power of the tools. For example, ncra can take several files as input. It is reasonable not to launch ncra on each file and then combine the result into a single file on a separate instruction but to feed all locally available files to ncra on APPLY phase to reduce the computation and I/O time. This is exactly the behavior triggered by the "all" quantile.

We leverage ncra to aggregate over the first array dimension using the "all" quantile. We utilize ncwa when several dimensions are to be aggregated, but without the "all" quantile.

Algorithm 2 Generic Distributed In Situ Array Aggregation

Input: $P, f_{aggr}, \mu, d_{aggr}$
Output: P_{aggr}
Require: $N > 1$

1: **for each** $key \in \{\text{GET-NEW-KEY}(d_{aggr}, p.key) : p \in P\}$ **do**
2: $\quad \mathbb{C} \leftarrow \{p : \text{GET-NEW-KEY}(d_{aggr}, p.key) = key \land p \in P\}$
3: $\quad Nodes \leftarrow \{c.node : c \in \mathbb{C}\}$
4: $\quad$ **for each** $node \in Nodes$ **do**
5: $\quad\quad \mathbb{P} \leftarrow \{p : p.node = node \land p \in \mathbb{C}\}$
6: $\quad\quad$ worker(*node*): APPLY ncra -y f_{aggr} $\mathbb{P}$:all \$interim   ▷ Delegation
7: $\quad\quad$ **if** $node \neq \mu(key)$ **then** worker(*node*): STREAM[$\mu(key)$] \$interim
8: $\quad$ worker($\mu(key)$): CONSUME[$|Nodes\setminus\{\mu(key)\}|$] \$interim
9: $\quad$ worker($\mu(key)$): FINALLY ncra -y f_{aggr} \$interim:all P_{aggr} ▷ Delegation

---

Figure 3 illustrates Algorithm 2. Subarrays with the same color reside on the same node, workers in parallel aggregate all local subarrays. The intermediate subarrays are gathered on one of the nodes which calculates the final result. The mapping $\mu : key \mapsto id$ returns the worker ID to which the partial aggregates must be sent.

Algorithm 2 is for $f_{aggr} \in \{max, min, sum\}$. Calculation of avg is reduced to calculating the sum and dividing each cell of the resulting array on $|A.d_1|$.

Lines 6 and 9 of Algorithm 2 are highlighted in gray to accent the work being delegated to an external tool.

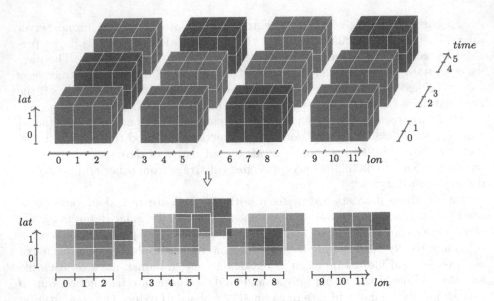

Fig. 3. Aggregation of system-level arrays.

4   Performance Evaluation

Microsoft Azure Cloud was taken for the experiments. Azure cluster creation, scaling up and down with given network parameters, number of virtual machines, etc. was fully automated using Java Azure SDK [22]. The latest version of Ubuntu Linux on which SciDB 16.9 runs is 14.04 LTS. We rented standard D2S V3 machines with 2 CPU cores, 8 GB RAM, 64 GB local SSD drive.

We selected band 1 from nine Landsat 8 scenes for path 190 and row 31 ($\approx$585 MB in total) such that the cloud cover percent for the majority of scenes is less than 20%. We evaluated SciDB version 16.9. We have written a Java program that converts GeoTIFF files to CSV files to feed the latter to SciDB. There are no official plugins or out-of-the-box ways to import an external file into SciDB 16.9. We aligned all scenes in UTM coordinates since they are slightly shifted relatively to each other and imported the scenes into a SciDB array with shape 9 × 7971 × 7941. We filled the cells with NULL values for areas in some scenes that appeared in the result of extension of that scenes during their alignment.

Import time of one Landsat 8 scene into SciDB takes about **40 min** on a cloud node, not local machine. Therefore, we imported all 9 scenes on a local computer, exported the resulting SciDB array into a file of proprietary SciDB binary format, and copied that file in the Cloud when needed (SciDB imports data from its proprietary format much faster). The selected Landsat 8 scenes are not big data. However, very long SciDB import time prevents testing SciDB on a larger data portion. Nevertheless, the data volume turned out to be sufficient for representative results.

SciDB is mostly written in C++, parameters used: 0 redundancy, 2 instances per machine, instances can allocate unlimited amount of memory, other settings are default. ChronosServer has 100% Java code, ran one worker per node, OracleJDK 1.8.0_111 64 bit, max heap size 978 MB (-Xmx). We used NCO and GDAL tools available from the standard Ubuntu 14.04 repository. NCO v4.4.2, last modified 2014/02/17. GDAL v1.10.1, released 2013/08/26.

We have evaluated cold query runs (a query is executed for the first time). Respective OS commands were issued to free `pagecache`, `dentries` and `inodes` each time before executing a cold query to prevent data caching at various OS levels. ChronosServer benefits from native OS caching and is much faster during hot runs when the same query is executed for the second time on the same data. There is no significant runtime difference between cold and hot SciDB runs.

For ChronosServer, a node contained $|P'|/M$ subarrays after "data cooking", where M is the number of cluster nodes. SciDB also distributes data uniformly.

Table 1 shows results for $f_{aggr} = min$. For other functions, ChronosServer reveals relatively the same performance. However, SciDB is up to 50% slower for $f_{aggr} = avg$. Note that ChronosServer is not equipped with an execution pool like ChronosDB. We anticipate that the performance of the presented aggregation queries will be about 2× faster when integrated into ChronosDB. Note also that the aggregation type that results into $A(lat, lon)$ is the most frequent as it is the aggregation over a time series.

Table 1. Results of performance evaluation, seconds

Aggregating to $A(lat, lon)$	8 nodes	16 nodes
ChronosServer	4.35	2.32
SciDB	87.73	44.87
Ratio, SciDB/ChronosServer	**20.17**	**19.34**
Aggregating to $A(time, lat)$	**8 nodes**	**16 nodes**
ChronosServer	4.25	2.29
SciDB	9.76	4.98
Ratio, SciDB/ChronosServer	**2.30**	**2.17**
Aggregating to $A(time, lon)$	**8 nodes**	**16 nodes**
ChronosServer	6.45	3.32
SciDB	18.45	9.52
Ratio, SciDB/ChronosServer	**2.86**	**2.87**
Aggregating to $A(time)$	**8 nodes**	**16 nodes**
ChronosServer	4.46	2.34
SciDB	8.49	4.38
Ratio, SciDB/ChronosServer	**1.90**	**1.87**

5   Related Work

Numerous techniques exist for remote sensing imagery processing. This work is novel because it is in the context of array DBMS research field. Four modern raster data management trends are relevant to this paper: industrial raster data models, formal array models and algebras, in situ data processing algorithms, and array DBMS. A recent comprehensive survey of the mentioned trends in their diverse aspects is in [21].

A recent survey on array models and algebras as well as industry standard data models is in [21]. Work [22] gives the peculiar features and merits of the ChronosServer data model. It is shown that the most popular array models and algebras can be mapped to Array Algebra [3]. Industry data models are also mappable to each other [8]. SciDB does not have a formal description of its data model. SciDB neither allows array dimensions to be of temporal or spatial types making it difficult or sometimes impossible to process many real-world datasets.

We performed the comparison only with SciDB [4] since it is the only freely available distributed array DBMS to date. The latest SciDB version does not operate in-situ and imports raster data only converted to CSV format – a very time-consuming, error-prone, and complex undertaking.

PostGIS/PostgreSQL [13] and RasDaMan [2] work on a single machine and allow registering out-database raster data in a file system. Enterprise RasDaMan version claims to be in-situ enabled and distributed but it is not freely available [14]. To the best of our knowledge, no in-depth, comprehensive performance comparison between SciDB and RasDaMan has been ever published. Intel released open source TileDB on 04/04/2016. It is yet not distributed neither in-situ enabled [11,27]. SciQL was an effort to extend MonetDB with functionality for processing multi-dimensional arrays [28]. SciQL does not provide in-situ raster processing. However, it has not yet finished nor its active development is seen so far. Commercial Oracle Spatial does not provide in-situ raster processing [10]. Commercial ArcGIS ImageServer claims in-situ raster processing with custom implementation of raster operations [1]. However, in a clustered deployment scenario all cluster nodes are recommended to hold copies of the same data or fetch data from a centralized storage. All commercial systems are not freely available.

Hadoop [6] and experimental SciDB streaming [26] allow to launch a command line tool, feed text or binary data into its standard input and ingest its standard output. Note two time-consuming data conversion phases in this case: data import into an internal database format and their conversion to other representation to be able to feed to an external software. In our approach we directly submit files to external executables without additional data conversion steps.

There are also other efforts for in-situ and distributed raster data processing: SAGA, SWAMP, SciHadoop, SciMate, Galileo, and others. However, none of them is an array DBMS. Most of them were not released or no longer actively maintained. These systems are reviewed in [19].

6 Conclusions

One of the core ideas behind our approach is to delegate major raster data processing work to feature-rich and highly optimized command line tools. The systems built using our approach are much faster than SciDB and provide much wider functionality. At the same time, the formal array model of ChronosServer maintains a high level of independence from the underlying raster file formats and the tools.

ChronosServer computes aggregates by up to 20× faster on Landsat 8 data than SciDB. ChronosServer data preprocessing stage takes about 5 min instead of several hours of SciDB import time not taking into account writing import code. Future work includes integrating the experimental aggregation techniques into ChronosDB. We anticipate them running even faster since they will leverage innovative ChronosDB execution capabilities.

Acknowledgments. This work was partially supported by Russian Science Foundation (grant №17-11-01052).

References

1. ArcGIS for server—Image Extension. http://www.esri.com/software/arcgis/arcgisserver/extensions/image-extension
2. Baumann, P., Dumitru, A.M., Merticariu, V.: The array database that is not a database: file based array query answering in RasDaMan. In: Nascimento, M.A., et al. (eds.) SSTD 2013. LNCS, vol. 8098, pp. 478–483. Springer, Heidelberg (2013). https://doi.org/10.1007/978-3-642-40235-7_32
3. Baumann, P., Holsten, S.: A comparative analysis of array models for databases. Int. J. Database Theory Appl. 5(1), 89–120 (2012)
4. Cudre-Mauroux, P., et al.: A demonstration of SciDB: a science-oriented DBMS. PVLDB 2(2), 1534–1537 (2009)
5. Earth on AWS. https://aws.amazon.com/earth/
6. Hadoop streaming. https://wiki.apache.org/hadoop/HadoopStreaming
7. Landsat apps. https://aws.amazon.com/blogs/aws/start-using-landsat-on-aws/
8. Nativi, S., Caron, J., Domenico, B., Bigagli, L.: Unidata's common data model mapping to the ISO 19123 data model. Earth Sci. Inform. 1, 59–78 (2008)
9. Newberry, R.G., Lupo, A.R., Jensen, A.D., Rodriges Zalipynis, R.A.: An analysis of the spring-to-summer transition in the West Central Plains for application to long range forecasting. Atmos. Clim. Sci. 6(3), 375–393 (2016)
10. Oracle spatial and graph. http://www.oracle.com/technetwork/database/options/spatialandgraph/overview/index.html
11. Papadopoulos, S., et al.: The TileDB array data storage manager. PVLDB 10(4), 349–360 (2016)
12. Planet Labs. https://www.planet.com/company/
13. PostGIS raster data management. http://postgis.net/docs/manual-2.2/using_raster_dataman.html
14. RasDaMan features. http://www.rasdaman.org/wiki/Features
15. Rodriges Zalipynis, R.A.: ChronosServer: real-time access to "native" multi-terabyte retrospective data warehouse by thousands of concurrent clients. Inform. Cybern. Comput. Eng. 14(188), 151–161 (2011)

16. Rodriges Zalipynis, R.A.: Efficient isolines construction method for visualization of gridded georeferenced data. Probl. Model. Des. Autom. **10**(197), 111–123 (2011)

17. Rodriges Zalipynis, R.A.: Representing Earth remote sensing data as time series. Syst. Anal. Environ. Soc. Sci. **2**(3), 135–145 (2012)

18. Rodriges Zalipynis, R.A.: Ecologic assessment of air pollution by nitrogen dioxide over the territory of Europe using Earth remote sensing data. Inform. Cybern. Comput. Eng. **1**(19), 126–130 (2014)

19. Rodriges Zalipynis, R.A.: ChronosServer: fast in situ processing of large multidimensional arrays with command line tools. In: Voevodin, V., Sobolev, S. (eds.) RuSCDays 2016. CCIS, vol. 687, pp. 27–40. Springer, Cham (2016). https://doi.org/10.1007/978-3-319-55669-7_3

20. Rodriges Zalipynis, R.A.: Array DBMS in environmental science: satellite sea surface height data in the cloud. In: 9th IEEE International Conference on Intelligent Data Acquisition and Advanced Computing Systems: Technology and Applications, IDAACS 2017, Bucharest, Romania, 21–23 September 2017, pp. 1062–1065. IEEE (2017). https://doi.org/10.1109/IDAACS.2017.8095248

21. Rodriges Zalipynis, R.A.: ChronosDB: distributed, file based, geospatial array DBMS. PVLDB **11**(10), 1247–1261 (2018). http://www.vldb.org/pvldb/vol11/p1247-zalipynis.pdf

22. Rodriges Zalipynis, R.A.: Distributed in situ processing of big raster data in the cloud. In: Petrenko, A.K., Voronkov, A. (eds.) PSI 2017. LNCS, vol. 10742, pp. 337–351. Springer, Cham (2018). https://doi.org/10.1007/978-3-319-74313-4_24

23. Rodriges Zalipynis, R.A., et al.: The Wikience: community data science. Concept and implementation. In: Informatics and Computer Technologies, pp. 113–117. DNTU (2011)

24. Rodriges Zalipynis, R.A., et al.: Retrospective satellite data in the cloud: an array DBMS approach. In: Voevodin, V., Sobolev, S. (eds.) RuSCDays 2017. Communications in Computer and Information Science, vol. 793, pp. 351–362. Springer, Cham (2017). https://doi.org/10.1007/978-3-319-71255-0_28

25. Rodriges Zalipynis, R.A., Pozdeev, E., Bryukhov, A.: Array DBMS and satellite imagery: towards big raster data in the cloud. In: van der Aalst, W.M.P., et al. (eds.) AIST 2017. LNCS, vol. 10716, pp. 267–279. Springer, Cham (2018). https://doi.org/10.1007/978-3-319-73013-4_25

26. SciDB streaming. https://github.com/Paradigm4/streaming

27. TileDB. http://istc-bigdata.org/tiledb/index.html

28. Zhang, Y., et al.: SciQL: bridging the gap between science and relational DBMS. In: IDEAS (2011)

Author Index